The Inside Story

DNA to RNA to Protein

PHOSPHO
PROTEIN
PHOSPHATASE
THWACK
P
ADP
P
PROTEIN KINASE
AMINO ACID
A
P
Golgi
A
RIBOSOME RECEPTOR
GDP
SNAP

The Inside Story

DNA to RNA to Protein

READINGS FROM *TRENDS IN BIOCHEMICAL SCIENCES*

EDITED BY

Jan A. Witkowski

Cold Spring Harbor Laboratory

Editor in Chief of *Trends in Biochemical Sciences*

The Inside Story: DNA to RNA to Protein

Publisher	John Inglis
Acquisition Editor	John Inglis
Development Manager	Jan Argentine
Project Coordinator	Maryliz Dickerson
Production Editor	Patricia Barker
Desktop Editor	Susan Schaefer
Production Manager	Denise Weiss
Cover Designer	Ed Atkeson

Cover illustration: Philip Leder (Harvard University) explains the inside story to colleagues at the 1969 Cold Spring Harbor Symposium entitled *The Mechanism of Protein Synthesis.*
Title page spread: Illustration by T.A. Bramley, "DNA Makes RNA Makes Protein."

Library of Congress Cataloging-in-Publication Data

The inside story : DNA to RNA to protein / Jan A. Witkowski, editor.
p. cm.
ISBN 0-87969-750-4 (pbk. : alk. paper)
1. Biochemistry--History. 2. Molecular biology--History. 3. DNA. 4. RNA. 5. Proteins--Synthesis. 6. Discoveries in science. I. Witkowski, J. A. (Jan Anthony), 1947-
QP511.I57 2005
572.8'09--dc22
2005007822

10 9 8 7 6 5 4 3 2 1

Author affiliations are those at time of original article publication. Correspondence can be addressed in care of Cold Spring Harbor Laboratory Press, 500 Sunnyside Blvd., Woodbury, NY 11797-2924.

All Cold Spring Harbor Laboratory Press publications may be ordered directly from Cold Spring Harbor Laboratory Press, 500 Sunnyside Blvd., Woodbury, N.Y. 11797-2924. Phone: 1-800-843-4388 in Continental U.S. and Canada. All other locations: (516) 422-4100. FAX: (516) 422-4097. E-mail: cshpress@cshl.edu. For a complete catalog of all Cold Spring Harbor Laboratory Press publications, visit our World Wide Web Site http://www.cshlpress.com./

Contents

Foreword

I was brought up to believe that finding out how to find things out was rather more important than what you actually found. In trying to educate young scientists, encouraging them to think about the kinds of approaches they might employ, what evidence they might seek, how they would know when they were on the right track, this seemed the right way to go about the business. Simply telling them how things were, that was the lazy way. In other words, knowing how we know what we know is at least as important, for a real scientist, as what is known. For a researcher, knowing what is not known is even more important. Scientific understanding is a foggy business, the clear patches few and far between and the limits of knowledge the blurrier the closer you approach. And it never gets any easier, finding things out.

This is why heroic stories of the past are so uplifting and such fun to read. Certain stories stick in the mind. Every biochemist should read Arthur Kornberg's book, *For Love of Enzymes:* his account of scouring Bethesda for the perfect potato is a beautiful story, pregnant with universal significance despite its humble specific objective, to find an abundant source of the enzyme nucleotide pyrophosphatase. How this led to the discovery of DNA polymerase becomes clear as the story unfolds. Or the hilarious account of André Lwoff's realization that ultraviolet light could induce the prophage—"Sir, I am entirely lysed." Some discoveries result from patient, well-planned journeys with one foot planted securely in front of the other, others from almost amateurish stumbling around in the right place at the right time. I once had the great privilege of listening to Bart Barrell recall the path that he and Fred Sanger had taken from sequencing tRNA to the development of the "dideoxy" method for DNA sequencing. Enzymes, techniques and methods developed along the way in a perfectly logical manner, but these sharp-eyed scientists often reused their tools in unexpected, unanticipated ways to great effect later in the struggle. Bart made it sound easy, and that was encouraging, as well as fascinating. You could trace a relatively straight path in retrospect, but at the time, nothing was quite certain. The stress is considerable.

So I love reading these stories, whether they concern matters I'm familiar with, or whether it's something I always wondered about. I recently came across the account of the discovery of the neutrino in a collection of Nobel Prize lectures, which explained that the original proposers of the existence of the neutrino had confidently asserted the impossibility of ever detecting one. It was left to Frederick Reines and Clyde Cowan, two young physicists working at Los Alamos to realize that they ought to be able to detect them. At first they thought that their detector would have to be near an atomic explosion, but this impractical scheme was refined when a new reactor was built at Savannah River, and with considerable ingenuity and much hard work they succeeded in detecting the first neutrinos. Very often, you would struggle to reproduce the work of the pioneers. Very often, too, once you see it, it's obvious. But you need to understand the actual nuts and bolts rather well. The more stories you read, the more you realize that advancing science is often a matter of thinking big but paying attention to the tiny details. And enjoying the details, as a gardener has to enjoy weeding as well as planting.

TIM HUNT
Cancer Research U.K.
Editor in Chief
Trends in Biochemical Sciences, 1992–2000

Preface

It was Bill Whelan, the first editor-in-chief of *Trends in Biochemical Sciences,* who decided to include historical articles in the journal. Inspired by the "100 Years Ago" column in *Scientific American,* Bill named the TiBS column "50 Years Ago," and began to solicit articles from prominent scientists and historians of science. Judging by the luminaries who accepted the invitation, he had struck a rich vein. The articles were both autobiographical and biographical, and ranged from brief anecdotes to serious contributions to the history of biochemistry. Of necessity, given the title of the column, the topics were drawn from what might now be regarded as the classical period of biochemistry, when much had still to be learned about the molecular constituents of the cell and how they functioned in synthetic and metabolic pathways. The column became an established part of TiBS, and the articles became longer and more scholarly.

In 1985, Judith Hall, then editor of TiBS, asked if I would like to take on "50 Years Ago." I had published several historical papers in TiBS and I leapt at the chance to promote historical studies of biochemistry. Judith and I decided to rename the column; selecting topics simply because they were 50 years old seemed rather arbitrary and, more important, I wanted to encourage papers on historical topics much closer to the present time. Judith suggested "Reflections on Biochemistry" as we wanted the authors not just to tell us what research they had done, but also to ruminate on its consequences. A further modification led to the current title: "Historical Reflections."

More recently, a conversation in the bar of Cold Spring Harbor Laboratory between me, Emma Wilson, editor of TiBS from 1999 to 2002, and John Inglis, Executive Director of Cold Spring Harbor Laboratory Press, produced the idea of assembling in book form the most interesting of the articles that were spread, largely unappreciated, across many years of red-bound volumes of TiBS. But we were amazed to find that there were over 260 articles! How should the selection be done? Should we try to cover the whole compass of biochemistry? Should we concentrate on autobiographical articles? Should we include only articles written by scientists—or even just Nobel laureates? When I came to look more closely, it was clear that the area the articles covered most comprehensively, where the most interesting selection could be made, was the Central Dogma, that is DNA, RNA, and protein synthesis. And the number of relevant articles was just right for the size of book we had in mind.

Fortuitously, the publication of this collection coincides with the 50th anniversary of the founding of the International Union of Biochemistry and Molecular Biology (IUBMB). TiBS came into being through the hard work of Bill Whelan and Bill Slater of the then-IUB, and Jack Franklin and Sandy Grimwade at Elsevier in Amsterdam. There were discussions in 1973, a contract was drawn up in 1974, and the first edition appeared in January 1976, with Whelan as editor-in-chief and Joan Morgan as editor. The partnership between the IUBMB and Elsevier continues to the present day, as the joint logos on the TiBS cover testify.

The history articles reprinted here are the work of many people. I thank the editors and editors-in-chief of TiBS, all of whom have believed that the history of our field is not only interesting and important, but also entertaining. At Cold Spring Harbor Laboratory Press, John Inglis initiated the project, and Maryliz Dickerson and Jan Argentine kept it on track. But it is the contributors to this book who most deserve our thanks, for reminding us of the human stories that often lie behind the polished sheen of the papers we read in the journals.

The General Idea

TIM HUNT

"Jim, you might say, had it first. DNA makes RNA makes protein. That became then the general idea." Thus did Francis Crick[1] explain to Horace Judson years later, long after he had written with such clarity and force on the subject of Protein Synthesis in the 1958 Symposium on "The Biological Replication of Macromolecules."[2] This article is celebrated for its prediction of the existence of tRNA (although by the time the article appeared in print, tRNA had been discovered), but it is chiefly worth reading and rereading, even today, for its enunciation of the two principles that together constitute the "General Idea." The first principle is the Sequence Hypothesis; the idea that the sequence of amino acids in proteins is specified by the sequence of bases in DNA and RNA. The second principle is the famous "Central Dogma"; not DNA makes RNA makes Protein, but the assertion that "Once information has passed into protein it cannot get out again." It isn't completely clear why one is a hypothesis and the other a dogma and the two together an idea. The Dogma stuck in some throats, mainly because it was called a dogma, with the heavy religious overtones.

Crick explains that calling it a dogma was a misunderstanding on his part; he thought the word stood for "an idea for which there was no reasonable evidence," blaming his "curious religious upbringing" for the error. But it probably wasn't that much of a mistake after all, for the Oxford Dictionary allows dogma to mean simply a principle, although the alternative "Arrogant declaration of opinion" is probably how most who were not molecular biologists took it, considering its never modest author. That is probably how they were meant to take it, too. It was the most important article of faith among the circle of biologists centered on Watson and Crick, and remained so for

From Hunt T. 1983. Introduction: The General Idea. In *DNA Makes RNA Makes Protein* (eds. T. Hunt, S. Prentis, and J. Tooze), pp. vii–xiv. Elsevier Biomedical Press, New York.

quite a long time, until the mechanism of protein synthesis became clear. Crick said that if you did not subscribe to the sequence hypothesis and the central dogma "you generally ended in the wilderness," although he did not offer alternative scenarios for public consumption, even though they probably played an important part in convincing him of the dogmatic status of the General Idea's second component.

In any case, the General Idea has proved completely correct (at least as stated in the 1958 article), and it is now difficult to recapture the atmosphere in 1970 which caused such glee in certain quarters following the discovery of reverse transcriptase. The importance of this enzyme was not that it "demolishes the hallowed idea of the molecular biologist that RNA cannot synthesize DNA"[3] (as we have seen, this wasn't the idea at all, let alone hallowed) nor yet that it solved the cancer problem (which it still hasn't), but because it opened the way for recombinant DNA technology to be applied to higher eukaryotes. It was a crucial tool, to be sure, but only a tool. The trouble was that Watson was very emphatic that RNA did not specify DNA synthesis in his justly famous and very influential textbook.[4] Even the 3rd edition says that "DNA makes RNA makes protein" *is* the Central Dogma, and adds that "RNA never acts as a template for DNA," although a couple of footnotes admit the existence of RNA tumour viruses and reverse transcriptase. It is curious and amusing to look back on this flurry; words like "hallowed," "shibboleth," "heresy," and "inspiration" joined "dogma" in the pages of *Nature.*[5] They were the last shots in the battle, as it turned out. Crick wrote a measured rejoinder,[6] and from then on it was possible to be a respectable Molecular Biologist. It is still difficult to understand why the General Idea aroused so much resentment among the more conservative biologists.

It is crucial to understand the General Idea. Together with the principle of Natural Selection it forms virtually the only solid ground biologists have to stand on. Some of the opposition to the Idea may have come from the constraints that it may appear to place on living systems; the discovery that they contain a highly structured logical core was perhaps surprising, and somewhat alien to biological thought, which tends to emphasize the variety, heterogeneity, and general illogicality of living systems. The strength of the Idea lies much more in what it forbids than what it permits. The Central Dogma necessarily eliminates Lamarckism in any shape or form, and the idea of DNA as a self-replicating store of information for the synthesis of proteins is wholly in accord with the Darwinian conception of life. As Howard writes, the DNA molecule itself "embodies in a pure chemical substance the three

Darwinian conditions for evolution: heredity from the accuracy of self-replication, multiplication from the fact of self-replication, and variation from the rare inaccuracies of replication."[7]

The General Idea has its limitations too, which in a way forms the subject matter of this book. The statement that DNA makes RNA makes protein is first and foremost logical, and stresses the one-way flow of information. The chemical and structural content of the Idea is low, and its physiological content zero. It fails to specify what proteins are for, and fails to mention the important fact that the regulation of the information flow requires proteins to interact with DNA. But the Central Dogma was strictly concerned with protein synthesis, not with the control of gene activity. Indeed, when the Central Dogma says that "information cannot escape from proteins" it sounds close to a denial of the very essence of living systems, for proteins are the very stuff of life. We wouldn't get very far if all we had was DNA. But it is at the level of proteins that simple logic fails, as the linear strings of information are converted into three-dimensional structures according to defined but complicated rules (which are still incompletely understood). And these structures have specific functional properties, which are quite impossible to deduce by simple examination of the structure.

While it is a challenging problem to predict the structure of a protein from its amino acid sequence, it would be impossible to tell what a protein *did* in a cell, even if every atom in its structure had been mapped with complete precision. It might be possible to guess, based on analogies with known enzymes, but given the sheer variety of potential substrates, and of the variety of possible transformations that might be performed on each of the substrates, it seems to be a task bordering on the logically impossible to say what a particular gene, and hence protein, is for based solely on its base (= amino acid) sequence. Likewise, it would be difficult to design an enzyme from scratch, if you were a real genetic engineer. What we actually do at present is to analyze. A gene is revealed by the effects of its loss or alteration. An enzyme can be discovered, purified, and sequenced. Its gene can be located, purified, and cloned. But it is difficult, if not impossible, to deduce the function of a given DNA sequence. It may be possible to rule out the possibility that it specifies a protein owing to the lack of extended open reading frames (although the existence of introns makes this difficult in the genomes of higher organisms). It is even more difficult to recognize a regulator of DNA, for these sequences are exempt entirely from the rules of the Central Dogma or the Sequence Hypothesis. They are presumably constrained only by the requirements of uniqueness (so that they can be addressed unambiguously) and of recognizability by certain proteins. We will return to this theme later.

At about the same time that the "transforming principle" was being identified as DNA, the slogan "One gene, One enzyme" was coined by the fungal geneticists. They did not proclaim it as a dogma, or even really label it a hypothesis; it seems to have served principally as a useful definition of a gene. It is very interesting that this definition did not arouse the same kind of bridling hostility among the intellectual community of the day that the Central Dogma provoked. Was it because the implications were less explicit, because nobody cared, or because the slogan was couched in physiological terms?

A gene was a heritable entity that had detectable effects in organisms and an enzyme likewise something you could assay by its function. At that time, there was no general acceptance that genes were made solely of DNA, so to translate "One gene, One enzyme" into the current equivalent "DNA makes Protein" would not have been possible in any case. Even if it had, there is a curious disparity between the logical chemical language of the Central Dogma and the functional description of One Gene, One Enzyme; a great deal is changed by the translation. What is gained in rigor is lost in physiological descriptiveness. It is an old conflict, between the observational biologists, the physiologists, and the biochemists, which is by no means over, as witnessed by Penman's current insistence on the importance of the cytoskeleton, for example.[8] Actually, according to François Jacob, even Monod had difficulty at first accepting that repression acted at the DNA level, that it could be as simple as that. His objections, which Jacob overcame in part by reference to toy trains, make interesting reading.[9]

In fact, neither of these catch-phrases really deals with our current concerns. We now take for granted that the General Idea is correct (though the detailed mechanisms remain in many cases to be worked out). What is needed is an extension, like this:

DNA → RNA → Protein → Cells → Organism

Why not? Is this not what we all believe? Today's Dogma? The trouble is, such a statement of faith (for such it surely is) acts to keep the vitalists away, but it does not serve any other profound or useful function. It does not define limits, or exclude possibilities. As soon as the sequence of amino acids in proteins is achieved, they fold and gain properties which are no simple function of that sequence and, of course, the proteins interact with each other and with the structures whose building they catalyze to generate a system endowed with life. The pleasing simplicity and rigor of the General Idea is left far behind as soon as this "messy and illogical"[7] realm is entered, where the laws of chemistry and physics are obeyed all right, but the system is more or less unfathomable because of the complexity of its interactions.

What is even worse is that the role of DNA goes well beyond simply specifying amino acid sequences. The instructions in the genetic material include what might be called analog information as well as the digital. It isn't enough to specify the structure of proteins; the amount and timing of their synthesis must be specified as well. This specification is not made by a simple digital code based on the sequence of bases in DNA; rather, DNA is recognized by proteins in much the same way that they recognize any other substrate, and proteins control the expression of the information contained in the linear strings of bases. Proteins do not read DNA as ribosomes read mRNA, using simple nucleic acid recognition rules. Proteins "feel" the shapes they find in the grooves, or look for sequence-determined, but not rigorously sequence-defined alterations in the backbone, like Z-DNA, for example. Evidently, proteins recognize the right sequence when they see it. At which point, the General Idea has to give up, because the proteins' trapped information is being used to control the flow of information from DNA. There is no more logic in that than in the recognition of glucose-6-phosphate by glucose-6-phosphate dehydrogenase. Certainly nothing to be dogmatic about.

The recent history of molecular biology has shown over and over again how strange are the ways of Nature, even when you understand the Central Dogma, the General Idea, whatever you call it. Not that it's a "considerable oversimplification." All the Idea ever sought to explain was how proteins were made, to account for the role of the genetic material in this process. As the articles in this book show, even fleshing out the General Idea with the specific details of its implementation has taken a great many people a great many years, and has revealed a wealth of unexpected beauty in the search for the Actual.[10] The search goes on, of course; it is still an open question whether development has a logic to it as simple as that of the lac operon. Brenner puts it as clearly and wittily as anyone,[11] asking whether there is such a thing as the "'Make a hand" program encoded in the genome. This is high-level stuff. We do not even know if there is a button in the genome which when pressed (metaphorically speaking) will give a red cell. It is going to be a lot of fun finding out. In effect, the hypothesis to be explored and tested was stated boldly (looking at it now, almost foolhardily) by Crick in the famous article, right at the beginning.[2] It seems just that he should have the last word. "I shall also argue that the main function of the genetic material is to control (not necessarily directly) the synthesis of proteins. There is a little direct evidence to support this, but to my mind the psychological drive behind this hypothesis is at the moment independent of such evidence. Once the central and unique role of proteins is admitted there seems little point in genes doing anything else."

REFERENCES

1. Judson H.F. 1979. *The Eighth Day of Creation*, Jonathan Cape, London.
2. Crick F.H.C. 1958. On protein synthesis. *Symp. Soc. Exp. Biol.* **12:** 138–163.
3. Anonymous. 1970. Après Temin, le Deluge. *Nature* **227:** 998–999.
4. Watson J.D. 1976. *Molecular Biology of the Gene*, 3rd edition. W.A. Benjamin, Menlo Park, California.
5. Anonymous. 1970. Central Dogma reversed. *Nature* **226:** 1198–1199.
6. Crick F.H.C. 1970. Central Dogma of molecular biology. *Nature* **227:** 561–563.
7. Howard J.C. 1982. *Darwin: A Very Short Introduction.* Oxford University Press.
8. Cervera M., Dreyfuss G., and Penman S. 1981. Messenger RNA is translated when associated with the cytoskeletal framework in normal and VSV-infected HeLa cells. *Cell* **23:** 113–120.
9. Jacob F. 1979. The switch. In *Origins of Molecular Biology* (eds. A. Lwoff and A. Ullman), Academic Press, New York.
10. Jacob F. 1982. *The Possible and the Actual.* Pantheon Press, New York.
11. Brenner S. 1981. Genes and development. In *Cellular Controls in Differentiation* (eds. C.W. Lloyd and D.A. Rees). Academic Press, London.

Albrecht Kossel

S. RAPOPORT

Institute for Physiological and Biological Chemistry, Humboldt University, Berlin

Professor Kossel in a very real sense was the founder of modern biochemistry.

A.P. Mathews[1]

Albrecht Kossel, Emeritus Professor of Physiology in the University of Heidelberg and Director of the Institute of Protein Investigation at Heidelberg, died on 5 July, 1927. This 50th anniversary of his death is a fitting occasion to pay tribute to his pioneering work.

He entered Hoppe-Seyler's laboratory in 1877 and there he found the inspiration for a life-long research program: the chemistry of the cell nucleus. Hoppe-Seyler had set out to find the chemical structures of living systems; his aim was to establish principles common to all types of protoplasm. This undertaking met with almost insurmountable difficulties. New prospects opened with the study of the cell nucleus, an "organ of the cell whose structure and function must be associated with the *general* processes of life" (Kossel). Miescher had defined "nuclein" by its content of phosphorus but nucleins were found in materials without nuclei such as egg yolk and casein. Purines, such as guanine, had been found but were not connected with nuclein. Uric acid had been known for a long time but its origin was completely unknown.

Albrecht Kossel (1853–1927). (Reproduced from Hoppe-Seyler's Z Physiol. Chem 169 [1927] by permission of Walter de Gruyter & Co., Berlin.)

Reprinted from *Trends in Biochemical Sciences* July 1977, 2(7): 163–164.

Within 20 years, Kossel brought order into the chaos. He defined the nucleic acids, discovered and proved the structure of their purine and pyrimidine bases, found their carbohydrates and predicted the types of chemical bonds in nucleotides. He showed that uric acid was their metabolic product and proved the endogenous synthesis of purines.

He then turned to the proteins of the nucleus. He recognized the protein nature of protamins and classified them according to their content of basic amino acids. He discovered and named the histones. Histidine was discovered in the course of this work and procedures were developed for determining the amino acid composition of proteins. These procedures, based on separations of phosphotungstates and silver salts, remained standard methods until the advent of chromatographic analysis.

Kossel's concept of the exposed functional groups of proteins showed remarkable insight. "I imagine the protein molecule in such a way that it may respond to a chemical attack with any of its characteristic groups. Similar to grapes on a vine, the protein molecule contains a large number of characteristic groups, like guanidine, imidazole, indole, etc. If special combinations are needed, they are present in receptive forms." He received the Nobel Prize for Medicine and Physiology in 1911.

Apart from his scientific research, Albrecht Kossel set another high standard by his integrity and by his fair and upright position during World War I. He was one of the few scientists who resisted the upsurge of ultra-nationalistic emotions during that conflict and refused to support the propaganda campaigns that endeavored to persuade his countrymen that the war-time standards of nutrition were adequate.

REFERENCE

1. Mathews A.P. 1927. *Science* **66:** 293.

Albrecht Kossel and the Discovery of Histones

DETLEF DOENECKE AND PETER KARLSON

Physiologisch-Chemisches Institut der Philipps-Universität, Marburg, FRG

In a paper dated 4 July 1884 Albrecht Kossel reported a "peptone-like component of the cell nucleus (Ueber einen peptonartigen Bestandtheil des Zellkerns)."[1] He proposed the name "histone" for a basic compound which he had isolated from goose erythrocytes by a rather simple procedure. The red blood cells were lysed in a mixture of water and ether and the erythrocyte nuclei were subsequently resuspended and washed in water. Finally, the nuclei were extracted with dilute hydrochloric acid and a precipitate was formed upon addition of sodium chloride to the acidic extract of the erythrocyte nuclei. The results of the chemical analysis of these nuclear extracts suggested to Kossel that the basic compounds he had isolated belonged to a class of substances (which had been described before as A-peptones, propeptones or albumoses), apart from classical proteins. In accepting this classification, Kossel did not yet consider the newly discovered compounds to fit all criteria of a protein as defined at that time. Kossel wrote: "I propose the name 'histone' for this substance." He did not explain why he chose this name. At first sight, the name histone may refer to biological tissues in general (which later turned out to be a valid suggestion). At the time of this publication, however, Kossel referred only to goose erythrocytes, which he would hardly have regarded as a tissue.

In the discussion of his data, Kossel suggested that the histones might be bound to nucleic acids, since he had observed that isolated histones "are soluble in neutral water, whereas the water, that had been used to wash the nuclear substance before the addition of the hydrochloric acid, did not con-

Reprinted from *Trends in Biochemical Sciences* September 1984, 9(9): 404–405.

tain any traces of histones." This apparent binding of histones to nucleic acids in the nuclei showed a striking similarity to the binding of protamines (discovered by Miescher in 1874[2] in salmon sperm) to deoxyribonucleic acid. In fact, Miescher had already found that immature testes of salmon did not contain protamine, but a protein-like substance (which was later found to be histone).

At the time of Miescher's and Kossel's initial discoveries, the basic amino acids were still unknown, and protamine, for example, had been defined by Miescher as a nitrogen-rich base. The situation changed when in 1886 Schulze and Steiger[3] discovered arginine and in 1889 Drechsel was the first to describe lysine.[4] This allowed Kossel to reinvestigate histones and protamines as to their amino acid composition. In 1896, he concluded in a paper "On the Basic Substances of the Cell Nucleus"[5] that lysine, arginine and a newly discovered amino acid, which he called histidine, were among the cleavage products of protamines.

With the increased knowledge about the structure of amino acids and proteins, the definition of histones as basic nuclear proteins became obvious and Kossel and his co-workers searched for histones and protamines in other systems. In 1894, Leon Lilienfeld, a co-worker of Kossel, introduced calf thymus as an experimental system and described the extraction of thymocyte histones in a paper "On the Chemistry of Leucocytes."[6] There, he summarized his own and Kossel's previous data: "the histone is a protein which is remarkably basic. This follows from the fact that it combines with hydrochloric acid to form a compound which is readily dissolved in water." Later, in his Nobel lecture,[7] Kossel wrote about the role of basic amino acids in histones and protamines: "the insertion of these nitrogen-containing groups into the protein molecule leads to the fact that very basic groups are present in a highly reactive state." He refers to avian erythrocyte nuclei and continues: "similar substances occur in a salt-like interaction with nucleic acids in the tissues of a wide variety of lower and higher animals."[7] In contrast to the work on protamines, which was mainly continued by Felix and his co-workers,[8] the interest in the chemistry of histones decreased after Kossel's time. It was again Felix who published a series of papers "On the Structure of Histones in the Thymus Gland,"[9] in which he described the chemical analysis of calf thymus histone as a defined species of basic proteins. This definition of histones as a separate (and rather homogenous) class of proteins, which was based on solubility criteria, had already been published in a classification of proteins proposed by an international nomenclature body in

1908.[10] There, histones are listed as a class of proteins along with albumins, globulins, glutelins and others.

The idea that histone was a homogenous type of protein predominated until the 1940s. Then, Stedman and Stedman[11] compared the histones, from a wide variety of species, for solubility under varied conditions. In addition to their own work, which established histones as a component of chromatin in general, Stedman and Stedman quoted the work of Mirsky and his associates, who had extracted histones from nucleoproteins before and who had discovered that histones were "components of threadlike structures which have been termed chromosomes."

At the 1947 Cold Spring Harbor Meeting, Mirsky reported on "chemical properties of isolated chromosomes"[12] and compared the histone content of chromatin of several tissues. In summary, he and his co-workers, Pollister and Ris,[13] state that chromosomes are composed of four fractions: "an abundance of deoxypentose nucleic acid, along with some pentose nucleic acid, and two clearly separable fractions, histone and residual protein."

The breakthrough in histone research came when Stedman and Stedman analysed the histones of several tissues from a variety of species by paper chromatography of their constituent amino acids.[11,14] They succeeded in separating two classes of histones which they called "main histones" and "subsidiary histones." The basis of this separation was a differential solubility in aqueous alcohol. The chromatographic analysis of the amino acids revealed that the "main histones" were rich in arginine, whereas the basic character of the subsidiary histones was due to a high lysine content. No tryptophan was found in either of these two fractions. Besides this successful attempt to fractionate histones, Stedman and Stedman drew another conclusion from their data: "the hypothesis that one of the physiological functions of histones is to act as gene supressors."[14]

With the advent of molecular biology and the progress in protein chemistry the interest in the structure and function of histones steadily increased. A wide variety of extraction procedures were proposed (for details, see Johns in Ref. 15 and earlier reviews, Refs. 16 and 17) and it became obvious that the two main groups which had been defined by Stedman and Stedman were again composed of individual protein species.

The introduction of ion-exchange chromatography,[18,19] electrophoresis,[20–23] N-terminal group analysis[24,25] and advanced extraction and precipitation methods[15] further helped to define individual histone fractions. This, finally, initiated sequencing studies and the first primary structure of a histone was published by DeLange et al. for histone H4 (then called F2A1) from calf thymus[26] and from pea embryos.[27]

This comparison showed for the first time that the primary structure of these histones was highly conserved during evolution. In the meantime, protein and DNA sequencing have revealed the primary structures of numerous histone species from several animal and plant cells types (for review, see Ref. 28). The discovery of the subunit structure of chromatin, the elucidation of the histone arrangement within these chromatin sub-unit cores and the work on higher order chromatin structures (for review, see Ref. 29) allowed, in the meantime, the assignment of structural roles to the five individual histone species, which are found throughout nature. H2A, H2B, H3 and H4 have been defined as core histones, since they form the histone octamer moiety of the nucleosomal core, whereas the very lysine-rich histone H1 is mainly involved in the formation of higher order chromatin structures.

The role of H1 and its variants is exemplified in avian erythrocyte nuclei. The condensed chromatin of these transcriptionally inactive nuclei contains an additional member of the H1 histone family, H5, which has a high amount of arginine residues, in addition to the high percentage of lysine, which is characteristic for histone H1. Thus, when Kossel selected goose erythrocytes ("the red blood cells of birds present very favourable conditions for the chemical investigation") as a system for his studies, he chose a system which provides a broad range of the structures and functions of the basic chromosomal proteins for which he proposed the name "histone" in 1884.

REFERENCES

1. Kossel A. 1884. *Hoppe-Seyler's Z. Physiol. Chem.* **8:** 511.
2. Miescher F. 1874. *Verh. Naturforsch. Ges. Basel* **6:** 138.
3. Schulze E. and Steiger E. 1886. *Hoppe-Seyler's Z. Physiol. Chem.* **11:** 43.
4. Drechsel E. 1889. *J. Prakt. Chem.* **39:** 425.
5. Kossel A. 1896. *Hoppe-Seyler's Z. Physiol. Chem.* **22:** 176.
6. Lilienfeld L. 1894. *Hoppe-Seyler's Z. Physiol. Chem.* **18:** 473.
7. Kossel A. 1912. Nobel Lecture 1910, Imprimerie Royale, Norstedt, Stockholm.
8. Felix K. 1960. *Adv. Protein Chem.* **15:** 1.
9. Felix K. and Rauch H. 1931. *Hoppe-Seyler's Z. Physiol. Chem.* **200:** 27.
10. Joint Recommendations. 1908. *J. Biol. Chem.* **4:** 48.
11. Stedman E. and Stedman E. 1951. *Phil. Trans. Royal Soc. B.* **235:** 565 (and earlier references cited therein).
12. Mirsky A.E. 1947. *Cold Spring Harbor Symp. Quant. Biol.* **12:** 143.

13. Pollister A.W. and Ris H. 1947. *Cold Spring Harbor Symp. Quant. Biol.* **12**: 147.
14. Stedman E. and Stedman E. 1950. *Nature* **166:** 780.
15. Phillips D.M.P. 1971. *Histones,* Plenum Press.
16. Allfrey V.G., Mirsky A.E., and Stern H. 1955. *Adv. Enzymol.* **16**: 411.
17. Butler J.A.V. and Davison P F. 1957. *Adv. Enzymol.* **18:** 161.
18. Crampton C.F., Moore S., and Stein W.H. 1955. *J. Biol. Chem.* **215**: 787.
19. Davison P.F. 1957. *Biochem. J.* **66:** 708.
20. Davison P. F., James D.W.F., Shooter K.V., and Butler J.A.V. 1954. *Biochim. Biophys. Acta* **15:** 415.
21. Cruft H.J., Mauritzen C.M., and Stedman E. 1954. *Nature* **174:** 580.
22. Khouvine Y., Gregoire J., and Zalta J.P. 1953. *Bull. Soc. Chim. Biol.* **35:** 244.
23. Neelin J.M. and Connell G.E. 1959. *Biochim. Biophys. Acta* **31:** 539.
24. Luck J.M., Cook H.A., Eldridge N.T., Haley M.I., Kupke D.W., and Rasmussen P.S. 1956. *Arch. Biochem. Biophys.* **65:** 449.
25. Phillips D.M.P. 1958. *Biochem. J.* **68:** 35.
26. DeLange R.J., Smith E.L., Fambrough D.M., and Bonner J. 1968. *Proc. Natl. Acad. Sci. USA* **61:** 1145.
27. DeLange R.J., Fambrough D.M., Smith E.L., and Bonner J. 1969. *J. Biol. Chem.* **244:** 5669.
28. Isenberg I. 1979. *Annu. Rev. Biochem.* **48:** 159.
29. Igo-Kemenes T., Hörz W., and Zachau H.G. 1982. *Annu. Rev. Biochem.* **51:** 89.

Early Nucleic Acid Chemistry

FRITZ SCHLENK

Department of Biological Sciences, University of Illinois at Chicago, Chicago, Illinois 60680, USA

The contemporary biochemist, perusing the publications of the past century, may easily be misled into some arrogance when sorting out the erroneous data and leads, the slow progress and the primitive reports of that early period. However, when recalling the limited techniques and the lack of facilities at the early stages of biochemistry, he will change his mind. The difficulties encountered by the investigators of nucleic acid structure were many. To begin with, it was difficult to ascertain the degree of uniformity of the starting material. There were no established methods for studying macromolecules, and no guidelines for the isolation of structural units. Fortunately for the identification of new compounds, reference material or closely related substances had already been synthesized by organic chemists in experiments that were usually unrelated to nucleic acid problems. Much information on purines already existed, and some pyrimidines had been synthesized. Emil Fischer's work on monosaccharides provided the basic information for the identification of ribose and deoxyribose. Thus, nucleic acid research of that period owes much to organic chemistry.

BEGINNINGS

The history of nucleic acid research begins with Friedrich Miescher's search for chromatin in pus. Miescher was trained in Basel, in the laboratory of his uncle, the renowned anatomist Wilhelm His, Sr. After receiving his medical degree in 1869, he joined F. Hoppe-Seyler in Tuebingen. Extraction of soiled bandages from infected wounds led to the isolation of material that he con-

Reprinted from *Trends in Biochemical Sciences* February 1988, 13(2): 67–69.

sidered to be a protein, more acid in nature than any earlier known protein. Miescher's publication was delayed, and when it appeared two years later in an irregular collection, *Hoppe-Seyler's Medizinisch-Chemische Untersuchungen,* his mentor had added two papers of other co-workers in which similar acid protein material from different sources was reported.[1]

This delay in publication was not unusual for the period. The practice of publishing research findings as they occurred was unknown prior to the establishment of the *Zeitschrift fuer Physiologische Chemie* in 1877. Its founder,[2] Felix Hoppe-Seyler, along with T. Schwann, was a pupil of the physiologist Johannes Mueller. At the University of Tuebingen, which was established around 1500, Hoppe-Seyler had research quarters in the picturesque old castle; a tablet at the entrance to a side wing commemorates his activity there from 1862 to 1872. Thereafter, he accepted a position at the University of Strasbourg.

Miescher returned to Basel in 1870, to start his academic career and to continue the investigation of chromatin. His study on "Protamine, a New Organic Base from Salmon Sperm"[3] achieved better uniformity of source material, extraction technique, results and speed of publication.

> Nuclein was found to be the principal fraction (48.7%); it is an albuminoid substance of acid property, rich in P (9.6%), and free of sulfur. Several similar compounds have been isolated earlier from various sources in an impure state. Now, for the first time, the possibility of its preparation in the pure state has been realized. Nuclein is combined with protamine; a mixture of several basic components may hardly be assumed.

The latter remark pertains to protamine-like compounds; Miescher was not aware of the presence of purines and pyrimidines. However, on his suggestion Jules Piccard,* the head of the chemistry department in Basel, did more work on nuclein and found 6–8% of the previously known purine bases guanine and hypoxanthine (sarkin) as components.[4] He suggested that "the presence of such significant quantities of these rare compounds in a readily accessible material may be of some interest to chemists and physiologists." The quality and speed of Piccard's publication suggests that he would have isolated all nucleic acid bases in the course of a few years, had he only continued the investigation of nuclein. Instead this task took several decades in the hands of others.

*Jules Piccard came from a renowned Swiss family of scientists; his twin sons, Jean-Felix and Auguste, later became famous for balloon excursions into the stratosphere and diving records by bathysphere to explore the depths of the oceans.

SLOW PROGRESS

The term "nucleic acid" was used first in 1889 by R. Altmann,[5] who studied the phosphate-containing material from thymus, egg yolk and salmon sperm; his results confirmed Miescher's observations. However, the term "nuclein" prevailed until the turn of the century. No attempt was made by Altmann to confirm Piccard's discovery of guanine. Likewise, A. Kossel in his numerous publications made only passing reference to Piccard's work. He may have felt that the latter had preempted his domain, because as late as in 1910, he wrote: "Guanine has been known for some time in various animal tissues, and was found, for example, in the spermatozoa of salmon by Piccard, although indeed this investigator had no suspicion that it had any genetic relationship with nuclein" (Ref. 6, p. 396).

The recognition of two distinct types of nucleic acid, DNA and RNA, occurred about 25 years after Miescher's discovery, and another 40 years passed before the first biological function was described. During this early period, the limited amount of nucleic acid research was mainly in the hands of investigators with a medical background. Organic chemists at that time did not take much interest in these ill-defined, almost untractable compounds; in their own field, the methods of organic synthesis provided easy access to new substances that could be purified readily and often had great commercial value.

Neither Miescher (1844–1895) nor Hoppe-Seyler (1825–1895) contributed much to the biochemistry of "nuclein" beyond their initial work. In 1871, Miescher was appointed to head the department of Physiology in Basel. He then devoted his efforts mainly to cytological problems, but this work was interrupted repeatedly by bouts of tuberculosis.[7]

A NEW APPROACH

Biochemical investigations of nucleic acids came to be one of the main activities of another of Hoppe-Seyler's students, Albrecht Kossel (1853–1927). Heidelberg was the principal site of his academic career, and he attracted many co-workers and guest investigators to his laboratory. E. Kennaway[8] has given us a charming set of recollections of his sojourn with Geheimrat Kossel, and of the commotion and torchlight parade of the students when the award of the Nobel Prize to Kossel was announced in 1910. Kossel was probably the first investigator to surmise that nuclein may be involved in growth and differentiation[9]:

> The assumption that nuclein is a resource, at the expense of which a starving organism lives, has to be refuted on the basis of all experiments. The quantity of nuclein changes little, whether the organism is starving or not.

The most valid basis for this conclusion seems to have been Miescher's observation on salmon sperm, which is formed at the expense of tissue reserves; the fish does not take nutriment during the period of spawning. Kossel studied this problem with a few starving chickens, but the quantitation of nucleic acid by the methods of that time was rather uncertain.

COMPONENTS IDENTIFIED

Progress in clarifying the structure and composition of nucleic acids was painfully slow. Guanine was known prior to its isolation from nuclein by Piccard[4]; it had been discovered already in 1846 by Unger[10] in guano, the bird excreta imported as fertilizer to Europe from the South American west coast. Guanosine was isolated by Schulze and Bosshard in 1886 from plant material.[11] A relation to nucleic acids was not suspected, but the identity was established in 1910 by comparison with the nucleoside that Levene had obtained from guanylic acid.[12] Adenine was isolated from thymus gland by Kossel[13]; he derived the name from the Greek term for gland: aden, adenos. The name purine was coined by Emil Fischer[14] to indicate the unaltered, pure nature of the basic ring system. Kossel and Neumann[15] discovered thymine and came to the conclusion that a carbohydrate group is present in thymus nucleic acid. Kossel's co-workers, Ascoli and Steudel, discovered cytosine and uracil. Pinner[16] synthesized pyrimidine and suggested its name in analogy to pyridine. Sometimes, degradation products of nucleic acid bases were obtained and, for a period, hypoxanthine and xanthine were believed to be nucleic acid components. Also, a variety of nucleic acids was assumed, each having only one kind of a base. It is impossible to enumerate all the misconceptions and the work necessary to undo them. Review articles were not customary at that time, but the Nobel Lecture of A. Kossel in 1910 gives an idea of all the vagaries of early nucleic acid biochemistry.[6]

Ribose and deoxyribose were the last principal nucleic acid components to be identified. The investigations of Emil Fischer and his school on carbohydrates provided the fundamental information.[17] Xylose and arabinose were known then as naturally occurring pentoses, and Fischer projected the configuration of the other two pentoses and suggested the names lyxose and ribose by rearrangement of some of the letters of xylose and arabinose, respectively.[18] How different would the development of nucleic acid biochemistry have been, if Fischer had made it one of his main enterprises!

The isolation of the carbohydrate of pentose nucleic acid was tried repeatedly, and its identity with arabinose, xylose, and lyxose was suggested by various investigators. Success in the identification of ribose[19] and deoxyribose[20,21] in 1909 and 1930, respectively, was achieved by P.A. Levene and his co-workers, mainly W.A. Jacobs, E.S. London, T. Mori, and S.R. Tipson. In both instances, the isolation of the nucleosides was a pre-requisite to provide the starting material. Here again, Levene did the pioneering work; the term nucleoside was coined by him for the reason that they link carbohydrate in a glycosidic union to the nucleic acid bases. His work with Tipson and with Stiller also led to the recognition of the furanoid structure and of positions 3 and 5 of the pentoses as the sites of esterification of phosphoric acid.

Levene's numerous contributions were summarized in collaboration with L.W. Bass in 1931, in *Nucleic Acids*,[24] the first monograph of consequence covering the entire field, with emphasis on the chemistry of nucleic acid constituents; of necessity, biological data were minimal and altogether speculative at that time. In all, Levene has done far more for chemical nucleic acid research than any of his predecessors.[25] His work is often underestimated by biochemists and biologists of a more recent period, who did not forgive him his erroneous concept of the "tetranucleotide structure" and his reservation about a possible macromolecular structure of nucleic acids. Several other leading specialists of that period, including Steudel and Feulgen, likewise favored the concept of a tetranucleotide structure, and nobody contested it. In his famous monograph (1931), Levene devoted only a few pages to "Nucleic Acids of a Higher Order." He summarized his opinion[24]:

> Thus, in conclusion, it must be admitted that judgement as to the existence of nucleic acids of a higher order should be postponed until the work is repeated on a larger scale. On the other hand, the presence of a ribopolynucleotide in the animal tissues must now be regarded as well-established.

In the late 1930s, however, Levene accepted the macromolecular structure of all nucleic acids, which by then had been firmly established by ultracentrifugation and dialysis experiments. Still, the repetitive occurrence of tetranucleotide units in these macromolecules persisted in the mind of many investigators of that period. J.A. Witkowski[26] recently has reviewed the reasons for such simplifying concepts of macromolecular structure of nucleic acids as well as proteins. Numerology, psychology and inadequate analytical data played a role.

Organic chemistry also played a key role in other developments of nucleic acid research. As described recently by J. Brachet,[27] the production of special stains and the development of specific color reactions greatly aided progress in nucleic acid cytochemistry. Thus, during the second quarter of this century, cell biologists showed an increasing interest in nucleic acids and lifted them from the status of biochemical oddities. A fortunate interplay among scientific disciplines resulted.

THE BEGINNINGS OF A NEW ERA

The first unequivocal demonstration of a specific biological activity of DNA was provided in 1944, by O.T. Avery, C.M. MacLeod, and M. McCarty.[28] They succeeded in demonstrating that the transforming principle isolated from smooth cultures of pathogenic *Pneumococcus* is a specific deoxyribonucleic acid. Recognition of their results, however, was not immediate. As described by M. McCarty,[29] one of the first comprehensive presentations of the data was given by him at an exclusive meeting of top scientists, among them seven Nobel Laureates, and a small group of younger scientists. Unfortunately, the proceedings of this conference in 1945, at the resort of Hershey, Pennsylvania, were not published. I remember the excellent presentation given by Maclyn McCarty; unfortunately, it did not arouse much excitement at the time. One of the Nestors of the group, Linderstrøm-Lang, restricted his summarizing remarks about progress essentially to the mysteries of proteins. He bemoaned their complexity and stated that the primary structure of proteins probably never could be resolved, and that synthetic efforts at best would lead to caricatures of the cellular products. Proteins preoccupied the affection of most investigators at that time to such an extent that their surmized role as carriers of genetic information was not readily abandoned.[29] With some delay, however, the DNA experiments were extended, and other transformations were found. DNA became very popular.

The effects of E. Chargaff to elucidate the base composition of DNA from various species led to the recognition of individuality and of the A–T and G–C equivalence; from there it was only a short step to the double helix. The recognition of various kinds of RNA was not long in coming; H.G. Khorana synthesized polynucleotides, and the establishment of the genetic code by M.W. Nirenberg was the crowning event.

The limelight now shifted to the ever increasing studies of the biological functions of DNA and RNA. However, chemistry continued to provide valuable contributions. The most important—now historical—account of this transition period is the compendium on *Nucleic Acids, Chemistry and Biology,* edited by E. Chargaff and J.N. Davidson.[30] In its 46 chapters nearly all contributors to the progress of that period are represented. A similar effort now (three decades later) would involve many hundred contributors and result in a treatise filling a small library.

REFERENCES

1. Miescher F. 1871. *Hoppe-Seyler's Med. Chem. Untersuchungen* **4:** 441–460.
2. Hoppe-Seyler F. 1877. *Z. Physiol. Chem.* **1:** 1–3.
3. Miescher F. 1874. *Chem. Ber.* **7:** 376–379.
4. Piccard J. 1874. *Chem. Ber.* **7:** 1714–1719.
5. Altmann R. 1889. *Arch. Anat. Physiol.* (Physiol. Abt.) 524–536.
6. Kossel A. 1910. *Nobel Lectures: Physiol. and Med.* 1901–1921, pp. 392–405, Elsevier.
7. Miescher F. (Centenary) 1944. *Helv. Physiol. Pharmacol. Acta, Suppl. II.*
8. Kennaway E. 1952. *Ann. Sci.* **8:** 393–397.
9. Kossel A. 1882. *Z. Physiol. Chem.* **7:** 7–22.
10. Unger B. 1846. *Ann. Chem. (Liebig)* **59:** 58–68.
11. Schulze E. and Bosshard E. 1886. *Z. Physiol. Chem.* **10:** 80–89.
12. Schulze E. 1910–1911. *Z. Physiol. Chem.* **70:** 143–151.
13. Kossel A. 1885. *Chem. Ber.* **18:** 1928–1930.
14. Fischer E. 1897. *Chem. Ber.* **30:** 2226–2254.
15. Kossel A. and Neumann A. 1894. *Chem. Ber.* **27:** 2215–2222.
16. Pinner A. 1885. *Chem. Ber.* **18:** 795–763.
17. Bergmann M., Schotte H., and Lechinsky W. 1922. *Chem. Ber.* **53:** 158–172.
18. Fischer E. and Piloty O. 1891. *Chem. Ber.* **24:** 4214–4225.
19. Levene P.A. and Jacobs W.A. 1909. *Chem. Ber.* **42:** 2703–2706.
20. Levene P.A., Mikeska L.A., and Mori T. 1930. *J. Biol. Chem.* **85:** 785–787.
21. Fruton J.S. 1979. *Trends Biochem. Sci.* **4:** 49–50.
22. Levene P.A. and Stiller E.T. 1934. *J. Biol. Chem.* **104:** 299–306.
23. Tipson R.S. 1945. *Adv. Carbohydr. Chem.* **1:** 193–245.
24. Levene P.A. and Bass L.W. 1931. *Nucleic Acids,* Chem. Catalog.
25. Tipson R.S. 1957. P.A. Levene (obituary). *Adv. Carbohydr. Chem.* **12:** 1–12.
26. Witkowski J.A. 1985. *Trends Biochem. Sci.* **10:** 139–141.
27. Brachet J. 1987. *Trends Biochem. Sci.* **12:** 244–246.
28. Avery O.T., MacLeod C.M., and McCarty M. 1944. *J. Exp. Med.* **79:** 137–158.
29. McCarty M. 1985. *The Transforming Principle,* Norton.
30. Chargaff E. and Davidson J.N., eds. 1955; 1960. *Nucleic Acids, Chemistry and Biology* (3 Vols), Academic Press.

P.A. Levene and 2-Deoxy-D-ribose

JOSEPH S. FRUTON

Yale University,
New Haven, Connecticut, USA

Among the many biochemical papers that appeared in 1929 under the authorship of Phoebus Aaron Theodore Levene (1869–1940)[1] were three reports that merit special recollection, because they led to the identification of 2-deoxy-D-ribose as the sugar component of "thymonucleic acid" (DNA). Two of them were with a visitor from the Soviet Union, Efim Semenovich London (1868–1939), and the third was with a Rockefeller Foundation Fellow from Japan, Takajiro Mori (b. 1897).

In February 1929, Levene and London reported briefly that digestion of thymus DNA with canine intestinal juice gave a guanine nucleoside, from which they obtained a crystalline sugar by mild acid hydrolysis.[2] The elemental analysis and color reactions of the nucleoside and the sugar indicated that the latter was a deoxypentose. Several months later, Levene and Mori[3] gave a fuller description of the sugar (which they named "thyminose") and concluded that it was not identical with either 2-deoxy-D-xylose or 2-deoxy-L-ribose. They could not establish the nature of thyminose and requested "the leaving of the solution of this part of the problem to this laboratory."[3, p. 806] By February 1930, an answer was available: an apparently pure sample of synthetic 2-deoxy-L-ribose now had the same optical rotation, but of opposite sign, as that of thyminose, which was clearly the D-enantiomer.[4] It was evident that the 2-deoxy-L-ribose described in the previous paper had not been optically homogeneous.

The importance of the identification of thyminose as a deoxypentose for the understanding of the structure of DNA was immediately evident. Levene and London reported their finding at the spring 1929 meeting of the National Academy of Sciences,[5] and reaffirmed the tetranucleotide structure of

Reprinted from *Trends in Biochemical Sciences* February 1979, 4(2): 49-50.

DNA that had been proposed in 1912 by Levene and Walter A. Jacobs[6] and in 1921 (in revised form) by Levene,[7] with the replacement of the presumed hexose of DNA by a deoxy pentose. The essential feature of this structure was its similarity to the proposed tetranucleotide structure of yeast nucleic acid (RNA), whose sugar unit Levene and Jacobs had identified during the years 1909–1911 as D-ribose, and for which they had demonstrated the sequence of linkage of the three components in the four nucleotides derived from RNA. The new structure of DNA was published in a later 1929 paper by Levene and London[8]; they described in some detail the isolation of the nucleosides of guanine, hypoxanthine (derived from adenine), thymine and cytosine from the digestion mixture. It is noteworthy that the yields appear to have been variable and extremely modest, "the maximal being 1.5 g of guanine nucleoside from 200.0 g of nucleic acid."[8, p. 797]

The nature of the sugar component of DNA had presented a vexing problem for about 35 years, after Albrecht Kossel and Albert Neumann[9] had shown that acid hydrolysis of thymus nucleic acid yields, among other products, levulinic acid ($CH_3COCH_2CH_2COOH$). From the work of Bernhard Tollens it had been known that this substance is readily yielded by hexoses, and for several decades thereafter it was widely believed that, whereas RNA contains a pentose, the sugar unit of DNA is a hexose. This view appeared to be supported by the work of Hermann Steudel in 1907, and in 1912 Levene and Jacobs had briefly reported the isolation of a guanine hexoside from a DNA hydrolysate.[10] There were puzzling features, however; for example, the glycosidic bonds in DNA were much more acid-labile than those in RNA, but synthetic purine hexosides were not hydrolysed more rapidly than purine pentosides.

In 1914, Robert Feulgen[11] attempted to solve the puzzle by suggesting that the sugar unit of DNA is glucal, a substance obtained by Emil Fischer in the previous year by the reduction of acetobromo-glucose with zinc dust. The glucal prepared by Fischer gave a positive Schiff reaction for aldehydes (restoration of the color of a fuchsin solution previously bleached by SO_2), as did a neutralized acid hydrolysate of DNA, in contrast to the behavior of similar hydrolysates of RNA. Several years later, however, Max Bergmann[12] showed that a pure preparation of glucal does not give the aldehyde reaction, and explained Fischer's result by showing that members of the glucal family are readily converted to the corresponding 2-deoxy sugars, which give a positive Schiff reaction.

Shortly after the publication of Bergmann's work, Feulgen[13] discussed the status of his glucal theory, without mentioning the conversion of glucals

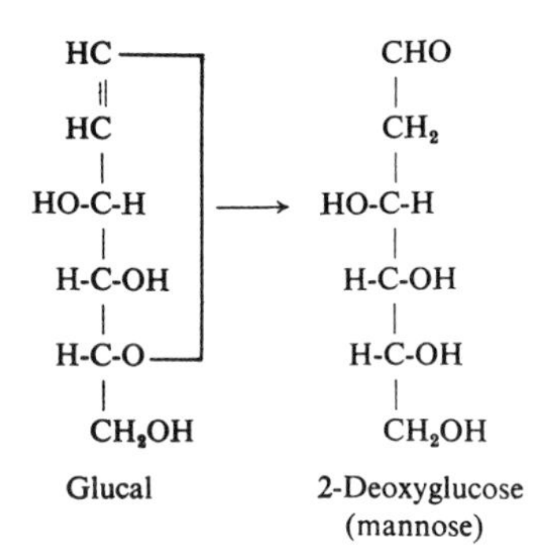

to deoxy sugars. Indeed, he was led to conclude from the available data on the sugar component of DNA that "thymonucleic acid must be constructed differently from yeast nucleic acid, and the great similarity assumed by Levene does not in fact exist."[13, p. 296]

Although Bergmann's work on glucal had eliminated it as a possible DNA component, he notably extended the knowledge about 2-deoxy sugars. Such sugars had been studied near the turn of the century by Heinrich Kiliani, who had isolated digitoxose (a 2-deoxy-5-methyl pentose) from digitalis glycosides. He also showed that the 2-deoxy sugars give a characteristic color reaction with ferric sulfate and a positive "pine-stick test," in which an acid solution of the sugar gives a succession of color changes. By 1929, therefore, when Levene and London had isolated thyminose, which also gave positive reactions in the color tests introduced by Schiff and Kiliani, Bergmann's work on 2-deoxy sugars was well known, and taken up by other chemists. It should be noted in passing that Bergmann's considerable contributions to carbohydrate chemistry during the 1920s have tended to be overlooked in the face of his later important work in the peptide field, especially the invention (with Leonidas Zervas, in 1932) of the "carbobenzoxy" method of peptide synthesis.

In Levene's laboratory, the study of 2-deoxy-D-ribose was later continued by Robert Stuart Tipson (b. 1907). After having shown that glycosides of D-ribose are furanosides, he demonstrated in 1935 that the 2-deoxy-D-ribosyl unit of thymidine is also a furanoside, by tritylation to yield the 5′-hydroxyl derivative. Thus, only the 3′ and 5′ hydroxyl groups of the sugar were available to form the phospho-diester bond in oligodeoxyribonucleotides. Much of the subsequent development of organic chemical studies on DNA were then carried out in England, where the decisive synthetic achievements of

Alexander Todd (now Lord Todd) completed a continuous, albeit tortuous, effort that had begun some 70 years before with the work of Kossel.

Levene's role in the DNA story has occasionally been presented in a negative light because of his reiteration of the tetranucleotide theory of nucleic acid structure, although some recent historians, notably Olby,[14] have attempted to redress the balance. In adhering to the idea that polynucleotides are aggregates of tetranucleotides, Levene reflected similar attitudes of the 1920s toward the constitution of proteins and polysaccharides. These views now lie in the dustbin of history, but so far as the structure of DNA is concerned, Levene's experimental work, first with Jacobs, and then with London, Mori and Tipson, laid foundations whose importance for the subsequent achievements deserves to be recalled.

REFERENCES

1. Van Slyke D.D. and Jacobs W.A. 1944. *Biog. Mem. Nat. Acad. Sci. U.S.A.* **23:** 75–126.
2. Levene P.A. and London E.S. 1929. *J. Biol. Chem.* **81:** 711–712.
3. Levene P.A. and Mori T. 1929. *J. Biol. Chem.* **83:** 803–816.
4. Levene P.A., Mikeska L.A., and Mori T. 1930. *J. Biol. Chem.* **85:** 785–787.
5. Levene P.A. and London E.S. 1929. *Science* **69:** 556.
6. Levene P.A. and Jacobs W.A. 1912. *J. Biol. Chem.* **12:** 411–420.
7. Levene P.A. 1921. *J. Biol. Chem.* **48:** 119–125.
8. Levene P.A. and London E.S. 1929. *J. Biol. Chem.* **83:** 793–802.
9. Kossel A. and Neumann A. 1894. *Ber. Chem. Ges.* **27:** 2215–2222.
10. Levene P.A. and Jacobs W.A. 1912. *J. Biol. Chem.* **12:** 377–379.
11. Feulgen R. 1914. *Z. Physiol. Chem.* **92:** 154–158.
12. Bergmann M. and Schotte H. 1921. *Ber. Chem. Ges.* **54:** 440–445.
13. Feulgen R. 1923. *Chemie und Physiologie der Nukleinstoffe,* Bornträger, Berlin.
14. Olby R. 1974. *The Path to the Double Helix,* Macmillan, London.

Schrödinger's *What is Life?*: A 50-Year Reflection

G. RICKEY WELCH

University of New Orleans,
New Orleans, Louisiana, USA

Erwin Schrödinger's *What is Life?* (Cambridge University Press, 1944) is one of the most talked about books in twentieth-century science. The book came under intense scrutiny at the centenary of Schrödinger's birth in 1987,[1] and was revisited at a recent symposium.[2] Discussion has focused largely on its impact on the early development of our understanding of chromosomal structure and of the concept of the "genetic code." Reflecting on the course of the scientific unity of physics and biology over the 50 years since its publication, one might ask, "Does the fixation on the historical role of *What is Life?* in the birth of molecular genetics represent a myopic view of the author's philosophical intent and shortchange the book's biophysical legacy? Were the book's ideas original? Was Schrödinger merely applying physics as a tool to analyse a biological object; or, rather, was he pondering an expanded natural philosophy that would provide a physicalist basis for biology?" The fiftieth anniversary of the publication of *What is Life?* serves as a fitting opportunity to (re)examine the significance of the biophysical issues raised by Schrödinger.

A BIOGRAPHICAL PRELUDE

Schrödinger was certainly not the only theoretician from the golden period of physics in the early twentieth century to delve into the science of biology.

Reprinted from *Trends in Biochemical Sciences* January 1995, 20(1): 45–48.

Niels Bohr, whose father was a physiologist, is a notable example; but of the pioneers from that germinal realm of relativity and quantum physics, I would suggest that Schrödinger had perhaps the deepest interest in the physicalist ramifications of the life sciences. His appreciation of biology can be traced to frequent childhood discussions with his father, whose avocation was botany. (*What is Life?* is dedicated to the memory of his parents.) Schrödinger read Darwin's *The Origin of Species* at an early age and became enthralled by the mechanistic view of the living world.

Schrödinger entered the University of Vienna in 1906, hoping to pursue physics under the great Ludwig Boltzmann. The untimely death of Boltzmann dictated a course of study for Schrödinger under the master's capable protégés. His closest friend from those days in Vienna was Franz ("Fränzel") Frimmel, a botany student with whom he had long late-night discussions about the meaning of life. Together, they read Richard Semon's 1904 book, *Die Mneme als erhaltendes Prinzip* (The Mneme as Conservative Principle), which left a lasting impression on the young Schrödinger. Semon was a pupil of Ernst Haeckel and the atomistic "plastidular" view of living systems. Semon's *Die Mneme* was derived from a psycho-lamarckian analogy between memory and inheritance. It is probably more than coincidence that Lamarck's *Philosophie Zoologique* was one of the few books Schrödinger retained from his father's library (the remainder having been sold owing to financial necessity).

From the writings, lectures and personal contacts of his later life as a world-famous physicist, it is evident that Schrödinger maintained a sideline interest in biology. Much of his thinking comes to a head in *What is Life?*, which was based on a series of public lectures at Trinity College in Dublin. The text is an amalgam of the statistical physics and thermodynamics of the author's turn-of-the-century training in Vienna, the quantum physics of his latter-day profession, and his early conviction towards biological order, evolution and heredity as deterministic Darwinian processes.

The Dublin lectures were very popular. However, the book elicited controversy even before its publication. In preparing the written version, Schrödinger added an epilogue "On Determinism and Free Will," which contained a denunciation of Western philosophy/religion and an espousal of Vedantic beliefs. Objection to this attack on the Church resulted in a collapse in the original plan for Cahill & Co. in Dublin to publish the book. Through a friend, the physical chemist Frederick Donnan, Schrödinger got the text published by Cambridge University Press; and thus did *What is Life?* enter the history of science. (For further biographical details, see Refs 3–5.)

FIGURE 1. A portrait of Schrödinger, taken by Lotte Meitner-Graf (c. 1935). The Professor is shown smoking his favorite Peterson pipe from Dublin. (Reproduced from Ref. 3 with permission from Cambridge University Press.)

WERE SCHRÖDINGER'S IDEAS ORIGINAL?

I do not intend here to review the many commentaries on *What is Life?* I would, however, like to address (and hopefully give some redress to) the question of the originality of his central ideas—a subject that has attracted much attention and that, some would argue, is the basis of the book's legacy. The majority of the book's content is, at face value, devoted to the molecular aspects of genetic action, as the author regarded the chromosome as the fundamental directing element in the generation of life's order. Accordingly, most of the commentaries on the book have concentrated on the genetic aspects of its historical status. Max Perutz,[6] for example, has professed cogently that *What is Life?* contains a considerable amount of textbook information on hereditary mechanisms and mutations, that the book's "backbone" is simply a paraphrase of a 1935 article by Timoféeff-Ressovsky, Zimmer and Delbrück[7] (entitled "The Nature of Genetic Mutations and the Structure of the Gene"), and that Schrödinger's lack of knowledge of the relevant literature invalidates the originality of his views. The legitimacy of such claims aside, many commentators [1,3–5,8] have argued that the book, with its elegantly written, although speculative, popularist style, was influential in attracting the early attention of physicists and chemists, as well as biologists, to the paramount biological problem of the molecular basis of gene action. James Watson,[9] for example, recalled "...I came back to the University of Chicago and spotted the tiny book *What is Life?* by the theoretical physicist Erwin Schrödinger. In that little gem, Schrödinger said that the essence of life was the gene. Up until then I was interested in birds. But then I thought, well, if the gene is the essence of life, I want to know more about it. And that was fateful because, otherwise, I would have spent my life studying birds, and no one would have heard of me."

Moving away from the oft-discussed role of *What is Life?* in the history of genetics, I should like to consider in some depth Schrödinger's application of statistical physics and thermodynamics to biology. His position was that "it is in relation to the statistical point of view that the structure of the vital parts of living organisms differs so entirely from that of any piece of matter that we physicists and chemists have ever handled...." Schrödinger illustrated his biological concern by reference to the so-called "$\sqrt{n}$ rule," whereby physicochemical laws manifest an inaccuracy with a relative error of the order of J/$\sqrt{n}$, where n is the number of molecules responsible for the process. As most physical laws deal with systems entailing very large numbers of particles, the error in predictability is usually insignificant. Whence, he wondered how the gene (or chromosome)—a *single* macromolecule—could determine dynamically "the very orderly and lawful events within a living organism."

Schrödinger's resolution to the paradox led him to propose that life's spatiotemporal order must be due, ultimately, to the inherent structural properties of the chromosome, which he conceptualized as a heat-stable "aperiodic crystal," involving a "miniature code" for a one-to-one correspondence with the organismic plan. Thus, he said, "we are here obviously faced with events whose regular and lawful unfolding is guided by a 'mechanism' entirely different from the 'probability mechanism' of physics." Schrödinger's longstanding romance with biology seems to reach a crescendo here, as he mused over such a "fantastic description, perhaps less becoming a scientist than a poet."

Schrödinger's concentration on the rationality of ordered cellular processes from the standpoint of statistical physics was incisive, but (at the risk of imparting a whiggish historiography*) we now see his analysis as more

*As phrased by Ernst Mayr,[10] "whiggish" is the term applied to that kind of biased, hindsighted history of science, "in which every scientist is judged by the extent of his [*sic*] contribution toward the establishment of our current interpretation of science."

appropriate to the molecular characterization of the *phenotype* rather than to that of the *genotype.* In Schrödinger's day, little was known empirically about the structural organization of cellular metabolism. Also, we should be reminded that biologists (especially in the early twentieth century and, to some extent, even today) tend to regard subcellular processes according to simplistic nineteenth-century physicochemical notions based on dilute, weak-electrolyte aqueous solutions and on gas-phase kinetics. Even so, early in the twentieth century a number of theoreticians from such areas as biochemistry, cell biology and developmental biology were already contemplating seriously the ideas of spatial order and microenvironmental deviation in the cellular machinery.[11–13] In particular, the crystal metaphor was being used at this time as a "field" concept by developmental biologists.[11]

An interesting parallel to Schrödinger's quest for understanding of the physical basis of cellular order, but on the phenotypic rather than the genotypic side, may be seen in an early paper entitled "Surface Structure in the Integration of Cell Activity," published in 1930 by the Oxford biochemist, Rudolph Peters.[14] Akin to Schrödinger's motivation, Peters felt that "the problem of greatest significance is that presented by the orderly, continuous and co-ordinated direction of chemical processes" in the living cell. With only limited cytological evidence available at the time, he argued for a mechanism of "directed chemical change" executed by enzyme populations associated with a reflexive proteinaceous network extending throughout the cell. Based on simple numerical calculations for the putative microcompartments posed by such ultrastructure, Peters pictured how fluctuations about some measured average value of cellular concentration can be of great kinetic consequence for the localized biochemical events. In a similar vein to Schrödinger, he asserted that "it is dangerous to apply thermodynamic arguments directly" to cellular processes. Instead, Peters suggested that "owing to the micro-heterogeneous nature of the system, surface effects take precedence over ordinary statistical, mass action relationships, and become in the ultimate limit responsible for the integration of the whole and therefore the direction of activities." Crossing paths with Schrödinger, Peters, taking "the 'gene' theory as a guide," speculated that, "if we isolate in thought the instant...at which the 'gene' first commences to exert its directive effect...," we should find that the protein-based cellular order begins with the "protamine" portion of the chromatin.

We now know that larger eukaryotic cells are, indeed, laced with a dense array of fibrous proteinaceous elements, as well as with spacious lipoprotein membrane surfaces; and a multitude of metabolic pathways has been found to be associated with this cytomatrix.[13] The theoretical issue of microenvironmental statistical variation in metabolic phenomena has continued to draw interest even in later years.[15] Perhaps the most succinct way of characterizing local biochemical flow processes in living cells is to refer to them in a modified "probabilistic" manner, as a diffusion under a bias, where the bias is rendered by the ambient structural setting.[13] Such a specific designation was, in fact, suggested in 1925 by the mathematical biologist Alfred Lotka.[16]

Schrödinger also examined the global problem of accounting for the organization of the living state within the context of the second law of thermodynamics. He maintained that such internal order must be "paid for" at the expense of degradation of substrates—"negative entropy" (or "negentropy")—from the surroundings. In the immediate aftermath of the publication of *What is Life?*, Schrödinger was criticized for not being explicit with the technical details of the nature of open systems and with the significance of Gibbs free energy in relation to negentropy. He retorted that the book was directed to a lay audience and that he was deliberately short on detail, although in a later edition he provided an addendum on this point. While Schrödinger lacked knowledge of some of the relevant biology, we can rest assured that, from his academic upbringing under the shadow of Boltzmann in Vienna, he was well aware of the overall thermodynamic nature of the matter.

Schrödinger's thermodynamic reasoning here was perspicacious. Although the basic principles had begun to appear in physics/chemistry textbooks[17] at the time, the subject of nonequilibrium thermodynamics was not well known, especially in the science of biology. Today, the thermodynamics of nonequilibrium processes has generated a well-defined scientific edifice that deals implicitly with the idea of negentropy as free-energy exchange across the boundary of open systems and internal dissipation.[18,19] Perhaps Schrödinger's view came from Boltzmann who, in the late nineteenth century, had lectured and written of living organisms as evading the second law of thermodynamics by feeding on negative entropy.[20] The bioenergetic significance of Boltzmann's depiction had been identified (and duly cited), long before *What is Life?*, by Lotka.[13] In 1940, Ludwig von Bertalanffy (respected by many as the father of "general systems theory"), independently had characterized living systems in virtually the same terms as Schrödinger.[21]

The central ideas in *What is Life?*, when considered individually, would not seem all that original. Nevertheless, Schrödinger's unique synthesis and physicalist insight gave notice, to a wide and diverse scientific audience, of a number of important, but relatively obscure, biophysical (as well as genetic) issues. This role, in itself, is sufficient to ensure a venerable position for the book in the history of science.

FROM IDEAS TO THOUGHTS

The true legacy of *What is Life?* is to be realized by reflecting not so much on the author's "ideas" but on the background "thoughts" that may have driven him to associate the ideas. At present, there is a communal feeling of unity (perhaps a *fin-de-siècle* euphoria?) amongst many scientists, especially those of a theoretical bent. As the twentieth century (and an old millennium) comes to a close, bookshops are replete with titles beginning with *The End of...,* from all fields of human endeavor. Many prominent physicists are now boldly asserting that a "theory of everything" (the so-called "TOE") is at hand. One has only to open the pages, for example, of Paul Davies' book *The New Physics* (Cambridge University Press, 1989), to feel the exciting holism in present-day physics and the urgency in its unification with biology.

These holistic times stimulate many thinkers to cross the barriers separating the individual intellectual disciplines and to engage in liberal analogical thinking between them. Schrödinger was a perceptive and inquisitive student of science during a similar period some 100 years ago. I agree with the biographical speculation of Walter Moore,[3] that *What is Life?* had its origins in the author's student days in Vienna, in those late-night biological discussions with Fränzel Frimmel. As the nineteenth century turned into the twentieth century, there was an abundance of books hailing the unity of science,[13] among them some by biologists such as Ernst Haeckel, of whom we know Schrödinger was aware.[3] Moreover, the physicist Boltzmann, Schrödinger's scientific idol from his university period, was a Darwin enthusiast and had given considerable notice to biological issues in the late nineteenth century.[20]

Moore[3] suggests that Schrödinger's incentive for interdisciplinary transcendence came from the *Naturphilosophie* spirit of Goethe, in whom he was well versed. The polymathic credo in Goethe's *On General Theory* tells all: "If you desire to reach out into the infinite, move in all directions in the finite.'"Goethe and Lamarck (whom, as noted above, Schrödinger also read) were two of the greatest universalists in the modern history of science.[22]

What is Life? is not really about genes, nor is it about entropy. Schrödinger, in fact, declares his purpose in the opening pages. After emphasizing "the keen longing for unified, all-embracing knowledge" and apologizing to the reader for his "second-hand and incomplete knowledge" of biology in the Preface, he states on the first page of the first chapter his focus, "at the risk of making fools of ourselves," on the fundamental question: "How can the events *in space and time* which take place within the spatial boundary of a living organism be accounted for by physics and chemistry?" (Schrödinger's italics). The true significance of *What is Life?* is to be understood within the realm of physicalism, a philosophical program of unification that accords a special relational privilege to physics.[23] Within this paradigm, the natural world is not analysed by merely reducing material substance to simple physical objects; rather, the theoretical character of space-time associations for processes to a physical format is reduced.

Schrödinger was, of course, caught in the beginning of the "molecular biology" era, which commenced in earnest shortly after the turn-of-the-century rediscovery of Mendel's work. Yet his concern was not with the nature of genes *per se;* rather, he wanted to know the causal, dynamical basis of "events" (such as structures, organization and hereditary transitions) in the living state. Schrödinger's famous "aperiodic crystal" metaphor for chromosomal structure should be interpreted in this dynamical sense. To a theoretical physicist, the idea of a "crystal" immediately conjures up mathematical symmetry groups. A hallmark of relativity and quantum physics in this century has been the elucidation of symmetry/asymmetry as the dynamical principle underlying all the forces of Nature.[24] Pierre Curie,[25] in the late 1800s, used the concepts of crystal symmetry groups in a dynamical manner (probably unknown to Schrödinger) that, today, provides the theoretical basis for our understanding of many kinds of coupled thermodynamic flow process in living cells.[13,18] While we cannot thank Schrödinger solely for the appearance of this paradigm in biology, the prescience of his analogical thinking is historically noteworthy.

In looking at the plethora of works on the theoretical foundation of the unification of biology and physics being written today, Schrödinger would surely smile contentedly. Not only should he be revered for his fundamental role in the physics revolution of the twentieth century, but he should be remembered in the history of science as one of the first pioneers from the early stage of that revolution to extend the physicalist implications ("albeit," as he said, "with secondhand and incomplete knowledge") to other areas.

Despite the phenomenal successes of theoretical physics and of molecular biology, we are, alas, no closer to answering the ultimate question *What is*

Life? than in Schrödinger's day. With time, the explanation becomes more elusive and metaphysical. Dare we infer that the author sensed the uncertainty himself, in adding the controversial epilogue? As voiced by Lotka,[16] a contemporary of Schrödinger, "(In) searching for the essential characteristics of *life,* those that should finally and conclusively distinguish the living from the nonliving, are we not just searching for the thing in nature that should correspond to a word in our vocabulary? Are we not hunting the Jabberwock?"

ACKNOWLEDGEMENT

I am grateful to D.B. Kell and J.A. Witkowski for helpful comments.

REFERENCES

1. Welch G.R. 1989. *BioEssays* **11:** 187–190.
2. Gnaiger E. et al. 1994. *What is Controlling Life?,* Innsbruck University Press.
3. Moore W. 1989. *Schrödinger: Life and Thought,* Cambridge University Press.
4. Yoxen E.J. 1979. *Hist. Sci.* **17:** 17–52.
5. Witkowski J.A. 1986. *Trends Biochem. Sci.* **11:** 266–268.
6. Perutz M.F. 1987. *Nature* **326:** 555–558.
7. Timoféeff-Ressovsky N.W., Zimmer K.G., and Delbrück M. 1935. *Nachrichten aus der Biologie der Gesellschaft der Wissenschaften Göttingen* **1:** 189–245.
8. Symonds N. 1986. *Q. Rev. Biol.* **61:** 221–226.
9. Watson J. 1993. *Science* **261:** 1812–1813.
10. Mayr E. 1982. *The Growth of Biological Thought,* Harvard University Press.
11. Haraway D.J. 1976. *Crystals, Fabrics, and Fields: Metaphors of Organicism in Twentieth-Century Developmental Biology,* Yale University Press.
12. Witkowski J.A. 1987. *Medical History* **31:** 247–268.
13. Welch G.R. 1992. *Prog. Biophys. Mol. Biol.* **57:** 71–128.
14. Peters R.A. 1930. *Trans. Faraday Soc.* **26:** 797–822.
15. Smeach S.C. and Gold H.J. 1975. *J. Theor. Biol.* **51:** 59–96.
16. Lotka A.J. 1925. *Elements of Physical Biology,* Williams & Wilkins.
17. De Donder T. and Van Rysselberghe P. 1936. *Thermodynamic Theory of Affinity,* Stanford University Press.
18. Caplan S.R. and Essig A. 1983. *Bioenergetics and Linear Nonequilibrium Thermodynamics,* Harvard University Press.
19. Westerhoff H.V. and Van Dam K. 1987. *Thermodynamics and Control of Biological Free-Energy Transduction,* Elsevier.
20. Broda E. 1983. *Ludwig Boltzmann: Man-Physicist-Philosopher,* Ox Bow Press.
21. Davidson M. 1983. *Uncommon Sense: The Life and Thought of Ludwig von Bertalanffy,* Tarcher.

22. Welch G.R. "Goethe's *Gestalt, Bildung,* and *Urphänomen* in Biology: A Twentieth-Century Physicalist View,' in *Goethe Scienziato* (ed. A. Grieco). Einaudi. (In press.)
23. Poland J. 1994. *Physicalism: The Philosophical Foundations,* Oxford University Press.
24. Pagels H.R. 1985. *Perfect Symmetry,* Simon & Schuster.
25. Curie P. 1908. *Oeuvres de Pierre Curie,* Gauthier-Villars.

Schrödinger's *What is Life?*: Entropy, Order and Hereditary Code-scripts

J.A. WITKOWSKI

Imperial Cancer Research Fund Laboratories, St. Bartholomew's Hospital, London, UK

There is a select group of books that those in intellectual circles profess to know but are unlikely to have read. Examples that come to mind are Darwin's *Origin of Species,* Wittgenstein's *Tractatus Logico-Philosophicus* and Joyce's *Finnegan's Wake* for those with scientific, philosophical and literary pretensions, respectively. For the biochemist and molecular biologist Erwin Schrödinger's little book *What is Life?* (abbreviated to *WIL*) belongs to the same genre. Widely thought to have been an important factor in the development of molecular biology in the late 1940s and early 1950s, it emphasized the idea of a code for transmission of heredity and it alerted physical scientists to the problems awaiting solution in biology (for detailed analyses and background information, see Refs 2–6).

What was it that Schrödinger wrote in 1943 that has played such an important role in modern biology? Why was Schrödinger of the wave equation interested in biology?

ERWIN SCHRÖDINGER

At the time that Schrödinger wrote *WIL* he had been director of the Institute for Advanced Studies in Dublin for two years.[7] The Institute had been established by Eammon de Valera, the Irish Prime Minister and mathematician, and was modelled on American Institutes like that at Princeton. Schrödinger

Reprinted from *Trends in Biochemical Sciences* June 1986, 11(6): 266–268.

had led a rather unsettled life for a number of years, and Dublin provided a haven of peace and quiet for him.

He had originally studied at the University of Vienna where his interest in statistical mechanics was aroused by Hasenohrl, who had succeeded Boltzmann as Professor of Physics. By 1921 Schrödinger's reputation was such that he was invited to become professor of theoretical physics in Zurich, a chair previously held by Einstein. It was in Zurich that Schrödinger produced his greatest work when, in a matter of a few months, he developed "wave mechanics." Apparently following up a suggestion by Einstein, Schrödinger demonstrated that particles like electrons could have a wave-like nature, and developed the famous Schrödinger wave equation that described the behaviour of the electron in the hydrogen atom. In 1927 he moved to Berlin where he spent six happy years in close association with giants of modern physics like Einstein, Planck, von Laue and Nernst.

By 1933 the political scene was far from happy when the Nazis came to power, for Schrödinger had not concealed his dislike of fascism. When he was given an opportunity to leave Berlin for a post in Oxford he accepted. Thus it was in Oxford, rather than Zurich or Berlin, that Schrödinger learned that he was to share the Nobel Prize with Paul Dirac. However, his stay in Oxford was not happy[8] and it may have clouded his political judgement, for, in 1936, he accepted a post in Graz, Austria. Two years later the Nazis invaded Austria and Schrödinger was forced to flee to Rome.

He moved to Dublin in 1939, although the Institute did not open until 1941, and stayed there until 1957. He worked on a variety of topics but he was chiefly concerned with attempts to derive a unified field theory for gravitational and electromagnetic fields. He did not succeed. In 1957 Schrödinger returned to Vienna and died there in 1961.

BIOLOGY AND THE PHYSICIST

Shortly after his arrival in Dublin, Schrödinger was faced with the task of giving a series of public lectures. He chose to discuss the phenomenon of life from the standpoint of a "naive physicist," and in particular the constraints a knowledge of statistical physics and quantum mechanics placed on such speculation.

Schrödinger's interest in the wider consequences of quantum physics had been aroused some ten years earlier when he had become involved in a discussion[9,10] over the relationship of the Heisenberg uncertainty principle to freewill and physical indeterminacy. At about the same time he also asked Karl Przibram for information about the gene,[4] and in 1935 he published a collection of essays one of which discussed thermodynamics in relation to living organisms.[11] From Schrödinger's notebooks and correspondence he seems to have begun work on *WIL* in earnest in 1942. His notebook listed three genetics' textbooks, Sherrington's Gifford Lectures, *Man on his Nature*,[12] and an article by Timoféeff-Ressovsky, Zimmer and Delbrück[13] on radiation and mutation. The latter contained a theoretical discussion by Delbrück on the nature of the gene and formed the basis for a large part of *WIL*.

WHAT IS LIFE?

I found three features of *WIL* particularly interesting. Firstly, very early on Schrödinger introduces the idea that "the most essential part of a living cell—the chromosome fibre—may suitably be called an aperiodic crystal" (*WIL*, p. 3) and that the chromosome "contains in some kind of code script the entire pattern of the individual's future development and of its functioning in the mature state" (*WIL*, p. 20). This is developed further in Sections 45 and 46. Here Schrödinger returns to the theme of the gene as an "aperiodic solid" and the code-script is expressed in the isomeric form that the gene can adopt. He uses morse code as an example of generating a very large number of code-words from relatively few symbols.

Secondly, Schrödinger wrote that his only motive for writing his book was to draw one general conclusion from his discussion of the mechanism of heredity. That general conclusion is startling: "it emerges that living matter, while not eluding the 'laws of physics' as established up to date, is likely to involve 'other laws of physics' hitherto unknown" (*WIL*, p. 66). However, Schrödinger is careful to make clear that he is not invoking vitalistic forces like the notorious "entelechy" which Driesch introduced to direct development of the embryo.[14] "The obvious inability of present day physics and chemistry to account for such events is no reason at all for doubting that they can be accounted for by those sciences" (*WIL*, p. 2). It is only because the physicist has so far not been concerned with the type of construction found in a living organism that present day laws of physics are adequate to describe living organisms.

The third feature that struck me is Schrödinger's discussion of this inadequacy of current physical laws to deal with living organisms. It is this theme that seems to underly the whole book. Schrödinger's earliest interest was in statistical dynamics[7] and in *WIL* he constantly emphasizes that the laws of physics are statistical, and they are only laws and exhibit regularity because they deal with enormous numbers of molecules. But, as Schrödinger goes on, processes in living organisms deal with *small* numbers of molecules and he takes the gene as an example. Delbrück estimated from the induction of mutations by X-rays that the size of the critical part of the gene is of the order of 1000 atoms.[13] How is it that life can be orderly when it is based on too few atoms for statistical considerations to apply? Schrödinger seems to offer two resolutions of this paradox. Firstly, given that the gene is an aperiodic crystal, the Heitler-London forces governing interatomic interactions in solids will apply to the gene and, since solids at room temperature are effectively at absolute zero, they can be treated as behaving dynamically rather than statistically.[3] Secondly, Schrödinger places great stress on the fact that organisms maintain order (and minimize their entropy) by feeding on negative entropy, making use of order already present in the environment (*WIL*, pp. 68–75). Karl Popper[8] has severely criticized this notion of "feeding on negative entropy." He pointed out that this could be said of any steam engine, oil-fired boiler or similar machine. They all may be said to be "continually sucking orderliness" from the environment (*WIL*, p. 75). Schrödinger called this the "order-from-order" principle (*WIL*, p. 86), a principle that is not "non-physical" or "super-physical" and is not alien to physics (*WIL*, pp. 81–86).

INFLUENCE OF *WIL*

I have already remarked that Schrödinger's discussion of the gene as an aperiodic crystal, and his references to a code-script and to new laws of physics had a significant influence on physicists in the post-war period. As Jacob[15] put it "Just to hear one of the leaders in quantum mechanics asking 'What is Life?' and then describing heredity in terms of molecular structures, interatomic bonds and thermodynamic stability was enough to fire the enthusiasm of certain young physicists and to bestow some sort of legitimacy on biology." Here are some examples:

> I had at about that time been deeply impressed by a little book written by the great Austrian physicist Erwin Schrödinger . . . The hereditary code-script? . . . "Chromosomes!" I exclaimed. "DNA, builder's craft! Let's work on the nose of Cleopatra!" *Chargaff*[16]

> Schrödinger's book had a very positive effect on me and got me, for the first time, interested in biological problems. *Wilkins cited in Ref. 3.*

> On those who came into the subject just after the 1939–1945 war Schrödinger's little book . . . seems to have been particularly influential . . . Schrödinger's book was very timely and attracted people who might otherwise not have entered biology at all. *Crick*[17]

> [After reading WIL] I became polarized towards finding out the secret of the gene. *Watson*[18]

Not everyone agrees that Schrödinger's theoretical discussion of the gene was important. Waddington,[19] whilst recognizing that Schrödinger's book "had a very great influence, both because of its elegance and intellectual quality" claimed that the part of *WIL* "that made the original impact is in fact a re-writing of the classical paper which we used to refer to as TZD," that is the paper by Timoféeff-Ressovsky, Zimmer and Delbrück.[13] Cohen[20] dismisses the supposed influence of *WIL.*

However, it seems to me that Olby[3] is right when he suggests that these molecular biologists who read *WIL* "found in it what they were looking for," for the discussions of gene, hereditary code-scripts and aperiodic crystals are few and are employed to illustrate the main theme of the paradox between the statistical laws of physics and organized life.

THERMODYNAMICS, ORDER AND LIFE

Schrödinger's resolution of this paradox has been discussed in detail by Olby[2,3] and Yoxen[4] from different perspectives. Yoxen shows that Schrödinger's contemporaries found the main theme of *WIL*—the apparent conflict between living organisms and the laws of thermodynamics—to be its most interesting feature. This apparent conflict was an important contemporary problem and had been recognized as such for a number of years prior to 1943. I have already referred to the essay published by Schrödinger in 1935 but it was not only physicists who were concerned with this problem. In his 1933 address to the British Association, Sir Frederick Gowland Hopkins, the eminent biochemist, used arguments strikingly similar to those of Schrödinger ten years later: "life has one attribute that is fundamental. Whenever and wherever it appears the steady increase of entropy displayed by all the rest of the universe is then and there arrested . . . every living unit is a transformer of energy."[21]

Other reviews and articles[22–24] made similar points and I think that Yoxen makes a good case that Schrödinger's book was not revolutionary but rather was a topical contribution to a contemporary problem. By the 1950s it had lost this topicality and by the 1960s the molecular biologists were looking back with nostalgia to a "historical object." At a rather different intellectual level, Schrödinger made an appearance as a professor discussing the thermodynamics of life in one[25] of George Gamow's series of *Mr Tompkins* books popularizing science!

One conclusion reached by Yoxen is that the molecular biologists "tend to make statements about the past that are heavily influenced by their present interests." These statements about the importance of *WIL* to the development of molecular genetics have obscured the contribution of *WIL* to an earlier problem of molecular biology, namely order at the molecular and cellular levels. This analysis of *WIL* appears to reconcile the contradiction between the main theme of the book and what we are led to expect of it by the pronouncements of Chargaff, Wilkins, etc. Nevertheless, to the physical scientists looking at the field of biology as outsiders, the problem of gene structure was one that caught their imagination.

CONCLUSION

Is there anything in *WIL* for the modern biochemist? I believe there is, but it is nothing technical, nothing that will help in understanding our contemporary problems. Firstly, it is a good read, written in a brief, succinct and elegant style that seems to be attained only rarely these days. Secondly, and more importantly, it is worth reading for the insight it gives of an attempt to think about the fundamental problems of biology at a time when the necessary data had yet to be obtained. Some may dismiss such an attempt as mere speculation not worthy of consideration at the time it was published, let alone now. I believe this to be quite mistaken and I shall finish with an opinion expressed by Ross G. Harrison,[26] the great experimental embryologist. Harrison was well known for his reluctance to theorize, yet he once wrote of a speculative suggestion by Needham: "This may be an advanced position to take at the present time, but then progress is made only by taking advanced positions."

REFERENCES

1. Schrödinger E. 1944. *What is Life?*, Cambridge University Press.
2. Olby R.C. 1971. *J. Hist. Biol.* **4:** 119–148.
3. Olby R.C. 1974. *The Path to the Double Helix,* Macmillan.
4. Yoxen E.J. 1979. *Hist. Sci.* **17:** 17–52.
5. Teich M. 1975. *Hist. Sci.* **13:** 264–283.
6. Judson H.F. 1979. *The Eighth Day of Creation,* Jonathan Cape.
7. Heitler W. 1961. *Biogr. Mem. Fellows R. Soc.* **7:** 221–228.
8. Popper K. 1976. *Unended Quest,* Fontana.
9. Donnan F.G. 1936. *Acta. Biotheōr.* **2:** 1–10.
10. Schrödinger E. 1936. *Nature* **138:** 13–14.
11. Schrödinger E. 1935. *Science and the Human Temperament,* Cambridge University Press.
12. Sherrington C.S. 1940. *Man on his Nature,* Cambridge University Press.
13. Timoféeff-Ressovsky N.W., Zimmer K.G., and Delbrück M. 1935. *Nachr. Ges. Wiss. Gottingen, Math.-Phys. Kl., Fachgruppe 6* **1:** 189–245.
14. Driesch H. 1907–1908. *The Science and Philosophy of the Organism,* Black.
15. Jacob F. 1974. *The Logic of Living Systems: A History of Heredity,* Allen Lane.
16. Chargaff E. 1978. *Heraclitean Fire. Sketches from a Life before Nature,* Rockefeller University Press.
17. Crick F.H.C. 1965. *Br. Med. Bull.* **21:** 183–186.
18. Watson J.D. 1966. *Phage and the Origins of Molecular Biology* (ed. J. Cairns et al.). Cold Spring Harbor Laboratory, Cold Spring Harbor, New York.
19. Waddington C.H. 1969. *Nature* **221:** 318–321.
20. Cohen S. 1975. *Science* **187:** 827–830.
21. Hopkins F.G. 1933. *Nature* **132:** 381–394.
22. Butler J.A.V. 1946. *Nature* **158:** 153–155.
23. Needham J. 1943. *Time: The Refreshing River,* Allen and Unwin.
24. Brillouin L. 1949. *Am. Sc.* **37:** 559–568.
25. Gamow G. 1953. *Mr Tompkins learns the Facts of Life,* Cambridge University Press.
26. Harrison R.G. 1936. *Collecting Net* **11:** 217–226. Reprinted in Wilens S., ed. 1969. *Organization and Development of the Embryo,* Yale University Press.

Finally, the Beginnings of Molecular Biology

SEYMOUR S. COHEN

State University of New York at Stony Brook, New York, USA

Warren Weaver coined the term "molecular biology" in 1938,[1] shortly after an extraordinary series of discoveries, which during 1934–1936 had pulled together the disciplines of structural and metabolic biochemistry, microbiology and genetics. Following the initial demonstration by J. Sumner, and later by J. Northrop and M. Kunitz, that enzymes were crystallizable protein molecules, W. Stanley, a young colleague of Northrop and Kunitz, had crystallized tobacco mosaic virus in 1935.[2] In 1936, F.C. Bawden and N.W. Pirie, in collaboration with the X-ray crystallographers J.D. Bernal and I. Fankuchen, had confirmed Stanley's result, clarified the meaning of "crystal" and discovered nucleic acid in the virus.[3] The concept of infectious pathological agents, previously familiar to the microbiologists as tiny cells, had been enlarged to include filterable entities, i.e. viruses, which now could be described as nucleoprotein molecules. Early commentators on these findings called attention to the similar reproductive capabilities of viruses and genes, and in 1937 E. Wollman added "La possibilité" d'"inoculer" des genes à des cellules ne nous semble pas pouvoir être exclue a *priori*."[4] Also, H.J. Muller praised the X-ray analyses of the virus as "bridging the gap between the structures of the chemist and those of the microbiologist and geneticist."[5]

In 1936, A. Mirsky, who had studied the structure of hemoglobin for a decade in collaboration with M. Anson, had decided to see what the crystallographers and physical chemists at Cal Tech were thinking about proteins. Mirsky and Anson had worked together in the laboratories of Barcroft at Cambridge, Henderson at Harvard and at the Rockefeller Institute. They had

Reprinted from *Trends in Biochemical Sciences* February 1986, 11(2): 92–93.

not only discovered the renaturation of denatured hemoglobin, but in 1934 had described the equilibrium between active native trypsin and inactive denatured trypsin. Mirsky's sabbatical year led to a classical paper, written with L. Pauling, formulating a general theory of protein structure and of folding and unfolding.[6] Meanwhile, Anson used the time to work on carboxypeptidase, which he had just crystallized in Northrop's laboratory.

John Northrop,[7] possessing a doctorate in chemistry, had joined Jacques Loeb at the Rockefeller Institute in 1915. Loeb, who had headed the Laboratory of General Physiology at that Institute since 1910, believed that chemistry and physics would explain many biological phenomena. In 1913, Loeb had hired Moses Kunitz as a research assistant and Kunitz became the skilled executor of Loeb's experiments on the binding of ions to proteins. Kunitz, who at the age of 22 had fled the Czar's police in 1909, attended college and graduate school at night until he earned his PhD in biochemistry in 1924, the year of Loeb's death. Northrop and Kunitz worked together on the physical chemistry of gelatin for many years, both in New York and in the Princeton branch of the Institute, founded as the Department of Animal Pathology. Northrop began his classic work on pepsin and crystallized the enzyme in 1929. Kunitz started his studies on pancreatic proteinases and in 1933 he and Northrop briefly described the isolation and crystallization of the inactive precursor of chymotrypsin. The isolation of crystalline chymotrypsinogen, chymotrypsin, trypsinogen, trypsin, trypsin inhibitor, and an inhibitor-trypsin complex were reported in detail in 1935[8] and 1936.[9] In later years Kunitz went on to crystallize ribonuclease, deoxyribonuclease, hexokinase, inorganic pyrophosphatase and a few other proteins. As we know, the pure nucleases proved to be essential in the demonstration of the genetic roles of the nucleic acids.

Wendell Stanley had graduated in 1926 from a Quaker college as a science major and football star, and seeking a job as a football coach had visited the University of Illinois, where he fell under the spell of the organic chemist, Roger Adams. Stanley also worked with his future wife and their joint paper with Adams in the *Journal of the American Chemical Society* in 1929 opened facing the summary of a paper of J.B. Sumner and D.B. Hand, describing the estimation of the isoelectric point of a protein as the pH of minimal solubility in salt solution.[10] Stanley never alluded to this paper, which presents a method he subsequently used to crystallize the virus, but we can ask if a memory of the procedure and principle was recalled some five years later. Stanley spent part of 1929 and 1930 with H. Wieland in Munich, and returned in 1930 to work at the Rockefeller Institute with W.J.B. Osterhout on ion transport in the giant plant cell, *Valonia.*

In 1932 the Princeton branch of the Institute became the Department of Animal and Plant Pathology, enlarging its activity by the recruitment of L.O. Kunkel to organize the new plant studies. Kunkel, who wished to develop work on the plant viruses, hired Stanley in that year to extend studies on virus isolation, pointing to promising results of C.G. Vinson and A.W. Petre on the separation and concentration of tobacco mosaic virus by protein precipitants. During 1933–1934, using a relatively new local lesion method, Stanley worked furiously to test the infectivity of fresh or partially purified extracts exposed to more than 100 reagents. The successes and methods of the group led by Northrop and Kunitz in the same Institute were a continuing source of encouragement in this period. In 1935 Stanley reported in *Science* that virus, precipitated with salt at low pH, could be further purified by precipitation of impurities with lead acetate at high pH. Virus was then absorbed on asbestos at low pH and eluted at high pH. The infectious agent was crystallized as small "needles" by the addition of acetic acid to a defined low pH in the presence of 20% saturated ammonium sulfate.[11] In 1936, the initial brief report was enlarged to present improvements in the isolation procedures and some analytical details.[12]

Stanley's report challenged the conventional wisdom on the form and composition of organisms possessing properties of inheritable duplication. In an important essay, N.W. Pirie noted the lack of agreement on the meaning of the word "living" and suggested that "it seems prudent to avoid the use of the word 'life' in any discussion about border-line systems." This warning was influential for several decades, until the inception of space programs and the search for "life" on Mars.

Stanley's report also stimulated intensive activity on the part of the English duo, F.C. Bawden, a plant pathologist and Pirie, his biochemical collaborator. In 1930, young Bawden became a research assistant of R.M. Salaman, head of the Potato Virus Research Station at Cambridge. In 1934, while working on problems of virus detection by serological methods, he met Pirie, who had been characterizing fractions of a bacterial pathogen, *Brucella melitensis.* The two men attempted the isolation of a potato virus proving particularly difficult to purify and in 1936 reported that the infectivity was sensitive to proteinases.

In that year also Bawden moved to the Rothamsted Experimental Station and this permitted him to explore viruses infecting plants other than potatoes. However, Bawden's interest in potatoes lasted throughout his career and he helped to develop work subsequently important in agricultural practice. By the end of 1936 Bawden and Pirie, working in Cambndge, were able to confirm, correct and extend Stanley's report in a joint paper in *Nature* with Bernel and Fankuchen.[13]

Although the infectious virus could be isolated as Stanley had described, it became clear in the *Nature* report that the "needles" did not possess three-dimensional regularity. The purified virus particles were relatively stiff rods of constant diameter, which were aligned during flow, and packed in hexagonal arrays to form needle-like "liquid" or "para" crystals. These tactoids were later analysed in great detail by Bernal and Fankuchen. The English workers also noted the spontaneous separation of purified concentrated suspensions into two phases, of which the bottom phase was spontaneously birefringent in polarized light, i.e. had crystallized in arrays seen in the smaller "needles."

Stanley's initial report in 1935 described a product containing 20% nitrogen but the first complete paper, in 1936 (accepted in November 1935), reported nitrogen contents obtained by two methods of 16.1–16.6%, values consistent with those of proteins and most nucleoproteins. Although Bawden and Pirie have each questioned the identity of Stanley's initial product, the high nitrogen content was corrected by the end of 1935, before the first paper by the English group (submitted 17 November 1936, and published on 19 December 1936). Furthermore, Stanley's material was analysed by an American crystallographic group, and the data published in November 1936 were described by Bawden et al. in a note added on 2 December 1936: "Wyckoff and Corey have published an X-ray study of the ammonium sulphate crystals of tobacco mosaic and aucuba protein. Their measurements of the intermolecular spacings obtained with unorientated material agree with ours, notably the lines they record at 11.0, 7.44, 5.44 and 3.7Å correspond to our measurement of the planes (0006, 9, 12, 18) respectively." Thus, the material obtained by Bawden and Pine and analysed by Bernal and Fankuchen was identical to an early preparation of the virus obtained by Stanley.

By 1936 Stanley had not found phosphorus or sulfur in the virus. The English paper records the discovery of 5% RNA in this virus, a finding they soon extended to other plant viruses. Also in 1936, M. Schlesinger had found materials reacting like DNA in bacterial viruses.[14] At first Stanley was unwilling to accept the result concerning the presence of phosphorus and RNA, which had been communicated to him by Pirie early in 1936. Stanley thought the RNA was a disposable contaminant until 1937–1938, when a collaborator, H.S. Loring, found phosphorus and RNA in organic combination with protein in several viruses. In 1939 Ross and Stanley included cysteine and sulfur among virus components.

In the next few years, until World War II altered research priorities and directions in England and the United States, many questions posed by these initial studies were explored. In this period Bawden and Pirie made many new significant observations on the different kinds of plant viruses. Until 1942 Stanley's group enlarged, to answer crucial questions of the size and composition of the virus particles and the division of labor of its components. In 1946 Sumner, Northrop and Stanley received the Nobel Prize jointly for their contributions to the crystallization of proteins, enzymes and viruses. Their work had not only demonstrated the validity of Loeb's belief in the applicability of chemistry and physics in the study of biological problems, but had also established model systems for the future study of protein structure, enzyme mechanism and genetic elements. The studies during 1934–1936 particularly were an initial step in the cross-fertilization of the specializations, as described by Muller. This has led to the synthesis of "biochemistry, genetics and structural chemistry in the pursuit of the molecular basis of the form, function and evolutionary descent of living things," the super-discipline commonly known as "molecular biology."

REFERENCES

1. Cohen S. 1984. *Trends Biochem. Sci.* **9:** 334–336.
2. Tooze J. 1979. *Trends Biochem. Sci.* **4:** 96.
3. Pirie N.W. 1984. *Trends Biochem. Sci.* **9:** 35–37.
4. Wollman E. 1937. *Bull. Inst. Pasteur* **35:** 1–10.
5. Muller H.J. 1939. *Nature* **144:** 813–816.
6. Mirsky A.E. and Pauling L. 1936. *Proc. Natl. Acad. Sci. USA* **22:** 439–447.
7. Herriott R.M. 1983. *Trends Biochem. Sci.* **8:** 296–297.
8. Kunitz M. and Northrop J.H. 1935. *J. Gen. Physiol.* **18:** 433–458.
9. Kunitz M. and Northrop J.H. 1936. *J. Gen. Physiol.* **19:** 991–1007.
10. Stanley W.M., Jay M.S., and Adams R. 1929. *J. Am. Chem. Soc.* **51:** 1261–1266.
11. Stanley W.M. 1935. *Science* **81:** 644–645.
12. Stanley W.M. 1936. *Phytopathology* **26:** 305–320.
13. Bawden F.C., Pirie N.W., Bernal J.D., and Fankuchen I. 1936. *Nature (London)* **138:** 1051–1052.
14. Schlesinger M. 1936. *Nature (London)* **138:** 508.

Biochemical Origins of Molecular Biology: A Discussion

ROBERT OLBY

Leeds University, UK

At the initiative of Professor Seymour Cohen a series of essays on the above subject appeared in *TIBS* in August–November of 1984. Now, two decades after the first signs emerged of disagreement and conflict between biochemists and molecular biologists, it is fitting to discuss the essays of Cohen, Lauffer, Putnam and Zamecnik and to ponder the motivation behind such contributions.

The aims of the series were stated explicitly: to "enlarge the limited historiography that has essentially excluded biochemical discovery from the stories of the founding of molecular biology." Taking McCormmach's precept against judging "the past from a superior present," Cohen's aim was "to provide a more complete sense of the common knowledge within which it became reasonable to focus on the nucleic acids and on DNA particularly as the genetic material, and on RNA and ribosomes as the intermediates in the synthesis of specific proteins."[1]

In these essays we are reminded that long before the Phage Group was formed in 1945 to study the process of the replication of bacterial virus in the infected bacterial cell, biochemists had made an intensive study of plant viruses. They had crystallized the infective agents, studied their constitution and shape, and investigated the effects of their chemical modification. The impact of these studies was immense; it brought about an interdisciplinary and international effort to establish the structure of the virus particle and the biological functions of the nucleic acid and protein moieties. When it is claimed with confidence that such interdisciplinary research only began in

Reprinted from *Trends in Biochemical Sciences* July 1986, 11(7): 303–305.

the 1950s (as had Francois Jacob[2]) we should spare a thought for the inter-war years when X-ray crystallographers, plant pathologists, biochemists and biophysicists collaborated in the attack on tobacco mosaic virus (TMV).

Professor Cohen has rightly pointed out the relevance of these TMV studies to the development of biochemistry, for just as Buchner's discovery of extracellular fermentation of sugar was recognized as a significant achievement upon which could be built a broad research programme into the enzyme control of metabolism, so Stanley's preparation of paracrystalline TMV possessing intensive infectivity formed the starting point for a programme of biochemical and biophysical research into these ultramicroscopic infective particles. Cohen's insight here helps to place the plant virus work in an adequate historical context. The study of TMV was for the 1930s what the study of the transforming principle was for the 1940s. Had not Astbury declared in 1939: "To the molecular biologist there can be no question but that the most thrilling discovery of the century is that of the nature of the tobacco mosaic virus: it is but a nucleoprotein."[3]

Another area where biochemists made crucial contributions was the study of protein synthesis. From the early studies of the 1930s through to the 1960s there has been continuous effort to penetrate the mystery of this remarkable process. Thus enzyme synthesis in *E. coli* following upon the presence of galactose substrate has been studied first as "enzyme adaptation" and subsequently as "enzyme induction." Equally, attempts to simulate protein synthesis starting with the reversal of catabolism and proceeding to amino acid incorporation in a tissue slice and eventually in the cell-free system, have stretched over several decades. Here again, as Professor Zamecnik shows in his informative essay, biochemists were to the fore, as one would expect of a process which is not explained merely by cracking the code. Its study, explained Zamecnik, "was rooted in the simple biochemical desire to understand how the energy barrier from free amino acid to peptide was overcome."[4]

To draw attention to these important areas of research and to underline the role of the biochemists in them is a worthy task, but inevitably with the merits of brevity come temptation to partisan and partial reconstruction. Among the important biochemical insights which were brought to bear upon the problems of viral replication and protein synthesis only those have been exhibited which subsequent research has confirmed. True, Zamecnik mentions the Bergmann attempt to show that proteolytic enzymes can synthesize as well as break down proteins. But he does not tell us that some biochemists hoped to account for the specific amino acid sequences in a protein in terms only of the specificity of the synthesizing enzymes involved. Nor does he tell us that such a mechanism, dubbed the "multi-enzyme theory" by Ditta Bartels,[5] continued to be promoted by biochemists as late as June 1953, when Campbell and Worth published their discussion on the "Biosynthesis of Proteins." They identified two streams of thought on the subject "one derived from the study of isolated enzyme systems and suggesting a stepwise coupling of many small peptide units; the other based on the study of genetic inheritance of protein specificity and preferring synthesis on templates, each template being specific for a single protein structure and probably identifiable with a gene." After pointing out the merits of the template theory they gave their objections and explained that "the conception of the gene is essentially an abstract idea and it may be a mistake to try to clothe this idea in a coat of nucleic acid or protein."[6]

As for the biochemists' study of viral replication, neither Cohen nor Lauffer tells the reader how biochemical insights were *first* applied to this subject. Yet as prestigious a biochemist as Northrop regarded phage particles as "autocatalytic enzymes." He believed he could achieve for phage what Buchner had done for zymase—strip it of its vitalistic associations. Host cells synthesized a "normal" inactive protein. "When the active virus or bacterophage is added, this inactive protein or 'prophage' is transformed by an autocatalytic reaction into more active phage," just as pepsinogen and trypsinogen are converted into pepsin and trypsin with the presence of the active form.[7] Nor was Northrop the only biochemist to support such an explanation. Krueger claimed to have shown that the growth curves for phage replication followed those for enzyme synthesis,[8] and Stanley had considered viruses as protein molecules only slightly more complex than enzymes.[9]

What is important about both the multi-enzyme theory of protein synthesis and the theory of viral autocatalytic enzymes is not whether they have proved to be right or wrong (it is doubtful that scientific theories can ever be unambiguously judged in this way) but whether they have stimulated empirical research. This they both achieved. For the historian they have the added merit of providing clues to the *thinking of* biochemists at that time. To dig them up for exhibition long after their decent burial is not in order to judge the biochemists who supported them, or to taunt the profession for entertaining them, but to recapture the conceptual context in which biochemists approached the study of these subjects.

After reading the series of articles on the biochemical origins of molecular biology one is made aware that behind them lurks a sense of aggrievement that biochemistry and biochemists have been unjustly treated, and that the tendency of such authors as Stent and Watson to undervalue their contributions when presenting the popular history of molecular biology has been all too effective. Do biochemists have just grounds for complaint? Surely they do. There existed a tradition of "molecular biology" in the 1930s dominated perhaps by biophysics, virology and genetics, but to which biochemistry made important contributions. This tradition assumed both a "structural continuity" from molecules to cells and also a clear causal relationship between structure and function. Viral particles, too, could be fitted into this continuum of structures, so that Stanley believed it was possible "to blend atomic theory, the germ theory, and the cell theory into a unified philosophy, the essence of which is structure or architecture." Astbury, to whom we owe possibly the best definition of molecular biology, had close links with biochemists; he collaborated with Kenneth Bailey and maintained an active correspondence with and frequently visited Chibnall. Although the proteins initially reigned supreme, the nucleoproteins later gained the ascendant. For the biochemists, the problems of protein structure were not confined to the chemical pathways of synthesis and breakdown, but concerned also the chemical basis to biological specificity. Because their study of the former was more successful than their study of the latter, it does not follow that biochemists ignored specificity, and as Cohen's own pioneer study of DNA synthesis in phage beautifully illustrates, the study of synthesis can throw light on the *source of* specificity.

If biochemists felt that they were part of the interdisciplinary activity of molecular biology in the inter-war years, what caused them to be neglected in the 1960s? Was it the publication of the *Journal of Molecular Biology*? Surely not, for this periodical did not undermine existing biochemical journals. What, then, caused the concern? We can suggest three events. One was the appointment of a Working Group on Molecular Biology by the UK Government's Advisory Council for Scientific Policy. This Group, chaired by John Kendrew, published its report in 1968.[10] The second event was the publication of the *Festschrift* to Max Delbrück by the Cold Spring Harbor Laboratory in 1966. This work, *Phage and the Origins of Molecular Biology,* became available in the UK in 1967.[11] The third event was the publication of J.D. Watson's *The Double Helix.*[12]

The publication of the Kendrew Report provoked a response from the Biochemical Society which set up a subcommittee chaired by Hans Krebs. Its report, published in 1969,[13] supported many of the recommendations of the Kendrew Report, but on the understanding that molecular biology was the legitimate territory of biochemistry. Clearly the Krebs Report offers evidence that biochemists saw the case made in the Kendrew Report for the support of molecular biology as a threat to the institutional development of biochemistry, a perception which is more explicit in the correspondence between the biochemists involved.

The publication of the Delbrück *Festschrift* caused Kendrew to conclude that molecular biology derived from two traditions—one, the Informational School of the Phage Group described in the book, and the other the Structural School tracing back to Astbury and omitted.[14] Biochemists did not respond directly to the book, but in 1968 Cohen criticized its claims in his *Virus Induced Enzymes.*[15] The third event, the publication of *The Double Helix* by J.D. Watson gave Professor Chargaff the opportunity to give his informed judgement not only of literary style and genre but also of the "new" science of molecular biology.[16] The tone of this amusing review marked the more extreme response of biochemists to the image of molecular biology which now began to receive the attention of the general public.

Yet what are disciplines? Are they not, as Robert Kohler claimed "political institutions that demarcate areas of academic territory, allocate privileges and responsibilities of expertise, and structure claims on resources?"[17] Hopkins had identified biochemistry with "the study of metabolism,"[18] but F.G. Young in 1946 called it "the study at the molecular and atomic level of the organization and function of biological systems,"[19] and Neuberger in 1960 confessed that having considered himself a biochemist for all of his working life, he still did not find it easy "to arrive at a very satisfactory definition," but he would call a biochemist: "a man who brings chemical techniques to bear on biological problems . . .." What he found so attractive about biochemistry was "its vagueness." It touched on so many subjects.[20] As for molecular biology, Kendrew admitted that this fashionable term was unfortunate, and in the Kendrew Report it was deliberately dropped and the phrase "Biology at the molecular level" adopted instead.

If the treatment of biochemistry has been unjust, at least it has served to stimulate those most directly involved to offer their perceptions of the development of the field. These participants' accounts have enriched the source material of the history of science, and if they appear to some degree motivated by partisan considerations, at least they may offset the partisan approach originating from an anti-biochemical persuasion, and contribute to an appreciation of molecular biology before it entered its popular phase in the late 1950s.

REFERENCES

1. Cohen S.S. 1984. *Trends Biochem. Sci.* **9:** 334–336.
2. Jacob F. 1970. *La Logique du vivant,* p. 269, Gallimard.
3. Astbury W.T. 1939. *Int. Conf. Genet.* **7:** 51.
4. Zamecnik P. 1984. *Trends Biochem. Sci.* **9:** 466.
5. Bartels D. 1983. *History and Philosophy of the Life Sciences* **5:** no. 2.
6. Campbell P.N. and Worth T.S. 1953. *Nature* **171:** 1000.
7. Northrop J. 1938. *J. Gen. Physiol.* **21:** 361.
8. Krueger A.P. 1937. *Science* **86:** 279–380.
9. Stanley O.W. 1935. *Science* **81:** 645; 1939. *Physiol. Rev.* **19:** 550.
10. Report of the Working Group on Molecular Biology. 1968. Cmnd 3675.
11. Cairns J., Stent G.S., and Watson J.D., eds. 1966. *Phage and the Origins of Molecular Biology,* Cold Spring Harbor Laboratory.
12. Watson J.D. 1968. *The Double Helix,* Atheneum.
13. Report of the Subcommittee of the Biochemical Society. 1969. *Biochemistry, "Molecular Biology" and Biological Sciences,* Biochemical Society.
14. Kendrew J.C. 1967. *Sci. Am.* **216:** 3, 141.
15. Cohen S.S. 1968. *Virus Induced Enzymes,* New York, p. 25, Columbia University Press.
16. Chargaff E. 1968. *Science* **159:** 1448–1449.
17. Kohler R.E. 1982. *From Medical Chemistry to Biochemistry,* p. 1, Cambridge University Press.
18. Hopkins F.G. 1914. *Annu. Rep. Progr. Chem.* **11:** 188–212.
19. Young F.G. 1946. Definition submitted to the Board of Studies in Biochemistry London University.
20. Neuberger A. 1960. *in The Organization and Financing of Research in Biochemistry and Allied Sciences in Great Britain* (ed. K.S. Dodgson), pp. 50–51, Biochemical Society.

The Biological Role of Nucleic Acids

PETER KARLSON

Philipps University,
Marburg (Lahn), FRG

One hundred years ago, Albrecht Kossel published an article[1] with the ambitious title "Zur Chemie des Zellkerns." He described a method for the quantitative determination of nucleic acid phosphorus in tissues, starting with 15 g of fresh weight tissue. Proteins were denatured with acid and tannin, phospholipids were extracted with ethanol and ether, and phosphoric acid determined in the ash by contemporary gravimetric methods. The progress in our methodology is evident—by the 1950s, Erwin Chargaff and his co-workers needed only micrograms for a phosphorus determination and a full analysis of the base ratios of a sample of nucleic acid; today, microgramms [sic] or perhaps nanograms suffice for a complete sequence analysis of a stretch of over a thousand base pairs.

On the basis of his analyses, Kossel compared the ratio nuclein phosphorus to total phosphorus in various tissues and found a positive correlation to the number of cell nuclei:

> The data are not very numerous at present, but they justify the anticipation that through the chemical analysis of nuclein substances a method will be available, that will allow us to give an equivalent of the mass of the cell nucleus in weight instead of an estimate according to its volume.*

In the next paragraph, he continues:

> It would be very valuable to recognize a relation of the nucleins to certain physiological functions, since this will allow us to draw conclusions on the function of the cell nucleus.†

*Das vorliegende Material ist freilich noch wenig reichhaltig, aber es berechtigt zu der Hoffnung, dass durch die chemischen Untersuchungen über die Nucleinsubstanzen das Prinzip einer Methode gegeben ist, welche gestattet, für die bisher nur dem Raume nach geschätzte Masse des Zellkerns einen Ausdruck in Gewichtszahlen zu erhalten.

†Eine Beziehung der Nucleine zu bestimmten physiologischen Funktionen zu erkennen, wäre umso wertvoller, da damit zugleich ein Licht auf die Funktion des Zellkerns geworfen wäre.

Reprinted from *Trends in Biochemical Sciences* August 1982, 7(8): 302–304.

He dismisses the idea that "nuclein" (a term denoting nucleoprotein) is a storage product on the basis of feeding experiments with chicken and pigeons, and continues:

> Morphological observations make it likely that we have to look in another direction for the physiological function of nuclein, namely in relation to the formation of new tissue.‡

As evidence he quotes observations of a famous botanist, Julius Sachs, on the size and density of nuclei in growing plants, and his own analyses of "nuclein" in embryonic and full-grown muscle.

Several aspects of this early paper are remarkable. In spite of the time-consuming and not very accurate analytical methods, Kossel relied very much on analytical figures, and he expressed the hope that the morphological estimation of the mass of the cell nuclei would be replaced, in due course, by the chemical determination of nucleic acids. This is exactly the practice in modern times. Moreover, he looked at the function of the cell nucleus, hoping to find a solution from the physiological chemistry. And he proceeds in this direction with the analysis of embryonic and adult muscle tissue.

Finally, the ease with which Kossel quoted evidence from the plant kingdom is exceptional, since at that time physiological chemistry was mostly confined to higher animals and man. This is presumably due to the fact that he was quite interested in botany, to the extent that he originally wanted to study botany instead of medicine.

Of course, it was premature to really hope for the clarification of the function of the cell nucleus by the methods then available and this remained so for another half a century, until Caspersson started his cytochemical analysis (see below).

In the years to come, Kossel and his co-workers contributed much to the chemistry of "nuclein" (= nucleoprotein). As early as 1884, he described a basic protein in the nuclei of the red blood cells of birds to which he assigned the name "histone."[2] Together with Steudel, he also isolated the bases of nucleic acid and elucidated their chemical structure. This work was aided by the simultaneous synthetic studies of Emil Fischer, which provided reference material. Further quantitative analyses of hydrolysates led Steudel[3] to suggest a formula for a copper salt of DNA, namely, representing a tetranucleotide.

‡Morphologische Befunde weisen bereits darauf hin, dass wir die physiologischen Funktionen des Nucleins wahrscheinlich in einer anderen Richtung zu suchen haben, nämlich in einer Beziehung zur Neubildung der Gewebe.

This was the birth of the tetranucleotide "dogma" that was to last about three decades. Thus, the molecule of nucleic acid appeared a fairly simple structure, and Steudel[4] wrote:

> . . . it cannot be seen at present how such a simple compound, the nucleic acid, that the synthetic chemist will certainly synthesize in the near future, should bear only a part of the fundamental functions of life of the spermatozoon.

This was certainly a step backwards. The tetranucleotide left no room for individuality of this constituent, in contrast to the proteins.

The elucidation of the chemical structure of nucleic acids is mainly due to the work of P. A. Levene between 1900 and 1930.[5] However, this is not the place for a full history of nucleic acid research[6]; suffice it to say that with many others, Levene strongly supported the tetranucleotide theory. For a long time, tetranucleotides were believed to be the units which, under certain conditions, could aggregate to form colloids (it was the dark age of biocolloidology, as Florkin[7] put it). Later, when the macromolecular nature of nucleic acids became evident, the tetranucleotide unit was still believed to be a kind of building block—a repeating unit, such as that seen in polysaccharides. Indeed, the analogy between carbohydrate-containing nucleic acids and structures like hyaluronic acid or the chondroitin sulfates seemed obvious.

Fifty years ago, around 1932, Caspersson re-discovered the strong u.v.-absorption of nucleic acids, and used this property in his cytophotochemical studies of the cell nucleus. For this purpose, he used a u.v. microscope designed by A. Köhler at Carl Zeiss (Jena) as early as 1904. The quantitative measurements were made first with a photographic and later with a photoelectric method. Fields as small as 0.1 μm in diameter could be measured with an accuracy of 1%. Of course, it took a tremendous number of single measurements for only a single cell, but the results were remarkable. The first comprehensive paper appeared in 1936.[8] Nucleic acids and proteins were measured in the nucleus *in vivo,* and changes in the nucleic acid content were seen during mitosis. Moreover, the giant chromosomes of *Drosophila* and other dipterans had just been (re)discovered in 1933. They furnished excellent biological material for studies of the kind Caspersson had in mind. In a review published in 1941,[9] Caspersson gave a diagram of the chemical composition of a chromosome that was often reproduced in textbooks, and is reproduced here (Fig. 1). The establishment of such a chart for the chromosomes was certainly a remarkable achievement, but Caspersson went even further and turned to the dynamics of protein and nucleic acid synthesis.

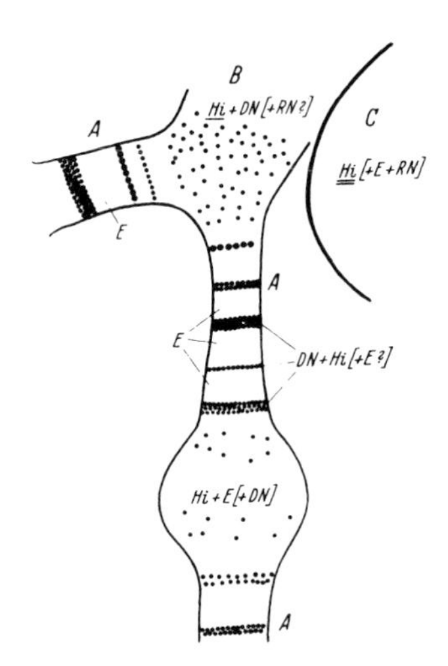

FIGURE 1. Scheme for the chemical composition of salivary gland nuclei. A Euchromatin, B heterochromalin, C Nucleolus, DN deoxyribose nucleotides, RN ribose nucleotides, E. Eiweiss (protein) of globulin absorption type, Hi protein of histone absorption type. Underlining designates large quantities, parenthesis insignificant quantities. (From Caspersson, Ref. 9.)

Studying the cytoplasm of pancreatic cells he found a high concentration of RNA in the ergastoplasm, a feature he related to the synthesis of protein:

> Thus, a connexion between the occurrence of ribose nucleotides in high concentration in the cytoplasm and the production of proteins at a high rate is very likely.

In a paper with Jack Schultz,[10] a relation between chromosomal proteins and nucleic acid synthesis was postulated. Investigating a special "variegated" race of *Drosophila,* with translocations of material from the Y-chromosome, they found higher nucleic acid contents and concluded:

> (These results) . . . suggest that the apparent genetically inert character of these (heterochromatic) regions is merely a consequence of their specialization in performing a function also performed by all other genes—namely, the synthesis of nucleic acid.

They then elaborate on the fact that other self-reproducing units—viruses and bacteriophages—also contain nucleic acid, and continue:

> It seems hence that the unique structure conditioning activity and self-reproduction, possibly by successive polymerization and depolymerization, may

> depend on the nucleic acid portion of the molecule. It may be that the property of a protein which allows it to reproduce itself is its ability to synthesize nucleic acid.

Indeed, Caspersson was very close to the discovery that the chemical nature of the genetic material is nucleic acid. What was needed, in retrospect, was a straightforward interpretation of his brilliant experimental results and measurements. But instead, he relied on the old experts, mainly on Einar Hammarsten, who had worked on nucleic acids for nearly two decades. Though Hammarsten had isolated high molecular, nearly undegraded DNA in form of "white fibers," he was also a colloidologist, and attributed the low osmotic pressure he had measured not to a high molecular mass, but to colloidal interactions with ions. For him, and also for Caspersson who worked for several years with Hammarsten, DNA was an aggregate of tetranucleotides:

> The deoxyribopolynucleotides are distinguished from the ribonucleotides by their strong tendency to polymerize into very long, uniform chains of molecules, which gives them an outstanding precondition to serve as support for cellular structures—properties that have already been postulated by E. Hammersten in 1924.[§]

Moreover, the view that genes had to be proteins was shared by all those who believed in genes as chemical entities, and Caspersson's following statement could well have been given by other authors around 1938:

> If genes are conceived as chemical substances, only one class of compounds need be given to which they can be reckoned as belonging, and that is the proteins in the wider sense, on account of the inexhaustible possibilities of variation which they offer. (Caspersson[8])

For these two reasons, Caspersson turned the argument he could have derived from this studies[10] upside down and speculated that the protein synthesized nucleic acid as a kind of support substance.

It is well known that the work of Avery et al. in 1944[11] finally gave the clue that nucleic acids were the true genetic material. From then on, it needed less than a decade until the double helix model came to light.

[§]Die Desoxyribosepolynucleotide zeichnen sich vor den Ribosenucleotiden durch ihre sehr starke Tendenz zur Polymerisation in besonders lange, streng einheitlich gebaute Molekülketten aus, was ihnen theoretisch ganz aussergewöhnliche Voraussetzungen dazu verleiht, die Unterlage für Zellstrukturen darzustellen—Eigenschaften, welche E. Hammersten bereits 1924 . . . postuliert hatte. (Caspersson[9])

REFERENCES

1. Kossel A. 1882. *Z. Physiol. Chem.* **7:** 7–22.
2. Kossel A. 1884. *Z. Physiol. Chem.* **8:** 511–515.
3. Steudel H. 1906. *Z. Physiol. Chem.* **49:** 406–409.
4. Steudel H. 1907. *Z. Physiol. Chem.* **53:** 14–18.
5. Levene P.A. and Bass L.W. 1931. *Nucleic Acids,* New York.
6. Olby R. 1974. *The Path to the Double Helix,* MacMillan, London.
7. Florkin M. 1972. In *Comprehensive Biochemistry* (ed. M. Florkin and E.H. Stolz), vol. 30, pp. 279–294.
8. Caspersson T. 1936. *Scand. Arch. Physiol.* **73:** Suppl.
9. Caspersson T. 1941. *Naturwiss.* **29:** 33–43.
10. Caspersson T. and Schultz J. 1938. *Nature (London)* **142:** 294–295.
11. Avery O.T., MacLeod C.M., and McCarty M. 1944. *J. Exp. Med.* **79:** 137–158.

Did a Tragic Accident Delay the Discovery of the Double Helical Structure of DNA?

KEITH L. MANCHESTER

University of the Witswatersrand, Johannesburg, South Africa

In their celebrated paper of 1953, Watson and Crick[1] described the structure of DNA as a double helix formed between two strands of DNA. Key to this model, in contrast to that proposed by Pauling and Corey,[2] was that the phosphates of the nucleotides faced outwards and were accessible to solvent, while the bases of the two strands faced inwards and hydrogen bonded with each other.

In the same issue of *Nature,* two more detailed papers appeared, one by Wilkins, Stokes and Wilson[3] and the other by Franklin and Gosling,[4] which set out the supporting crystallographic data. One of the references in the paper by Franklin and Gosling was to an article by Gulland and Jordan in the 1947 issue of the Cold Spring Harbor Symposium on Quantitative Biology. Although the precise citation is incorrect, Franklin and Gosling[4] credit Gulland and his colleagues with having shown that in aqueous solution the carbonyl and amino groups of the bases in DNA are inaccessible, whereas the phosphate groups are exposed, a finding that was to prove of crucial significance. Who was Gulland, and how near was he in 1947 to the discovery of the double helix?

INVESTIGATIONS AND INVESTIGATORS, 1938–1947

In 1947, John Masson Gulland held the Sir Jesse Boot Chair of Chemistry at University College, Nottingham, a position to which he had been appointed

Reprinted from *Trends in Biochemical Sciences* March 1995, 20(3): 126–128.

in 1936. Born in 1898, he was a graduate of Edinburgh and subsequently had worked with Sir Robert Robinson at St Andrews and Manchester, then with W.H. Perkin at Oxford. There he taught chemistry to biological and medical students and therefore became associated with Sir Rudolph Peters. From Oxford he moved to London to the Lister Institute of Preventive Medicine before his appointment to Nottingham. His work on nucleoprotein led to his election in 1945 to a Fellowship of the Royal Society. He was undoubtedly a distinguished chemist/biochemist, and by 1947 his bibliography ran to over 90 publications.

Gulland's views in 1947 on the structure of DNA are contained in the paper he presented at the Cold Spring Harbor Symposium.[5] Electrometric (pH) titration, which was a technique that had played an important role in the previous 30 years in nucleic acid analysis, revealed that the amount of secondary phosphoryl dissociation, if present, was extremely small, a finding that suggested an unbranched polynucleotide, as originally proposed[6] by Levene and Simms in 1926. When calf-thymus DNA that had been carefully prepared under conditions in which the pH did not vary significantly from 7 was titrated with acid or alkali, "No groups were titrated at first between pH 5 and pH 11 but outside those limits there occurred a rapid liberation of groups titrating in the ranges pH 2–6 and pH 9–12 respectively. On back-titration either with acid from pH 12 or with alkali from pH 2.5, a curve was obtained which was different from that representing the initial titration. Subsequent titration of the solution followed the second or back-titration curve, which is thus the true titration curve of the polynucleotide. It was concluded that the groups titrated between pH 8 and pH 12 were the hydroxyl groups of guanine and thymine, and that those titrating in the range pH 2.5–6.3 were the amino groups of guanine, adenine and cytosine." The simplest explanation of the effect described was that in the original DNA a linkage existed between hydroxyl and amino groups in the form of hydrogen bonding. The elegance of these results (Fig. 1), obtained with such simple equipment, is impressive.

Gulland further concluded that "For steric reasons this hydrogen bonding cannot exist between the amino and hydroxyl groups of the same nitrogenous radical, but must occur between nucleotides of the same or adjacent covalent chains." Studies of the viscosities of solutions of DNA[9] showed them to be strongly birefringent and extremely viscous. "The initial high viscosity of the solution at pH 7 was fully maintained as the pH varied between 5.6 and 10.9, but in solutions more acid than pH 5.6 or more alkaline than

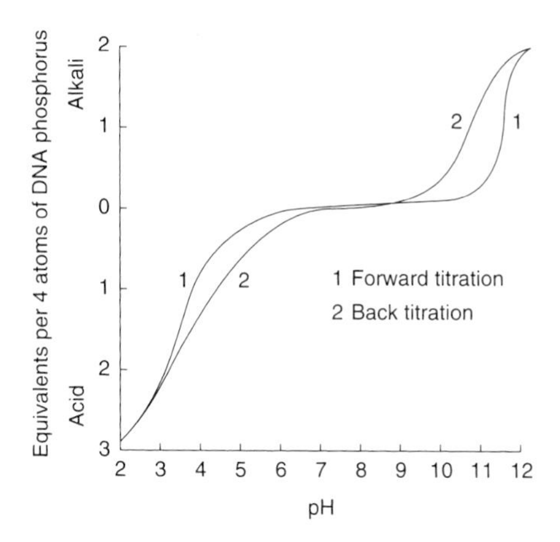

FIGURE 1. Dissociation curve of desoxypentose nucleic acid of calf thymus. (Reproduced from Ref. 7. This is a simplification of data presented in Ref. 8.)

pH 10.9, the viscosity fell sharply to a very low value. These critical values of pH coincided with those observed for the liberation of acid and basic groups in the titration." Figure 2 should be compared with Fig. 1. The fall in viscosity and streaming birefringence was obviously closely associated with the

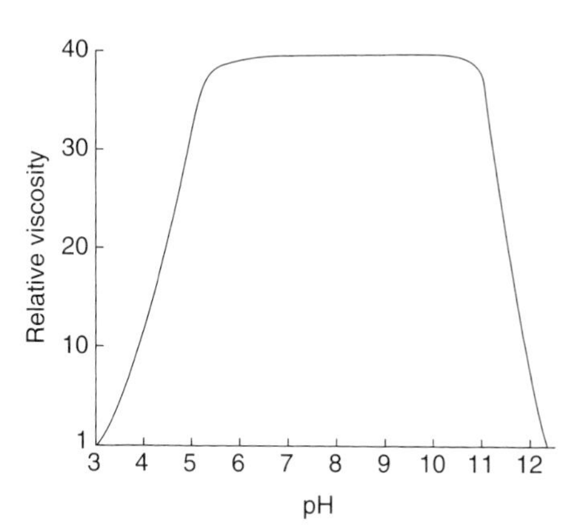

FIGURE 2. Viscosity of thymus desoxypentose nucleic acid solutions. (Reproduced from Ref. 7. This is a simplification of data presented in Ref. 9.)

fission of hydrogen bonds, and it was thus evident that at least a large proportion of the hydrogen bonds united nucleotides in neighbouring polynucleotide chains.

At around the same time, the crystallographer William Thomas Astbury was also searching for the structure of nucleic acids, using a different approach. Gulland would, of course, have been familiar with the work of Astbury, so it is necessary to look at Astbury's views in 1946/7 on the structure of DNA. Astbury was born in the same year as Gulland. After graduating from Cambridge he went to work with Sir William Bragg, first at University College London, then at the Royal Institution, before moving in 1928 to the University of Leeds as Lecturer in Textile Physics, with the task of setting up a Textile Physics Research Laboratory. He was elected a Fellow of the Royal Society in 1940 and filled the chair of Biomolecular Structure in the Department of the same name created for him at Leeds. Amusingly, Astbury later wrote,[10] "the title I wanted was 'Molecular Biology,' the name I myself had first propagated, but the committee thought it was asking too much to describe me as any sort of biologist."

In a lecture[11] given in 1946, Astbury considered the most significant fact about the nucleic acids to be their high density, and if the nucleotides could be stacked on top of one another, or in some similar arrangement, then he calculated that the spacing of the bases was about 3.1 Å. He suggested that the bases must be almost flat, because of their double-bonded nature, resulting in a structure "like a great pile of plates." However, if this was not the case, then the arrangement must be dimensionally equivalent, perhaps with "nucleotides of neighbouring molecules being *closely* interleaved in a surprisingly regular fashion" (Astbury's italics). Astbury, however, thought this latter possibility highly unlikely and favoured the pile of plates.

Even earlier, Signer, Caspersson and Hammarsten[12] had concluded from the optical anisotropy of solutions of DNA that the purine and pyrimidine rings lay in planes perpendicular to the longitudinal axis of the molecule. Moreover, the sodium salt of DNA had the form of thin rods, approximately 300 times longer than they were wide, with an M_r between 500 000 and 1 000 000. Pollister (quoted in Ref. 5) in 1948 remarked that "It has often been suggested that the hierarchy of coils (seen under the microscope in chromosomes) actually extends downward to a 'molecular spiral'." With hindsight, it seems that there were several clues to the structure of DNA almost a decade before the publication of Watson and Crick's paper. What factors might have been responsible for the apparent delay in progress?

DID THE DOUBLE HELIX MODEL MAKE A DELAYED ENTRANCE?

Bernal,[10] in his *Memoirs of Fellows of the Royal Society,* published in 1963, considered that Astbury, in the period leading up to 1953, missed the actual structure of DNA for two reasons, first because the X-ray analytical procedure for helical structures was not developed, and second because the strategic double-coil model of DNA could hardly have been guessed at before the establishment of the chemical equalities between the quantities of the purines and the pyrimidines by Chargaff.[13] This seems to be a fair assessment. Astbury's X-ray photographs were of the difficult to interpret A-form of DNA, and indeed, it was only in 1951/2 that Stokes, Cochran and Crick[14] developed the theory for analysis of helices (although this appears not to have been a particularly formidable task). Nevertheless, if only as a consequence of his and Gulland's joint participation in the first Symposium of the Society for Experimental Biology (entitled Nucleic Acid), held in Cambridge in 1946, Astbury could have realized by putting his own and Gulland's work together that the DNA structure in solution was made up of at least two polynucleotide chains possessing a regularity of structure and interacting in such a way as to hide the amino and hydroxyl groups of the bases, but not the phosphates, from the solvent. It is curious that in his paper[11] to the Society for Experimental Biology, Astbury does not make reference to Gulland's work, despite the fact that Gulland made two presentations[7,15] to the meeting bearing on the structure of nucleic acids. It may be worth noting that Gulland talked in terms of polynucleotide chains, whereas Astbury talked of columns and used terms such as "fibre axis" and "spiral-wise," and of the need to build models—a very prescient perception, but he made no mention of a helix.

Bernal[10] also asked the question of Astbury: "With all his abilities and intuition, why did some of the principal discoveries in the field, nevertheless, elude him?" Bernal comments that Astbury was more isolated than Linus Pauling in his scientific thinking. MacArthur,[16] a long time collaborator, felt that Astbury's contribution lay in the questions he asked and when—stimulus rather than detailed answer. It was also unfortunate that Astbury seemed to be unduly mesmerized by numbers. For example, regarding[17] the 3.4 Å spacing of superimposed nucleotides that he had proposed, he thought that the most significant point of the spacing was that it was practically identical to the spacing between sidechains of amino acid residues in an extended polypeptide. However, if it was unfortunate that the physicist and crystallographer Astbury missed the point of Gulland's titration results, how much

FIGURE 3. Resonance forms of the amidine (**a**) and amide (**b**) segments of adenine and thymine, respectively. (Reproduced, with permission, from Ref. 18.)

more remarkable that their significance should have been overlooked by Pauling.[2]

Since Gulland was certainly familiar with the work of Astbury, how did Gulland fail to connect his own findings of the masking of the amino and hydroxyl groups of the bases with the close, regular interleaving speculated on by Astbury? It would surely soon have been realized that, for spatial reasons, purines had to link with pyrimidines, although it could not immediately have been speculated that adenine links with thymine and guanine with cytosine. Although in his Cold Spring Harbor paper Gulland suggests a maximum of two hydrogen bonds for every four phosphorus atoms, with his experience of heterocyclic chemistry it seems likely that he would soon have realized that the delocalization of electrons to give the resonance forms of the amidine and amide segments of adenine and thymine, respectively (Fig. 3), would allow pairs of hydrogen bonds between interacting nucleotide bases. This would depend upon, and therefore suggest, a certain geometry that later became apparent from the crystallographic evidence.

How Gulland's views might have developed we shall never know, for he died tragically as the result of a railway accident in October 1947 (Fig. 4)—an event that was bound to bring to an end the work of his laboratory. In the same month, however, he had resigned his Chair to take up the Directorship of Research for the Institute of Brewing, a change that might well have distracted his attentions from DNA. Dare one also suggest that the interest he developed in organization and administration,[19] which detracted from his experimental work, may be partly responsible? But it is important to remember that the 1940s were a very difficult time for esoteric scientific research, and it is impressive that the experiments reported in Figs. 1 and 2 could be carried out during the difficult conditions of the war years or immediately after.

FIGURE 4. Professor Gulland was killed on 26 October 1947 in the last major accident experienced by the old LNER (London North Eastern Railway) just before its assimilation into British Rail on 1 January 1948. Partly because of the failure of a driver to read a notice about work on the lines, the 11.15 express from Edinburgh to Kings Cross was derailed at about 12.45 as the train passed through Goswick, 6 miles south of Berwick. Three coaches crashed down an embankment and 28 people, including Gulland, lost their lives. (Photograph from *The Illustrated London News*.)

Perhaps Gulland and collaborators really had no chance for, as Crick and Watson[20] admit, without data provided by Wilkins and Franklin on the crystalline A-form and paracrystalline B-form structures before publication "the formulation of our structure would have been most unlikely, if not impossible," and models such as those proposed by Furberg,[21] in which successive bases point out from the central column in opposite directions to provide a 6.8 Å spacing into which bases of neighbouring molecules might interleaf, could have been a serious possibility. As with Astbury, in 1947 Gulland and collaborators were unaware of the more precise quantitive relationships of nucleotide composition in DNA as established by Chargaff[13] in the early 1950s. None the less, it is remarkable to find that in 1943 Mirsky[22] had claimed that DNA from both animal and plant sources had equimolar quantities of purines and pyrimidines—"a definite restriction in possible variations among the deoxynucleic acids." Of course we now see the Watson and Crick publication as a major breakthrough, but in 1953 it was but an hypothesis, as Watson and Crick recognized,[1] and even in 1958, the second edition of the textbook by Fruton and Simmonds[23] described the double helix model as "an ingenious speculation."

If Gulland failed to arrive at the concept of the double helix for no other reason than his untimely death, his work, none the less, appears to have been pivotal in the formulation of the structure of DNA in 1953. It was partly on the basis of Gulland's work that Rosalind Franklin became convinced that the phosphate groups lay on the outside of the DNA "structural unit," and it appears[24] that it was a rereading of Gulland and Jordan's papers that convinced Watson that hydrogen bonds formed between bases within the same molecule. It was this realization that appears to have forced Watson to check systematically all the hydrogen bonding possibilities between the bases, leading to the chance observation (though we know that chance favours the prepared mind) of the identical shape of the adenine–thymine pair to that of the guanine–cytosine pair—and this was the breakthrough.

It seems in many ways surprising that, although mention was made of Gulland's work by Franklin and Gosling,[4] no mention of Gulland is to be found in the principal Watson and Crick papers of the time,[1,20–25,27] despite the central importance of hydrogen bonding in their proposal. Reference to Gulland's titometric work was made by Furberg,[21] but Pauling and Corey[2] were willing to admit acquaintance only with Gulland's earlier publications on the chemistry of the nucleotides. As stated by Wilkins in his Nobel lecture,[26] speaking of the work of Avery, MacLeod and McCarty, showing that bacteria could be transformed genetically by DNA, the work, "though published in 1944, seemed even in 1946 almost unknown, or if known, its significance was often belittled." This seems also, at least in part, to have been the case for Gulland's seminal papers of 1947. Had he lived, what might have been the outcome?

REFERENCES

1. Watson J.D. and Crick F.H.C. 1953. *Nature* **171:** 737–738.
2. Pauling L. and Corey R.B. 1953. *Proc. Natl Acad. Sci. USA* **39:** 84–97.
3. Wilkins M.H.F., Stokes A.R., and Wilson H.R. 1953. *Nature* **171:** 738–740.
4. Franklin R.E. and Gosling R.G. 1953 *Nature* **171:** 740–741.
5. Gulland J.M. 1948. *Cold Spring Harbor Symp. Quant. Biol.* **12:** 95–101.
6. Levene P.A. and Simms H.S. 1926. *J. Biol. Chem.* **70:** 327–341.
7. Gulland J.M. and Jordan D.O. 1947. *Symp. Soc. Exptl. Biol.* **1:** 56–65.
8. Gulland J.M., Jordan D.O., and Taylor H.F.W. 1947. *J. Chem. Soc.* 1131–1141.
9. Creath J.M., Gulland J.M., and Jordan D.O. 1947. *J. Chem. Soc.* 1141–1145.

10. Bernal J.D. 1963. *Biographical Memoirs of Fellows of the Royal Society* Vol. 9, pp. 1–35, Royal Society.
11. Astbury W.T. 1947. *Symp. Soc. Exptl. Biol.* **1:** 66–76.
12. Signer R., Caspersson T., and Hammarsten E. 1938. *Nature* **141:** 122.
13. Chargaff E. 1951. *J. Cellular Comp. Physiol.* **38:** (suppl. 1) 41–59.
14. Cochran W. and Crick F.H.C. 1952. *Nature* **169:** 234–235.
15. Gulland J.M. 1947. *Symp. Soc. Exptl. Biol.* **1:** 1–14.
16. MacArthur I. 1961. *Nature* **191:** 331–332.
17. Astbury W.T. 1952. *The Harvey Lectures,* Series XLVI, pp. 3–44, Charles C. Thomas.
18. Abeles R.H., Frey P.A., and Jencks W.P. 1992. *Biochemistry,* p. 334, Jones and Bartlett.
19. Cook J.W. 1947. *Nature* **160:** 702–703.
20. Crick F.H.C. and Watson J.D. 1954. *Proc. R. Soc. London Ser. A* **223:** 80–96.
21. Furberg S. 1952. *Acta Chemica Scand.* **6:** 634–640.
22. Mirsky A.E. 1943. *Adv. Enzymol.* **3:** 1–34.
23. Fruton J.S. and Simmonds S. 1958. *General Biochemistry* (2nd edn), p. 200, Wiley.
24. Watson J.D. 1968. *The Double Helix,* Weidenfeld and Nicolson.
25. Watson J.D. and Crick F.H.C. 1953. *Nature* **171:** 964–967.
26. Wilkins M.H.F. 1977. *Nobel Lectures in Molecular Biology 1933–1975,* pp. 147–175. Elsevier.

The Double Helix and All That

H.R. WILSON

University of Stirling, Stirling, UK

It is generally recognized that the discovery of the double-helix structure of DNA by James Watson and Francis Crick is one of the outstanding scientific achievements of this century. That by itself would be sufficient for it to merit interest, but this was greatly intensified by the publication of James Watson's *The Double-Helix,* giving a personal account of the discovery.[1] Since then many others have written about the discovery, with varying degrees of objectivity. It is interesting to analyse the processes involved in any scientific discovery, but it has proved to be exceptionally fascinating in this case. The combination of the importance of the discovery of the double helix, the events leading up to it, and the personalities involved seems to have all the ingredients of a drama, as was evident in the television drama documentary, *Life Story,*[2] recently shown in the UK.

When Watson and Crick announced their discovery in April 1953 I had been working for six months at King's College, London, with Maurice Wilkins (who shared the 1962 Nobel Prize for Medicine and Physiology with Crick and Watson), and I am often asked what I remember of the events of that period. What follows is a brief account of this.

In the Spring of 1952 I was nearing the end of my PhD studies at the University College of North Wales, Bangor, and wondering what I would do after the summer. My research involved X-ray diffraction studies of the effect of stress on metals and I was keen to change from solid-state physics to biophysics where I thought there were more interesting and relevant problems to be solved. I discussed this with my supervisor, Professor Edwin Owen, who suggested that I should apply for a University of Wales Fellowship and that I should "write to Perutz in Cambridge or Randall in London." He went on to

Reprinted from *Trends in Biochemical Sciences* July 1988, 13(7): 275–278.

say that J.T. Randall (later Sir John Randall) had set up a Biophysics Unit within the Wheatsone Physics Laboratory at King's College, London, and he drew my attention to Randall's paper "An Experiment in Biophysics" in *Proceedings of the Royal Society.*[3] At that time I happened to be reading Ritchie Calder's book *Profile of Science,*[4] which included a description of Randall's war-time work in developing the cavity magnetron and I was greatly impressed by the fact that some Americans described the device as "the most valuable cargo ever brought to the shores of America" when it was taken across the Atlantic by Sir Henry Tizard. Hence I decided to write to King's.

I was first interviewed by Randall. He was softly spoken and mild mannered and he put me at my ease straight away by saying that he'd noted that I came from the small seaside town of Nefyn, in North Wales, where he had spent his honeymoon! After questioning me about my research and my reasons for wanting to move into biophysics, I was introduced to Maurice Wilkins who showed me round the laboratories. He took me to see Alex Stokes and then Rosalind Franklin, with both of whom I spent some time. I was familiar with Stokes's work on X-ray diffraction line broadening, which was relevant to my PhD work, and I was most apprehensive that he might ask me awkward questions. However, my worries on that account were unfounded! In discussion with Franklin I was reassured to learn that she too had come into biophysics following X-ray diffraction studies in another area, namely that of amorphous carbon.

Wilkins was very understanding about my wishes to move into biophysics and mentioned his own background in solid-state physics and in the Manhattan Project. I asked him about reading material in biophysics and he mentioned Schrödinger's book *What is Life?,*[5] a copy of which I had browsed through in a bookshop that morning and for which I returned after I left King's that day!

My application for a University of Wales Fellowship was successful and I received an invitation from Wilkins to join him to work on X-ray diffraction studies of DNA, nucleoproteins and cell nuclei. I started at King's in September 1952.

BACKGROUND

It was only after I moved to King's that I began to understand the situation regarding the DNA structural studies that were going on there. I discovered that Rosalind Franklin and Raymond Gosling were working together on the structure of DNA, but not in collaboration with Wilkins. Moreover, they had the best DNA preparation. This was a preparation of calf thymus NaDNA that had been given to Wilkins some two years earlier by Rudolf Signer, of Bern, and from gels of which material Wilkins was able to draw thin, uniform fibres showing sharp extinction between crossed polarizers. Gosling and Wilkins had obtained X-ray diffraction photographs from these fibres indicating a high degree of crystallinity,[6] and were a great improvement on those obtained earlier by W.T. Astbury and Florence Bell in their pioneering studies of DNA.[7] They achieved this by passing hydrogen through water and then into the X-ray camera so that the fibres were kept in a moist atmosphere during the exposure.

I did not know at the time why Franklin and Gosling were not collaborating with Wilkins nor why the Signer DNA was now exclusively theirs for structural studies. I was very much aware, however, of the tension between Wilkins and Franklin which mainly manifested itself in the two of them having as little to do with each other as possible—helped by the fact that Franklin's office was in a different part of the building, two floors above Wilkins's. However, all the X-ray equipment was in one room and we shared a photographic dark-room, so I would have frequent contact with either Franklin or Gosling. I did not appreciate then what had brought about this state of affairs between Wilkins and Franklin, but it did seem strange to me that Wilkins, who was responsible for initiating the DNA X-ray studies at King's and to whom the DNA preparation had been given by Signer, was now excluded from its study. There was obviously more to it than merely a clash of personalities but I did not fully understand the position until I read Robert Olby's book *The Path to the Double-Helix*.[8] Whilst Wilkins had assumed that there would be collaboration between Franklin and himself, she had been told by Randall, before she started at King's, that only she and Gosling would be involved in the X-ray diffraction studies. Thus the seeds of future discord had been sown at the very beginning.

With the Signer DNA, Franklin and Gosling discovered that by increasing the relative humidity even higher than that required to produce the crystalline form previously discovered by Wilkins and Gosling, the structure changed to a paracrystalline form. They called the crystalline form A-DNA and the paracrystalline form B-DNA, and they showed that the A $\leftrightarrows$ B transformation could occur with change in relative humidity.[9]

Although Wilkins had handed over the Signer calf thymus DNA to Franklin and Gosling, he had other DNA preparations from *E. coli,* wheat germ and pig thymus provided by Erwin Chargaff. However, none of these preparations showed evidence of a crystalline (A-type) structure, and were all paracrystalline, giving B-type X-ray diffraction patterns, but of poorer quality than those obtained by Franklin and Gosling from the Signer preparation. Although this must have been very frustrating for him, Wilkins was very scrupulous about keeping to the agreement that allowed only Franklin and Gosling to work on the Signer preparation. When, as part of our comparative studies of DNA and nucleoprotein from the same source, I suggested that I should ask Franklin for some Signer DNA in order to compare calf thymus DNA with calf thymus nucleoprotein he strictly forbade me to do so.

In addition to his interest in DNA itself, which is chemically extracted material, Wilkins was also interested in its structure within the living cell, where it is normally combined with protein. He had obtained X-ray diffraction patterns from oriented squid sperm that were remarkably similar to those from paracrystalline DNA. A short paper on this was written in November 1952 and I remember seeing a draft in which the authors were Wilkins, Randall, Franklin and Gosling, but when it appeared in print the only authors were Wilkins and Randall,[10] with acknowledgements to Gosling for taking some of the X-ray photographs and to Franklin for discussion. The earlier draft left me wondering whether the two groups were about to merge, but obviously this was not to be!

EXPERIMENTAL

Our first studies in the autumn of 1952 involved a comparison, at different humidities, between the intermolecular separation in pig thymus and wheatgerm DNA with those in squid sperm (see Fig. 1 for examples of X-ray diffraction patterns). At low relative humidity the intermolecular separation was greater in the nucleoprotein (nucleoprotamine) of the sperm head than in DNA, but whilst DNA swells markedly with increase in water content, with a corresponding marked increase in molecular separation, the nucleoprotein swells much less and is insoluble in water. The observations were consistent with the phosphate groups being on the outside of the DNA molecule, as had also been concluded by Franklin and Gosling,[9] and with the protein wrapping itself around the DNA in the sperm head. We also showed that synthetic nucleoprotamine gave similar diffraction patterns to those of the intact sperm.

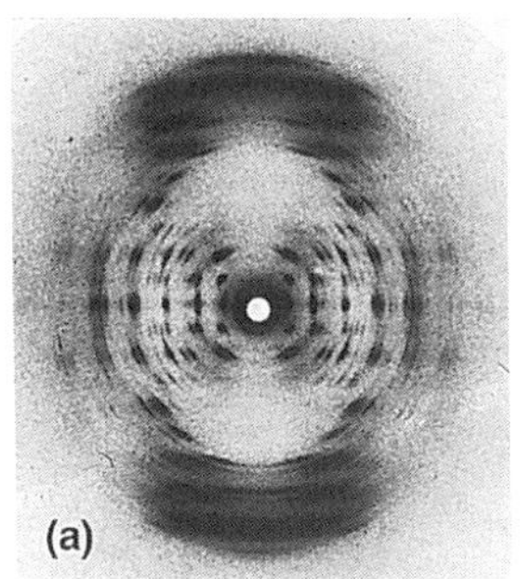

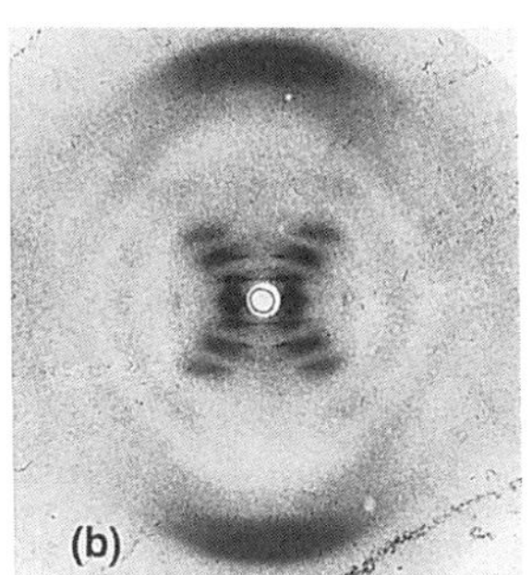

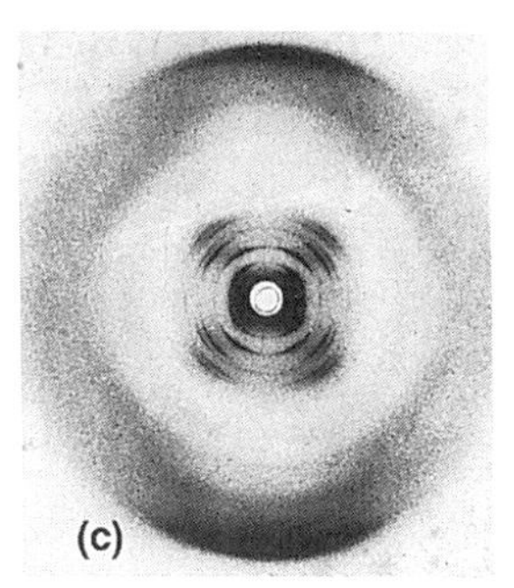

FIGURE 1. X-ray diffraction photographs from my Fellowship report (1954). (a) A-DNA (75% relative humidity) (b) B-DNA (92% relative humidity) (c) *Sepia* sperm (98% *relative humidity).*

Although the squid sperm specimens had not suffered any chemical treatment, they had been dried during preparation, and Wilkins wanted to be certain that the drying process had not affected the *in vivo* structure. For this reason, in January 1953, we extended the studies to those of undried preparations of live trout sperm. These confirmed that the sperm diffraction pattern was not an artefact of the drying process. We made similar studies on fresh rat thymus cells and on wet bacteriophage preparations. The diffraction patterns from bacteriophage were similar to those of DNA rather than to those of nucleoprotein, as would be expected if the DNA was packaged within a protein coat.

Wilkins considered comparative studies of DNAs from different sources to be very important and had already shown that they had similar paracrystalline structures. In January 1953 we started further comparative studies. The first few preparations showed only the usual paracrystalline structure but in February we obtained X-ray diffraction photographs from bacterial transforming principle, prepared by Harriet Ephrussi-Taylor in Paris, that indicated a crystalline, or A-type structure. Wilkins was particularly interested in the bacterial transforming principle, not merely because it was another source of DNA, but also because Avery, MacLeod and McCarty's studies had demonstrated that it was the genetic material of the bacterium.[11] Hence, not only was the transforming principle the first non-calf-thymus DNA to show crystallinity, but it was also material that was biologically active. Like the sperm diffraction studies, it showed that the structure being investigated was similar to that *in vivo* and was not a chemical artefact. Shortly afterwards we also obtained A-type diffraction patterns from human

DNA (from the leucocytes of a patient with chronic myeloid leukemia), and from a mouse sarcoma DNA. These were prepared by Leonard Hamilton and Ralph Barclay at the Sloan-Kettering Institute for Cancer Research, New York. Thus we had a number of DNA preparations, from different sources, that gave as good X-ray diffraction patterns as those from the Signer calf thymus preparation, and with which we could make detailed structural studies. Ironically, this was just when Watson and Crick announced their double-helix model for DNA!

DNA STRUCTURE

When I started at King's I was aware of Wilkins' view that DNA had a helical structure and I knew that he had encouraged Stokes to work out the theory of diffraction by a helical structure independently of Cochran, Crick and Vand.[12] I read Sven Furberg's PhD thesis and made notes of his two helical models of possible DNA structures (Furberg, S. [1949] PhD thesis, University of London), although I knew that a single chain did not give the correct density. I learned that Watson and Crick at Cambridge and Bruce Fraser at King's had built three-stranded helical models of DNA but that both were unsatisfactory. I was not aware of the details of these models except for the fact that the phosphate groups were on the inside and the bases on the outside in Watson and Crick's model, but the reverse was the case in Fraser's model.

Mere inspection of the paracrystalline (B-DNA) X-ray diffraction pattern suggested a helical structure, but the same could not be said about the A-DNA pattern. Nevertheless, the simplest and most reasonable assumption to make (although not necessarily the correct one) was that if DNA was helical in one form it was also helical in the other. However, Franklin and Gosling claimed that they had experimental evidence that in A-DNA the molecule was not helical and this clouded the issue for some time. It was a non-helical model of A-DNA that Franklin described in a colloquium at King's at the end of January 1953. In the notes I made at her talk there is no reference to B-DNA and I was not aware of her views on its structure at the time. I only learned of these later with the publication of Aaron Klug's paper[13,14] which is based not only on published work but also on Franklin's notebooks. Unfortunately, the situation that had developed at King's whereby there were two groups working on the same problem meant that there was a sensitivity in the relationship between them. This made it difficult to have a frank and meaningful dialogue about DNA structure. I believe that if we had obtained the A-type diffraction patterns from the other DNAs a few months earlier the situation would have been very different. It would have enabled the two groups to compare data and there would have been a firm basis for critical discussion. Certainly the longest discussion I had with Franklin was after we obtained A-type diffraction photographs from transforming principle. She expressed surprise at this result and I believe that until then she considered the A-form in calf thymus DNA to be due to a particular purine/pyrimidine base sequence, as discussed by her and Gosling.[9] My discussion with Franklin must have taken place a few days before she left King's for Birkbeck. Although she used to return to King's occasionally to use the densitometer to measure her virus diffraction patterns, I do not remember any detailed discussions about DNA during those visits.

Wilkins was very excited by the evidence that DNAs from different sources had similar crystalline structures. Even before then he was convinced that Chargaff's chemical analyses of the base ratios in DNAs[15] were of structural significance and it must have been after discussion with him in January 1953 that I made a note "to read about different base ratios for different nucleic acids." Other notes that I made in February refer to a talk Wilkins gave to a Biological Society when he described Avery, MacLeod and McCarty's studies on bacterial transformation,[11] together with the Hershey and Chase experiment on bacteriophage action,[16] as evidence that DNA is the genetic material. He stressed that what is important in biology is the process of self duplication of individual molecules, and he discussed the concept of complementarity. This must have been just at the time when Watson and Crick were discovering the complementary nature of their double-helix model, which explained the replication of DNA. This was a consequence of the proposed specific pairing of bases, which explained the physical basis of Chargaff's and also G.R. Wyatt's base ratio results,[17] and was the most important feature of the double-helix model.

I think that the day I first met Watson proved to be very significant in the DNA structure story! It was a few days after Franklin's colloquium that I came upon someone wandering along one of the subterranean corridors at King's looking for Wilkins, and he introduced himself as Jim Watson. I believe it may have been the fateful day that, according to *The Double Helix,* he had a confrontation with Franklin and when Wilkins showed him the print of the B-DNA photograph that so excited him.

THE DOUBLE HELIX

I remember vividly the sense of excitement and also of disappointment when, in March 1953, we heard that Watson and Crick had built a double-helix model of DNA that not only explained the base ratio results and seemed to be in agreement with the X-ray diffraction pattern of B-DNA, but also had important biological implications. It was the last feature of the model that accounted for the excitement and the disappointment was because we had not discovered it ourselves! However, there was no time to mope because Wilkins, Stokes and I were busily engaged in preparing a paper for *Nature,* which was to be one of two papers from King's that were to accompany Watson and Crick's paper announcing their discovery. (The second paper was to be by Franklin and Gosling.) All three appeared in *Nature* on April 25, 1953.[18–20] On 30 May Watson and Crick published a second paper discussing the genetical implications of the model.[21]

Wilkins must have been the first person from King's to see the double-helix model. My first view of it was when I went up to Cambridge with Wilkins, Bill Seeds and Geoffrey Brown, and where Crick explained it all so elegantly. I remember that I bought a book to read on the train and although I can not be sure I like to think that it was Howard Spring's *Fame is the Spur!*

POST DOUBLE-HELIX

The task of rigorously testing the double helix against the X-ray diffraction data began immediately at King's. Seeds started model building and Wilkins designed higher resolution X-ray cameras to obtain improved diffraction data. Hamilton, in New York, started preparing DNAs from a variety of sources, in the form of many different salts, and would collaborate in much future work. Franklin had by this time left King's to go to Birkbeck, but she and Gosling, from an analysis of the X-ray diffraction data, showed that A-DNA also had a double-helix structure,[22] and Wilkins, Seeds, Stokes and I did the same using a combination of X-ray diffraction and model building.[23]

Wilkin's group was joined, at various times, by Clive Hooper, Bob Langridge, Max Feughelman, Don Marvin, Geoffrey Zubay, Mike Spencer, Watson Fuller and Struther Arnott, and under his direction and leadership the structures of A-DNA, B-DNA and a third form, C-DNA, were refined, and the correctness of the Watson–Crick double helix firmly established. More extensive structural studies of nucleoproteins were also made.

I stayed at King's until the summer of 1957, when I left to take up a lectureship at Queen's College, Dundee, which was then part of St Andrews University. When I visited Dundee for interview I had dinner with Raymond Gosling, who was then on the staff, but was shortly to leave to take up an appointment in the West Indies. Amongst other things, we discussed the current situation regarding DNA structure, but since this was before Crick, Watson and Wilkins were awarded the Nobel Prize and before the publication of Watson's *The Double Helix,* our discussion was less animated than it would have been if it had taken place after those two events!

REFERENCES

1. Watson, James D. 1968. *The Double Helix.* Atheneum, and Weidenfeld and Nicolson.
2. Nicholson W., 27 April 1987, *Life Story,* BBC2 Horizon Programme.
3. Randall J.T. 1951. *Proc. R. Soc., A.* **208:** 1–24.
4. Calder R. 1951. *Profile of Science.* Allen and Unwin.
5. Schrodinger E. 1944. *What is Life?* Cambridge University Press.
6. Wilkins M.H.F., Gosling R.G., and Seeds W.E. 1951. *Nature* **167:** 759–760.
7. Astbury W.T. and Bell F.O. 1938. *Nature* **141:** 747–748.
8. Olby R. 1974. *The Path to the Double Helix.* Macmillan.
9. Franklin R.E. and Gosling R.G. 1953. *Acta Crystallogr.* **6:** 673–677.
10. Wilkins M.H.F. and Randall J.T. 1953. *Biochim. Biophys. Acta* **10:** 192–193.
11. Avery O.T., MacLeod C.M., and McCarty M. 1944. *J. Exp. Med.* **79:** 137–158.
12. Cochran W., Crick F.H.C., and Vand V. 1952. *Acta Crystallogr.* **5:** 581–586.
13. Klug A. 1968. *Nature* **219:** 808–810.
14. Klug A. 1968. *Nature* **219:** 843–844.
15. Chargaff E. 1950. *Experientia* **6:** 201–209.
16. Hershey A.D. and Chase M. 1952. *J. Gen. Physiol.* **36:** 39–56.
17. Wyatt G.R. 1951. *Biochem. J.* **48:** 584–590.
18. Watson J.D. and Crick F.H.C. 1953. *Nature* **171:** 737–738.
19. Wilkins M.H.F., Stokes A.R., and Wilson H.R. 1953. *Nature* **171:** 738–740.
20. Franklin R.E. and Gosling R.G. 1953. *Nature* **171:** 740–741.
21. Watson J.D. and Crick F.H.C. 1953. *Nature* **171:** 964–967.
22. Franklin R.E. and Gosling R.G. 1953. *Nature* **172:** 156–157.
23. Wilkins M.H.F., Seeds W.E., Stokes A.R., and Wilson H.R. 1953. *Nature* **172:** 759–762.

Light on a Dark Lady

ANNE PIPER

8 Lower Common, London, UK

Rosalind Franklin was a scientist at a time when the parameters of science were changing fast and there were new ideas and new methodology to be explored; however, this was also a time when there was still considerable prejudice against women scientists.

I have long felt that there is a need for more to be told of Rosalind: more of the woman herself and more of her work. What is generally known of Rosalind has been gleaned from Jim Watson's book *The Double Helix*.[1] Actually, she spent only two years of her short life at King's College London working on DNA; she also made a significant contribution to science through her research in other fields. Her fame, however, has been built around the work on DNA that she did at King's.

I was a friend of hers for a number of years and feel that I would like to pay tribute both to her ability and to her courage. This article, which was originally presented as a lecture to the Wimbledon Literary and Scientific Society, is dedicated to her.

As I prepared this article, I became more and more aware of what a very small part I had played in her life, although I probably knew her as long and as well as anyone apart from her close family. At times we would see a lot of each other; at others, circumstances would force us apart. However, even when we were closest, I soon came to realize that she was a very private and strictly "compartmentalized" person.

ROSALIND FRANKLIN'S BACKGROUND

I was eleven when I started at St Paul's Girls' School in 1931; Rosalind came in 1932. We became friends early on in our school life. We were both always

Reprinted from *Trends In Biochemical Sciences* April 1998, 23(4): 151–154.

FIGURE 1. Rosalind Franklin. Photographed by Vittorio Luzzati in the diningroom of the Cabane des Evettes in the French Alps (ca. 1950–1951). (Reproduced with permission of The National Portrait Gallery, London.)

in the first division for Maths, French, Latin and Science, but however good my exam marks were, Rosalind's would always be higher—except for one particular occasion well remembered by me! We both enjoyed sport and were members of most school teams until our second year of work for the Higher School Certificate. Rosalind studied Physics, Chemistry, and Pure and Applied Mathematics—I abandoned Chemistry.

Rosalind's background and upbringing were very different from anything that I had been used to. Her maternal grandfather, Jacob Cohen, won a scholarship in mathematics to University College London, at the age of thirteen, where he graduated with first class honours in Maths and Classics. He read for the bar and combined his work as a barrister with that of his professorial duties in Political Economy at University College; he was the first Jewish professor appointed to an English university. Rosalind's father was Ellis Franklin—a man whose great-grandfather Abraham was the first member of the Franklin family to come to England from Breslau, where Abraham's father was a Rabbi. Abraham settled in London in 1763, where he soon established himself as a Merchant Banker and prospered in his career. Throughout his life, he devoted much time, money and energy to a very wide range of charitable schemes. His was an orthodox Jewish family with a strong liberal tradition.

In 1938, the Franklin family took their usual summer holiday, renting two large houses in St David's, and invited me to stay with them there. I was overawed to find that, as well as hiring local help for the "rough," they had taken three of their maids with them.

Throughout the 1930s, Hitler's persecution of the Jews had increased. The Franklins spent much time working with the German/Jewish Refugee Committee in Woburn Square, and it was there that Rosalind and I spent many hours during the weekends and in our school holidays helping to sort out endless papers, and trying to impart some kind of order in dealing with the vast numbers of heartrending pleas for help.

In the Franklin family, there were a number of powerful and influential women who had made their mark by public work. They had concentrated their efforts on the disadvantaged members of the community and were at the forefront in seeing that women should have an education that would enable them to take their rightful place in society. Ellis Franklin had always assumed that his two daughters would follow the family tradition and use their time, energy and talents for the direct benefit of the community, rather than take up a career. This then is what he expected of his powerful, able and formidable daughter. I use the word "formidable" deliberately, for although when relaxed Rosalind was far from that, she was one of those very able people of great sensitivity who tend to mask their shyness with a brusque, abrupt manner. She never suffered fools gladly!

CAMBRIDGE

Rosalind was determined to get to Cambridge as soon as possible and succeeded in gaining a place at Newnham a year earlier than was then normal. I went to Girton a year later, but during our university years Rosalind and I did not see so much of each other.

Rosalind worked exceptionally hard throughout her undergraduate years; however, in Finals, she did not fulfil the expectations of her tutors, her friends or her family, and just failed to get a First. I suspect that she answered the questions on the papers in an unusual and original way that did not commend itself to the examiners. She was a prime example of those with good Seconds who achieve much more than those with good Firsts.

Rosalind considered her disappointing finals results to be a failure; I don't imagine that she had ever before encountered academic failure. She had always set herself extremely high standards and could not bear falling short of them. This did not always make her easy to work or live with. She was, however, offered a research grant and stayed on for a year in Cambridge to do research on gas-phase chromatography, under the supervision of Ronald Norrish; in 1945, her PhD thesis was accepted.

During that year, Rosalind lived in and helped to run a hostel that had been established by a French refugee, Adrienne Weil. The two became close friends and Adrienne—a scientist—helped Rosalind to emerge from her formal family background and traditional schooling at St Paul's. Adrienne supported her in the struggle to break away from the pattern expected of Rosalind by her parents. Rosalind was steadily becoming increasingly uneasy with the political outlook her family adopted and was very ready to move further toward a socialist standpoint. In the clever, lively Adrienne she found the friend, companion and guide that she needed and, subconsciously, had been looking for. With Adrienne's support and friendship, Rosalind was able to crystallize both her political ideas and her religious beliefs.

THE WAR YEARS

Rosalind and I left Cambridge in 1942 and were directed into war work. Rosalind joined BCURA (the British Coal Utilization Research Association), where she did research on the structure of carbons; I went to work for BAC (the Bristol Aircraft Corporation) and later joined the staff of Bristol University as an Assistant Lecturer in Applied Mathematics.

I don't think that either of us had much leave during the latter years of the war, but we did manage to fit in two walking holidays—one in Snowdonia and one in the Peak District. I particularly remember the Welsh holiday. One day we had made our way up Snowdon to the snow-covered Crib Goch ridge, when the mist set in. I was thankful for that as I was no longer able to see the steep drop on either side. I just managed to make my way along the ridge—driven more by my fear of Rosalind's tongue than of falling over the edge.

PARIS

In 1947, things changed for both of us. In August of that year, I got married, resigned from the University and came to live in London. With the war over by the spring of 1946, Rosalind had decided that she should leave BCURA.

She had written a note to Adrienne, who was back working in Paris, asking her to keep a lookout for someone anxious for the services of "a physical chemist who knows very little physical chemistry, but quite a lot about holes in coal." A friend of Adrienne's, Marcel Mathieu, a distinguished scientist, held a responsible position in a government agency that supported and controlled a great part of French scientific research. He had emerged from his wartime resistance activities with a reputation as a hero. Adrienne knew that Mathieu was to attend a conference on carbon research in London in 1946 at which Rosalind was to present a paper. They met and immediately took to each other. Obviously, here was someone that Rosalind could admire and respect, so she had no need to show her defensive side. Mathieu found her charming and was able to make a shrewd estimate of her outstanding scientific ability. She was appointed "chercheur" at the Laboratoire Central des Services Chimique de l'Etat in February 1947.

Rosalind spent four years in Paris, living for the most part in very cramped quarters. She had a single room in a flat belonging to a Professor's widow, where she was allowed use of the bathroom once a week; otherwise, she had to make do with a tin basin placed behind a screen. This single room made a strange contrast to the spacious, double-fronted house in Pembridge Place in which she had been brought up. Suffice it to say, she was very happy during her time in France. The cramped flat seemed to augment rather than limit her freedom and personal development. Her work went well; she was valued for herself; and so she blossomed in many ways.

My husband, Michael, and I stayed with Rosalind in her cramped quarters in Paris in 1948. Later, in 1951, with our two children, we set off from London in an old London taxi and camped out for six weeks in a derelict farmhouse in the Dordogne, at St Léon sur Vezere. Unfortunately, Joanna, our daughter, developed chickenpox just before she came. I remember sending a telegram to Rosalind: "Joanna Varicelle (Chickenpox) A bientôt."

Undaunted however, Rosalind did come to stay with us, and we spent a very happy and relaxed time together. One day, leaving the children with Michael, she and I drove over to Montignac and went into the Lascaux caves—now no longer open to the public. Another day, we went to Rocamadour, leaving Ben, then aged about six months, with *Grand'mère*, who owned and lived in half of the house that we were renting. Joanna won Rosalind's admiration by saying as we reached the top step, having climbed a long way up from the car-park to the village, "Can we do it all again?" Rosalind flourished in the freedom of a holiday like that. She was kind, generous, relaxed and always at her best with children. That is how I remember her. From St Léon, she returned to London, having had a break in Paris en route.

By then, a happy and rewarding period of Rosalind's life was over. She had gained much from her time at the Laboratoire Central: her research work had gone well; she had felt her contribution was appreciated; and she got on well with colleagues, who esteemed her for her intellectual prowess, her experimental skill and her ability to work with single-minded concentration on whatever she was doing. She did not experience any sex prejudice in Paris—had Madame Curie set the pattern? She had also been free to live as she wished, to expand her extensive culinary skills, to acquire a French dress sense, to enjoy herself in mixed company and so shed many of the restricting inhibitions that had surrounded her.

It was shortly after Rosalind's return to England that my life was disrupted and I changed from being one of a two-parent family with two children to being a single parent of four. I was very busy and had little time or energy for anything other than keeping going. During that period, I seemed to have had no conscience about asking for help from others, and I was able to find friends and relations who appeared not only willing but anxious to have my children to stay while I had a brief holiday. In 1953, Rosalind and I went together by train (the "Tauern Express" travelling from Victoria) to Athens, changing at Munich and Ljubljana. In Ljubljana, we stopped for a truncated 24-hour break and stayed with Duysan, a scientific colleague of Rosalind. I remember him saying to me, "she makes my clockwork tick." After a brief stay in Greece, Rosalind went on to Israel for a hitchhiking holiday and I returned home. Later in 1955, Rosalind and I went on a cycling tour in the Brittany–Normandy area. This was another high spot for me, in what was a grey and tough period of my life, and I am sure I was not sensitive to her needs. Despite our other preoccupations, our friendship was stalwart enough to endure and we had a really good holiday—although she was back to jumping on me. This time, it was for replying to the question of what did I do, by saying that I was a "femme de ménage." Nevertheless, that holiday was another time when we enjoyed being together and felt free from the constraints and criticisms of our daily work.

KING'S COLLEGE

Paris, for Rosalind, had been a happy and productive period. She had seemed ideally suited, both in temperament and in intellect, to continue working in

France. However, the Laboratoire was a government funded institution and it was unlikely that she, an alien—there was no European Union in the 1950s—could have had a permanent post or a fulfilling career there. She knew that, in order to further her career, she must move, and that she should extend her area of research. She was primarily an X-ray crystallographer, and work in that field had grown from being concerned with metallurgy and mineralogy into the field of biology. She certainly was greatly attracted to the new challenge of using the X-ray diffraction technique on biological substances and would therefore have been especially tempted by the possibility of carrying on her research at King's College London, which was then at the forefront of such work. This career benefit would weigh more than any personal inclination. She returned to England, but at this point things went wrong. She was horrified to find, when she arrived at King's, that the Senior Common Room (SCR) was out of bounds to all women—whatever their status. This was a restriction that she could not take lightly and her forceful reaction can hardly have endeared her to her colleagues.

Rosalind had been appointed to her post by Sir John Randall, Professor of Physics and Director of the MRC Biophysics Research Unit, to do her own research and to set up and expand a new X-ray diffraction unit. She understood that the X-ray diffraction unit was to be her unit. It was most unfortunate that, on the day of her appointment, Maurice Wilkins was on holiday. Wilkins was the second in command in the laboratory and was already working on DNA structure. At the time Rosalind was interviewed, Raymond Gosling (Wilkins' PhD student) had been asked to join Randall and was told that he would be Rosalind's student in future. Wilkins returned to find Rosalind installed and with Raymond Gosling working with her as her PhD student.

It was only years later that Wilkins saw Rosalind's letter of appointment, which set out, in a rather ill-defined way, the nature of her responsibility. The latter never seems to have been made clear either to Rosalind or to the senior staff at King's. This cannot have made things easy for anyone. She came to King's on what she believed to be the understanding that she was to be working in her own area of research, but that she could expect to enjoy cross-fertilization of ideas and discussions with colleagues who might perhaps use a very different line of approach from her own. She was not the only one to have misunderstood what was being required of her.

It is pointless now to argue how or why this misunderstanding first arose, although it does appear that lack of open communication may have been a contributory factor. Suffice it to say, it resulted in much unhappiness and frustration. It is only necessary to read Jim Watson's book *The Double Helix* to realize that time and energy were wasted in cross-purposes with crossed lines (see Box 1).

Neither Rosalind nor I had much time for leisured meetings, but I do remember that she spoke to me about the rebuff she had felt in being excluded from the SCR. This veto might appear trivial to some; to her it appeared just stupid. It led her to think that King's did not take women seriously; her reaction would have been to be on the defensive and to hide the hurt she felt with aggression. Once she had decided this, she would have appeared unapproachable and become submerged in her work.

Box 1. Setting the record straight

In the last two paragraphs of the epilogue to the *Double Helix,* James Watson speaks of those whom he had mentioned:

> All of those people, should they so desire, can indicate events and details they remember differently. But there is one unfortunate exception. In 1958, Rosalind Franklin died at the early age of thirty-seven. Since my initial impressions of her, both scientific and personal (as recorded in the early pages of this book), were often wrong, I want to say something here about her achievements. The X-ray work she did at King's is increasingly regarded as superb.
>
> The sorting out of the A and B forms, by itself, would have made her reputation; even better was her 1952 demonstration using Patterson superposition methods, that the phosphate groups must be on the outside of the DNA molecule. Later, when she moved to Bernal's lab, she took up work on tobacco mosaic virus and quickly extended our qualitative ideas about helical construction into a precise quantitative picture, definitely establishing the essential helical parameters locating the ribonucleic chain halfway out from the central axis. Because I was then teaching in the States, I did not see her as often as did Francis (Crick), to whom she frequently came for advice or when she had done something very pretty, to be sure he agreed with her reasoning. By then all traces of our early bickering were forgotten, and we both came to appreciate greatly her personal honesty and generosity, realising years too late the struggles that the intelligent woman faces to be accepted by a scientific world which often regards women as mere diversions from serious thinking. Rosalind's exemplary courage and integrity were apparent to all when, knowing she was mortally ill, she did not complain but continued working on a high level until a few weeks before her death.[1]

THE DOUBLE HELIX

Scientifically speaking, Rosalind had come to England at the right moment. Research on heredity was on the verge of taking a great leap forward; much work had already been done. It appeared that determining the actual structure of the DNA molecule would be fundamental, although at this stage the connection between DNA and heredity was only a theory lacking any form of proof. It was not clear whether further research should be based on X-ray diffraction work alone or combine such an approach with model building, or indeed, involve X-ray work at all.

During the first two months of 1953, the pace hotted up. Rosalind had already discovered that there were two forms of the helix, which she had christened A and B. She worked on the A-form using the laborious Patterson Technique—a lengthy and tedious process of mathematical analysis of the diffraction patterns generated. I remember our being in her flat one evening and churning away on a hand calculator to produce some of the results. She would show me her patterns with great pride, but I never had the remotest idea of the enormous importance of the work she was doing, nor did I appreciate the great significance of her achievements. At the same time, she had continued to take X-ray diffraction photographs and had an excellent photo of the B-form, which in January she handed to Wilkins. Two days later, she wrote what was to be her last paper at King's. A further two days passed before a paper by Linus Pauling on DNA arrived in Watson and Crick's laboratory in Cambridge. This paper demonstrated that Pauling was within easy reach of the solution but had it wrong. Watson decided to travel posthaste to King's, in order to consult with his friend Wilkins. However, when he arrived, Wilkins was not immediately available and Watson looked in on Rosalind. What exactly happened then is not entirely clear, but apparently they almost came to blows before Wilkins arrived and took Watson away. It was then that Wilkins, in good faith, showed Watson the photo that Rosalind had taken. It appears that Rosalind had given it to him for his own use and did not expect him to hand it to those she considered the opposition. The showing of the photograph further exacerbated the situation. It was particularly destructive, as Rosalind did not even know that it had happened.

Rosalind's draft paper on 17 March 1953 outlined her conviction that the B-form was helical and comprised two coaxial chains (the double helix). By this time, Watson and Crick had already reached the same conclusion and were all ready to publish.[2] Watson and Crick had set out to discover the secret of life using the technique of model building, while Rosalind's approach was based on her X-ray diffraction patterns of both the A-form and B-form of the sodium salt of DNA.[3]

TOBACCO MOSAIC VIRUS

By the time the Watson and Crick's paper on the Double Helix[2] was published in *Nature* on 25 April 1953, Rosalind had already left King's for a new post at Birkbeck College. Professor J. D. Bernal, who headed the Birkbeck lab, was someone whom Rosalind could respect as a scientist, but she would have found his dogmatic political views unattractive and her brand of socialism was a far cry from his old-fashioned communism. However, she must have found his active support of women students, and his eagerness to promote them, encouraging and endearing.

On her arrival at Birkbeck, Rosalind continued the work on the tobacco mosaic virus that Bernal had started in 1935. She worked in a small fifth-floor, attic-type room in an old house in Torrington Square. I remember visiting her in this small lab, where, although cramped, she was happy doing her research—even though the X-ray camera was in the semi-basement. Perhaps, as with the cramped conditions in the Paris flat, the physical conditions did not bother her that much, provided she was at peace with her work, with her colleagues and with herself.

Rosalind was in charge of a team working on the tobacco mosaic virus and she quickly infected those working with her with her enthusiasm and drive. In the four and a half years she was at Birbeck, she produced 17 papers (three published after her death), which gives an indication of the prodigious amount of research she carried out during that period. This would be a remarkable achievement under the best of circumstances, and was astounding when one realizes that for much of that time she was seriously ill and was very well aware of the prognosis for an inoperable cancer. Aaron Klug had come to Birkbeck in 1954. In Aaron, Rosalind found an ideal working partner and it was with him that the majority of her published papers were written. After she died, Klug took over the post as the head of the virus structure research group.

It must have been during the summer of 1957, when she already knew that she was terminally ill, that Rosalind prepared and took my four young children and myself for a river picnic on the Thames near Hurley. She had brought ice-cream with her, and the main memory my children have of this occasion is the behaviour of the dry ice that she had brought. They were all fascinated as she threw it out and they watched with awe as it zoomed about, apparently steaming around on the surface of the water. The inclusion of the dry ice in the picnic basket demonstrated again her kindness and thoughtfulness for others, especially children. She shared the children's delight as the ice whizzed about.

Rosalind died in April 1958; what she had achieved, her courage and her determination impressed all who knew her.

TRIBUTES

In 1962, the Nobel Prize for Medicine or Physiology was awarded to Francis H.C. Crick, James D. Watson and Maurice H.F. Wilkins. The three men, almost a decade earlier, had worked together, merging data from chemistry, physics and biology, to solve the structure of DNA—Crick and Watson building a hypothetical model that would conform in all its parts to what Wilkin's X-ray pictures had already shown of the molecule. The interplay of ideas, temperaments and circumstances was an especially fortunate one, since the result was something that, in Watson's words, was too pretty not to be true: the Double Helix.

Rosalind could not have received even a share in the Nobel Prize in 1962, as it cannot be awarded posthumously. There are, however, friends and colleagues who think it sad, perhaps unjust, that so little mention was made of her achievements either then or later. There is a blue plaque outside her flat at the corner of the Fulham Road and Drayton Gardens, and Maurice Wilkins showed me a plaque on the wall at King's commemorating all those who were involved in the DNA work: the four names—Rosalind's included—fit nicely in round the rim.

In paying tribute to a friend with whom I had a long association, I hope that what I have told you has brought to life one of the most able women scientists of our generation.

REFERENCES

1. Watson J.D. 1968. *The Double Helix,* Atheneum.
2. Watson J.D. and Crick F.H.C. 1953. *Nature* **171:** 737–738.
3. Klug A. 1968. *Nature* **219:** 808–810; 843–844.

The DNA Replication Problem, 1953–1958

FREDERIC L. HOLMES

Yale University School of Medicine, New Haven, Connecticut, USA

Recent discussions regarding the elusiveness of the locations and boundaries of genes within the genome accept the fundamental proposition that genes are composed of sequences of DNA. That basic knowledge entered the biological sciences with dramatic suddenness when Watson and Crick proposed their model for the structure of deoxyribonucleic acid (DNA). As Evelyn Fox Keller put it, "Before 1953, genes had been abstract hypothetical units; in that year they became concrete, knowable entities. The achievement of James D. Watson and Francis Crick permitted the identification of genes as sequences of DNA and offered a solution to the mystery of genetic replication."[1]

Since 1953, the enormous success of the Watson–Crick structure has retrospectively obscured the fact that the solution it offered to the mystery of genetic replication was initially problematic itself. This was so because the mode of replication that Watson and Crick proposed, whereby the two strands of DNA separated and acted as templates for the formation of the new strands, posed formidable difficulties. To Max Delbrück, one of the most prominent of the early molecular biologists, the scheme was too implausible to accept. For several years, the "replication problem" dominated discussion of the structure and function of DNA. Some enthusiasts quickly proclaimed that DNA accounted for a range of phenomena, including gene specificity and gene reproduction, for example[2]; however, more sceptical biochemists felt that this was still merely speculation.[3]

Reprinted from *Trends in Biochemical Sciences* March 1998, 23(3): 117–120.

In April 1953, Delbrück, who had received a description of the DNA structure from Watson, wrote back about a problem he foresaw arising from the relationship between the helical structure proposed for DNA and the inferred replication model:

> If we understand your model correctly it implies that the two threads are wound around each other plectonemically...[that is, in a helix in which the two strands are interlocked so that they cannot be separated without uncoiling]....For a DNA molecule of MW 3,000,000 there would be about 500 turns around each other. These would have to be untwiddled to separate the threads.

Delbrück suggested that the helix might contract to a superhelical state in which the threads rearranged themselves as a paranemic coil (i.e. in such a way that the two threads can be pulled apart sideways without interlocking). He wrote:

> One must postulate that the DNA opens up in some manner....In the structure you describe this...is opposed both by the two hydrogen bonds per nucleotide and by the interlocking of the helices, and it becomes a very important consideration to find a way out of this dilemma, or to think of a modification of the structure that does not involve interlocking.*

Watson rejected this suggestion on the grounds that such a mechanism would require successive segments of the DNA to be wound in opposite directions. Watson and Crick's model building had taught them that only a "right-handed" helix could be constructed; however, Watson did acknowledge that he and Crick were "without ideas" as to how the strands could unwind.*

Despite their lack of ideas on the subject, Watson and Crick were not deterred from discussing the process of the exact self-duplication of DNA in their second *Nature* article on DNA structure.[4] Their optimism did not persuade Delbrück. When he saw a copy of their manuscript, he wrote to Watson, telling him that he would be willing to bet that their complementarity idea was correct, but that the plectonemic coiling of the chains was radically wrong. His reason was that he could not see how such a chain could be untangled. In his response to Delbrück, Watson acknowledged that they were unhappy about the plectonemic coiling problem, but believed that the X-ray evidence and stereochemical considerations must take precedence over such complications.*

When he accepted Delbrück's invitation to present a paper on the structure of DNA at a Cold Spring Harbor Symposium on viruses, held in 1953,

*The personal communications cited here are from the Max Delbrück Collection, California Institute of Technology Archives.

Watson must have decided that he would have to meet Delbrück's objection head on. In his discussion, he acknowledged that the necessity for the two complementary chains to unwind was a fundamental problem if the two chains were indeed interlaced as in their model. Nevertheless, he rejected the option of paranemic coiling, or of complementary coiling in opposite directions, as incompatible with their model. Furthermore, he asserted that to have paranemic coiling with two regular simple helices going around the same axis was impossible—a point that can only be clearly grasped by studying the models. The difficulty of untwisting was formidable enough that they had asked themselves whether another complementary, but non-helical, structure could maintain the necessary regularity. They had considered a flat ribbon model; however, because it could not explain the strong equatorial reflections in the X-ray diagrams of DNA, they were not enthusiastic about it.[5] Although Watson did not mention Delbrück in his discussion, it is obvious from a comparison of his paper with the correspondence between them during the preceding weeks that he had composed a public answer to the objections that Delbrück had pressed privately on him. While admitting that they had not yet found a way out of the dilemmas Delbrück's objections posed, Watson and Crick made no concessions to them. The issue of untwisting was just another of the many problems they were confident could be resolved when more information became available.

Delbrück was probably persuaded by this time that his plectonemic coiling ideas could not solve the replication problem. He did not, however, abandon his resistance to the notion that the two chains could separate from one another in the zipper-like fashion that Watson and Crick had imagined. In May 1954, Delbrück set forth a cogent theoretical analysis of the problems. He described three ways in which the daughter duplexes might separate: (1) by slipping past each other longitudinally; (2) by unwinding of the two duplexes from each other; (3) by breaking and reuniting. Abruptly dismissing the first two as "too inelegant to be efficient," he proposed a mechanism by which the third possibility might occur. In this scheme, as replication proceeded synchronously along the two chains of a DNA duplex, the old chains were broken at each half turn of the helix and rejoined such that one terminal of each was connected to the open end of one of the new chains of the same polarity (i.e. a chain running in the same direction as defined by the 5′–3′ phosphodiester linkages). The other terminals would meanwhile become the open ends for the continuation of the replication process[6] (see Fig. 1).

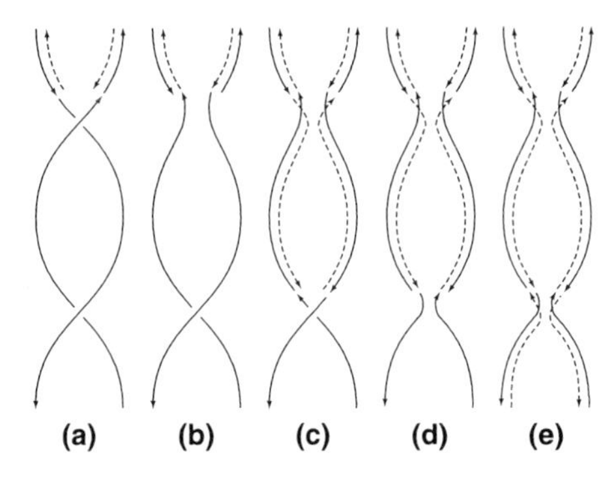

FIGURE 1. Max Delbrück's hypothesis for DNA replication, reproduced from Ref. 6: "Resolution of an interlock in a replicating duplex by breaking both old chains at each half-turn of the helix and rejoining the lower terminals of the breaks to the open ends of equal polarity of the new chains. Lateral view. (a) Location of first pair of breaks. (b) Rejoining of lower terminals of breaks. (c) Location of second pair of breaks. (d) Rejoining of lower terminals. Parental chains are represented by solid lines; new chains, by dashed lines. At the overlaps the lower chains are dotted."

Although Delbrück's scheme was abstract, it did offer a potentially testable consequence. Delbrück illustrated the possible alternative duplexes that would result from successive cycles of replication of a labelled parent duplex in an unlabelled pool of precursors (Fig. 2), according to both his own model and one in which breakage and reunion did not occur. He wrote:

> that the chains of the daughter duplexes consist of alternating sections of parental and assimilated nucleotides, each section with an average length of five nucleotides. If a labelled duplex replicates repeatedly at the expense of an unlabelled pool, then, according to this model, the label will be statistically distributed to the daughter duplexes at each successive replication. Without the breaks and reunions the distribution of label would occur only at the first replication. At each subsequent replication one daughter duplex would receive all the label, the other none.[6]

Delbrück's paper appeared to those interested in the genetic implications of the double helix as a provocative challenge. A number of scientists, most of them physicists, sought to resolve the unwinding dilemma at an abstract level, by means of topographical models. In 1954, John Platt suggested that the twisted strands of a rope do not need to be unravelled from one end to separate them. He suggested that it was simpler and energetically easier to make a "transfertwist," which involves pulling the strands apart in the middle and letting each strand twist about itself.[7] George Gamow proposed that

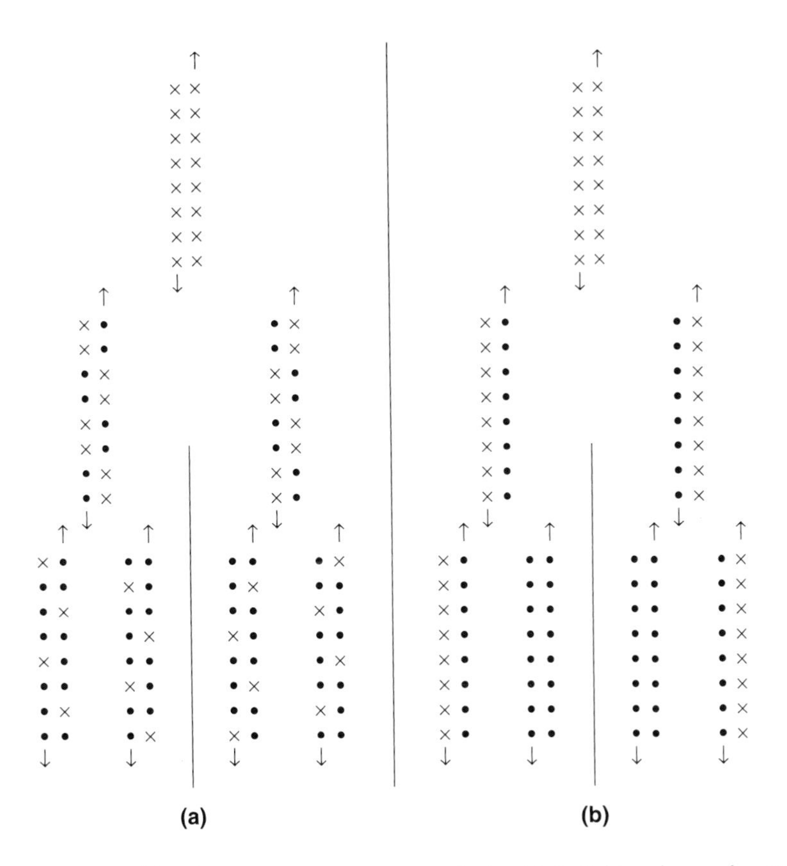

FIGURE 2. Testable alternative consequences of Max Delbrück's scheme for DNA replication as compared with a model in which breakage and reunion do not occur. Reproduced from Ref. 6. Delbrück described the figure: "Distribution of labeled parental chains to daughter-duplexes in two successive cycles of replication. x - labeled parental chains; • - unlabeled chain material assimilated from pool of precursors during replication. (a) With breaks and rejoins as postulated in the theory here presented. (b) Without breaks. For simplicity it is assumed that the breaks in (a) occur at exactly every second link and that the break points during the second replication are intermediate between those of the first replication."

separation of two helices can take place without any breaks if the long helical molecule is coiled into a spiral possessing the same repetition period as the original helix before division (G. Gamow, unpublished).

In 1955, further speculative solutions were advanced. Niels Arley proposed that a protein molecule built up within the DNA molecule could in turn serve as a template around which a replica of the original DNA molecule could be formed.[8] David Bloch offered the view that the histone molecules associated with DNA in chromosomes could act as a supporting framework that would allow the polynucleotide strands to rotate 180°, so that the bases would face outward for replication.[9] None of these schemes impressed Watson and Crick,[10] but their publication in leading journals indicates the urgency with which the replication problem was viewed during this period.

While he was at Caltech, in 1954–1955, Watson took Delbrück's challenge seriously enough that he tried to build an alterative DNA model that would obviate the need for unwinding. He and Delbrück both became suddenly enthused when Watson constructed a four-stranded DNA model possessing a ribbon structure rather than a helical one. The chains were held together by phosphate linkages and the bases were facing outwards. Watson described it as a beautiful structure. The new model offered an explanation for his earlier observation that half of the labeled DNA is lost in the transfer from parental to progeny phage. It did, however, postulate a phosphate triester linkage for which there was no experimental evidence and, consequently, the idea was soon abandoned.*

Developing topological sketches and alternative scale models to evade the unwinding dilemma were mental exercises; the most serious challenge Delbrück could deliver was to test his replication scheme experimentally, against that imagined by Watson and Crick, by following the distribution of labelled duplex material into daughter duplexes. At the time, Delbrück and others expected that the method most readily applicable to the problem would be to incorporate the radioactive isotope ^{32}P into phage DNA. The person considered most likely to make progress on the problem was Gunther Stent at Berkeley, who had already begun to explore the distribution of phage DNA in progeny molecules by using this marker. Because the ^{32}P isotope decayed relatively rapidly, however, one could not directly account for the isotope incorporated into the parental DNA by finding it in progeny DNA. Stent and his co-workers, who included the brilliant immunologist Niels Jerne, attempted to reach conclusions indirectly, by statistical analyses of the survival curves of parental and progeny phage progressively inactivated by the decay of the ^{32}P isotope contained in their DNA. Although Stent and Jerne could show that most of the transferred phosphorus resides in a minority of the progeny population, they could not draw any conclusions specific enough to adjudicate on the mechanism of DNA replication.[11] When similar methods were applied to bacterial replication the following year, Stent and his associate Clarence Fuerst showed that phosphorus atoms in "parental and newly assimilated DNA become intermingled with daughter nuclei," but they could not characterize the pattern of distribution involved with sufficient clarity to distinguish between the different proposed mechanism.[12]

At the University of Michigan, Cyrus Levinthal devised a method for tracing the distribution of ^{32}P-labelled DNA more directly, using an electron-sensitive photographic emulsion that could make visible the radioactive decay occurring within a single virus particle or DNA molecule. With this technique, he hoped to determine how DNA molecules are duplicated in the bacteriophage. Levinthal found that the labelled parental DNA was not randomly distributed, but concentrated in relatively few progeny. He too, however, had to acknowledge that the prediction made by the Watson–Crick model could not be verified by these experiments.[13]

By mid-1956, those biologists most concerned with the replication question were becoming aware that an experimental answer might prove very difficult. At the same time, they were trying to define with greater clarity what the theoretical possibilities were. Levinthal, for example, postulated three types of model: template-type, dispersive and complementary-type replication.[13] He also provided diagrams, similar to those of Delbrück, to represent the way in which radioactive label would be distributed according to each type (see Fig. 3).

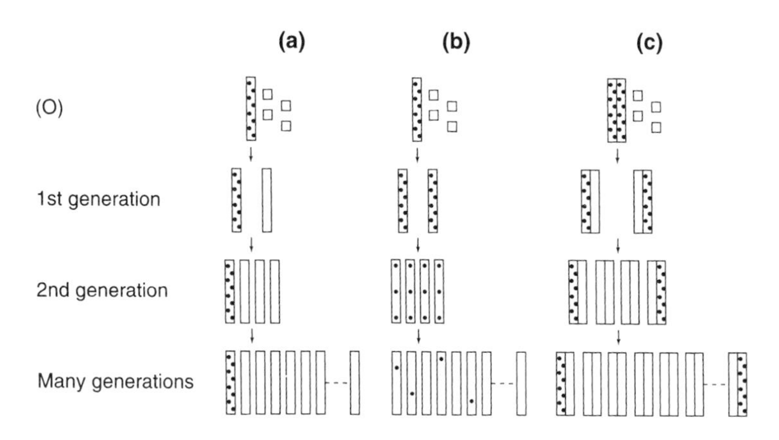

FIGURE 3. "[Levinthal's] three types of model for the replication process. The dots represent radioactive label, and the open squares represent the nonradioactive subunits used to build the new structure. (O) is the original labeled molecule. (a) is template-type replication, which leaves the label in one molecule; (b) is a dispersive type of replication...; and (c) is a complementary type." Reproduced from Ref. 13.

Delbrück and Stent reviewed the subject in detail in a paper—"On the Mechanism of DNA Replication"—presented at a symposium at Johns Hopkins University in June 1956. By contrast to the spare, self-assured style of Delbrück's first paper on the problem, he and Stent here balanced the experimental evidence and alternative models cautiously. They designated three classes into which the various proposed mechanisms could be divided: "conservative," "semi-conservative" and "dispersive" replication. They warned that an unambiguous interpretation of experimental results might be very hard to make if the processes of genetic recombination, or the successive replications that might occur in a single phage reproductive cycle, turned out to complicate or mask the effects of DNA duplication. Nevertheless, they expressed confidence that the whole issue would be resolved before too long. In keeping with the ethos of the phage group, of which Delbrück was a charter member, they believed that the decisive experiment would be carried out with bacteriophage.[14]

The decisive experiment that Delbrück and Stent predicted was performed just 16 months later, almost under Delbrück's own gaze at Caltech. It was, however, achieved by a method different from that he had anticipated, using an organism different from the one he had favored, and its outcome ruled out the scheme of DNA replication that he had once proposed. This experiment, designed and carried out by Matthew Meselson and Franklin Stahl, quickly became one of the classics of the emerging field of molecular biology. The fundamental difference between Meselson and Stahl's approach and the one that had been expected to solve the problem was that, instead of using a radioactive label to follow the distribution of parental atoms among the progeny, Meselson had the idea of incorporating a label that would impart an increased density into the parental DNA, and separating progeny DNA molecules of differing densities by centrifuging them in a solution in which a density gradient would be set up. After some unsuccessful efforts to carry out such an experiment in phage, using DNA labelled with 5-bromouracil, they shifted to bacteria and DNA labelled with ^{15}N. Meselson and Stahl grew the bacteria in a medium in which ^{15}N was the sole source of nutrient nitrogen and then transferred the organisms abruptly to a medium containing abundant ^{14}N. At time intervals corresponding to the generation time of the bacteria, they performed the following steps: they lysed the organisms to isolate their DNA; placed the DNA in a CsCl solution; spun the solution in an analytical ultra-centrifuge; and detected the position of the DNA in the gradient by taking ultra-violet absorption photographs.[15]

On photographs portraying the nitrogen isotope content of bacterial DNA at the time of the shift in medium, one sharp band, representing "heavy DNA," appeared. On those corresponding to the first generation after the shift, a second sharp band, representing a hybrid DNA—half heavy and half light—gradually replaced the heavy band. Finally, on photographs corresponding to subsequent generations, a third band, representing "light" DNA, appeared and became gradually more intense. Meanwhile, the hybrid band became less intense, but did not disappear. The result was "clean as a whistle," as Meselson wrote to Watson immediately after the first successful experiment.* The conclusions they drew from this result are represented most succinctly and simply in a schematic representation that they included in their paper[15] (see Fig. 4).

Although Meselson and Stahl were not deliberately copying the schematic diagram in Delbrück's earlier paper (compare Figs 3a and 4), the resemblance between the two representations symbolizes the degree to which the Meselson–Stahl experiment constituted the resolution of the replication debate initiated by Delbrück. To show that their results matched exactly the expectations of the Watson–Crick model for DNA duplication, Meselson and Stahl compared the mechanism Watson and Crick had proposed with a schematic representation of their own experiment: the similarity was immediately obvious[15] (see Fig. 5).

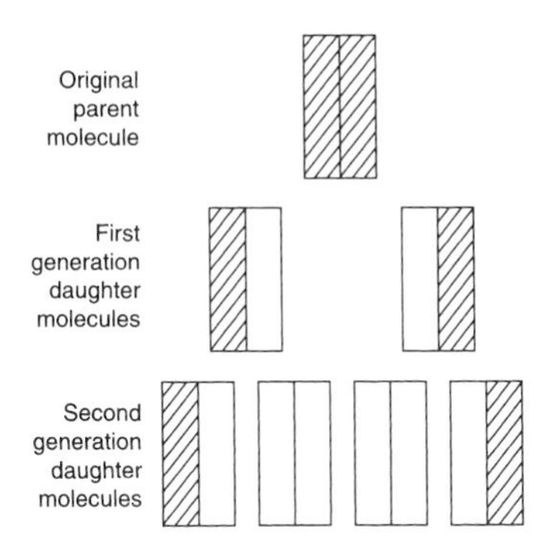

FIGURE 4. Schematic representation of conclusions drawn from the results of Meselson and Stahl's density labelling experiment, reproduced from Ref. 15. The original legend read: "The nitrogen of each molecule is divided equally between two subunits. Following duplication, each daughter molecule receives one of these. The subunits are conserved through successive duplications."

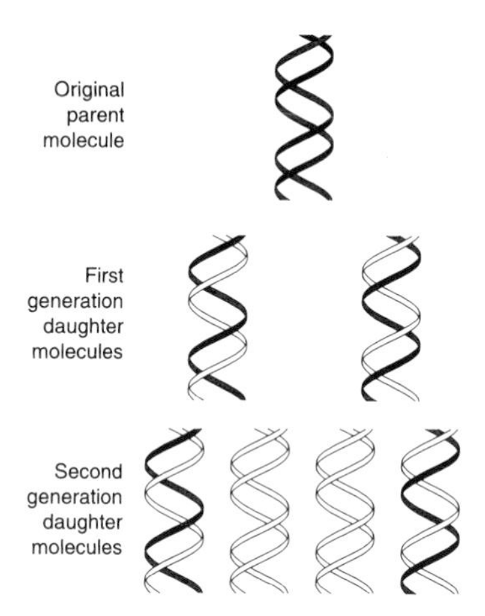

FIGURE 5. Illustration of the mechanism of DNA duplication proposed by Watson and Crick, reproduced from Ref. 15. Each daughter molecule contains one of the parental chains (black) paired with one new chain (white). Upon continued duplication, the two original parent chains remain intact. Two molecules each containing one parental chain will, therefore, always be found.

By the time Meselson and Stahl's paper appeared, their experiment, news of which had circulated through the molecular biology community by personal letters, had already become a landmark. It was described from the beginning as a beautiful experiment not only because of the ingenuity of its design, and the simplicity and clarity of its result, but because it came at just the right moment to provide powerful momentum for the acceptance of the structure and genetic implications of the double helix. In a letter to Meselson in June 1958, Maurice Wilkins wrote that the results of this experiment had instilled real confidence in the Watson and Crick duplication hypothesis—a view that rapidly became widespread.*

The Meselson–Stahl experiment did not resolve the difficulties Delbrück had initially raised about understanding how two chains interlocked in hundreds of turns could unwind. It only provided, what Watson called in 1965 in the first edition of his textbook *Molecular Biology of the Gene*, "solid evidence in favor of DNA strand separation."[16] Little more was known about how the strands actually come apart than in 1953. Yet, by contrast to that time, when Watson and Crick had admitted that the difficulty was formidable though not insuperable, Watson could write in 1965: "Nor does the need to untwist the DNA molecule to separate the two intertwined strands represent a real problem."[17] What the Meselson–Stahl experiment did was to give others the same assurance that Watson and Crick had displayed from the beginning, that somehow this problem would, in due time, find its solution.

REFERENCES

1. Keller E.F. 1992. In *The Code of Codes* (ed. D.J. Kevles and L. Hood), p. 266, Harvard University Press.
2. California Institue of Technology, Annual Report, 1954–1955, p. 67.
3. Fruton J.S. and Simmon S. 1958. *General Biochemistry* (2nd edn), pp. 200–201, John Wiley.
4. Watson J.D. and Crick F.H.C. 1953. *Nature* **171:** 964–967.
5. Watson J.D. and Crick F.H.C. 1953. *Cold Spring Harbor Symp. Quant. Biol.* **18:** 123–129.
6. Delbrück M. 1954. *Proc. Natl. Acad. Sci. U.S.A.* **40:** 783–788.
7. Platt J.R. 1955. *Proc. Natl. Acad. Sci. U.S.A.* **41:** 181.
8. Arley N. 1955. *Nature* **176:** 465–466.
9. Bloch D.P. 1955. *Proc. Natl. Acad. Sci. U.S.A.* **41:** 1058–1059.
10. Crick F.H.C. 1957. In *A Symposium on the Biochemical Basis of Heredity* (ed. W.D. McElroy and B. Glass), p. 536, Johns Hopkins University Press.
11. Stent G.S. and Jerne N.K. 1955. *Proc. Natl. Acad. Sci. U.S.A.* **41**: 704–709.
12. Fuerst C.R. and Stent G.S. 1956. *J. Gen. Physiol.* **40:** 73–90.
13. Levinthal C. 1956. *Proc. Natl. Acad. Sci. U.S.A.* **42:** 394–404.
14. Delbrück M. and Stent G.S. 1957. In *A Symposium on the Chemical Basis of Heredity*, pp. 699–736, Johns Hopkins University Press.
15. Meselson M. and Stahl F.W. 1958. *Proc. Natl. Acad. Sci. U.S.A.* **44:** 671–682.
16. Watson J.D. 1965. *Molecular Biology of the Gene*, p. 271, W.A. Benjamin.
17. Watson J.D. 1965. *Molecular Biology of the Gene*, p. 268, W.A. Benjamin.

Tritium-labeled Thymidine and Early Insights into DNA Replication and Chromosome Structure

J. HERBERT TAYLOR

The Florida State University, Tallahassee, Florida USA

The year was 1955 and I was planning an experiment that I hoped would reveal how the DNA in chromosomes was organized and test the hypothesis of semiconservative replication recently proposed by Watson and Crick.[1] I knew how to prepare chromosomes for autoradiography from experiments I had initiated at the Oak Ridge National Laboratory before moving to Columbia University in 1951. I had labeled DNA with ^{32}P and ^{14}C by growing plants or seedlings in the presence of these radioactive isotopes and then prepared autoradiographs of the squashed cells. The β-particles emitted when these isotopes decay have enough energy to move many microns in photographic emulsions so the location of the radioactive atoms cannot be determined with high precision. However, I had learned that a radioisotope of hydrogen, tritium, emitted a very low energy β-particle that would move less than a micron in photographic emulsions and would probably allow one to determine which of two chromosomes lying close together in an autoradiographic preparation had incorporated the isotope.

I knew I could label DNA selectively with thymidine during replication and that it would remain there and be distributed to the daughter chromosomes in subsequent cell divisions. Thymidine had been labeled with ^{14}C and shown to be used selectively for making one of the four nucleotides in DNA.[2] Could I prepare thymidine labeled with tritium that would be incorporated into the new DNA and remain there when the cells were removed and

Reprinted from *Trends in Biochemical Sciences* November 1997, 22(11): 447–450.

allowed to proceed through subsequent cell cycles? If so, I hoped to study the distribution of the radioactive DNA in individual chromosomes using autoradiography.

My plan was to grow bean seedlings for eight hours in a solution containing tritium-labeled thymidine, rinse them and transfer them to a fresh solution to grow for another 8–36 hours. By preparing autoradiographs after eight hours or less, I would have cells that had replicated their chromosomes only once in labeled thymidine and thus have the labeled DNA in their chromosomes along with the original unlabeled DNA. To destroy the spindle and make it easier to spread and flatten the cells with individual chromosomes separated, the plan was to transfer the seedlings to a solution with colchicine after removing them from the thymidine. I knew that plant cells would continue to cycle in the presence of colchicine, but the chromosomes would remain in one nucleus. If I found cells with twice the normal number of chromosomes, I would be looking at cells that had passed through one cycle of replication since the one in which they were labeled with the radioactive thymidine. Fortunately, Howard and Pelc[3] had recently used ^{32}P to label DNA and shown that the cell cycle in bean roots (*Vicia faba*) was about 24 hours. The cycle consists of a G1 stage of about eight hours after a division when DNA is not replicated, then an S phase of eight hours in which all chromosomes are replicated, and finally G2, another eight hours before division, which only requires an hour or so.

The problem that remained unsolved was the labeling of thymidine at a high specific activity. I had learned that exchange from tritium water in the presence of a platinum catalyst was one possible way, but the specific activity was very low in the experiment reported. In the summer of 1956, I joined Philip Woods, a former postdoctoral fellow at Columbia University and then in the Biology Department at the Brookhaven National Laboratory, to try to label chromosomal DNA with tritium. Soon after arriving, we learned that Walter L. (Pete) Hughes planned to label thymidine at a high specific activity, with the objective of labeling and killing cancer cells. We went to see him and explained our plans for labeling and following the distribution of DNA in chromosomes. He agreed to share the labeled thymidine, which he hoped to prepare. Within a few weeks he brought me his first batch and I carried out my first set of experiments. I did not know how long I would have to expose the autoradiographs to accumulate enough grains to detect labeled chromosomes, but I developed some of the autoradiographs in two weeks and reserved the remainder of the slides for later development. I was disappointed to find no labeled cells.

However, Pete Hughes told me that he had prepared a second batch of thymidine of higher specific activity. I repeated the experiments and stored the autoradiographs in the refrigerator for about two or three weeks. When these were developed, I saw that some chromosomes and many interphase nuclei were labeled so that grains appeared in the photographic emulsion that remained attached to the glass slides on which the cells were squashed. The resolution of the site of decay of tritium was as good or better than I had expected. Chromosomes in cells that reached division within eight hours after removal from thymidine had both chromatids labeled. However, after a longer time, when I found cells with twice the normal number of chromosomes, I saw that one chromatid of each chromosome was labeled, while the sister chromatid was free of labeled DNA. It soon became clear that chromosomes must consist of two subunits, which are separated during replication. A labeled subunit is synthesized along each, and at the following division both new chromosomes, which often remain attached at the centromere in colchicine-treated cells, appeared to be labeled because the subunits are too close together to be resolved. However, the two subunits remain intact and are segregated at the next replication. At this second replication, only unlabeled DNA is synthesized along each subunit. Therefore, only one of the two chromatids (new chromosomes) has a labeled subunit made during the previous replication in tritium-labeled thymidine. It registers on the photographic emulsion, but the sister chromatid does not. I was both pleased and surprised that the results were so clear and followed the prediction of semiconservative replication of DNA. Of course, I was aware that the chromosome and the DNA double helix were orders of magnitude different in size. Why a whole chromosome would segregate as predicted for the DNA helix would puzzle us for years to come.

I showed the cells to some of my colleagues in the Biology Department, but soon many others came to see the results and the remainder of that day and the next was largely devoted to show-and-tell sessions. We soon noted that in many instances, chromatids were labeled for only part of the length, but in each case the sister chromatid was labeled in the complementary segment. It appeared that a reciprocal exchange (sister chromatid exchange) had occurred at some time during or following replication. In the *Vicia* root cells, the exchanges were frequent enough to be annoying when one wished to demonstrate regular semiconservative distribution of DNA. However, I was pleased that I could detect the exchanges for one of my objectives in these experiments was to see if I could use autoradiography to study physical exchanges correlated with genetic crossing over during meiosis.

There was so much interest in my demonstration of semiconservative distribution of DNA at the chromosomal level that I realized I would have to defer studies of crossing over for a while. The summer period for my work at Brookhaven was drawing to a close and I would soon return to my home base at Columbia University for the fall term. In the meantime I attended the AIBS meeting in Storrs, CT, USA, where I was to present a short paper on a different topic from which I had submitted an abstract. The chairman of the session asked me to present instead our recent work on DNA and he agreed to extend the time. It was a warm September afternoon and because rumors had spread about our experiments, the room was overflowing and many who could not get in listened at the open windows.

A few days later an international meeting on DNA was held in Tokyo, which I did not attend. However, several people from Brookhaven and some from the Storrs meeting attended and carried word of our experiment to Tokyo, and soon our results were relayed around the world. A few weeks later, after I returned to Columbia, Francis Crick called to ask if I would supply a photograph for him to use in an article he was writing for *Scientific American.* In the meantime, at Columbia, my friend and mentor, Franz Schrader, was alarmed that my experiments were known around the world, but no publication or even an abstract had appeared in print. Even though I was convinced that our experimental results were clear and correct, the photographic evidence was not as good as I wanted. Nevertheless, I put together what was available and Professor Schrader communicated it to the *Proc. Natl. Acad. Sci. U. S. A.* for quick publication.[4]

I immediately began experiments with another plant I had available, *Bellevalia romona* of the lily family. It has only four pairs of large chromosomes, three of which are easily distinguishable by their morphology. One problem was that I had to grow roots from bulbs and I had only a few bulbs, each of which produced only a few roots. In spite of these technical difficulties, I obtained autoradiographs superior to those of *Vicia* and there appeared to be fewer sister chromatid exchanges in *Bellevalia* chromosomes.

The fall term was a busy one with visitors often coming by the lab. The most impressive group was Francis Crick, Max Delbruck and Alexander Rich who arrived unannounced and asked, no demanded, to see the evidence for semiconservative replication. Fortunately I had a cell of a *Bellevalia* root at the second division after the labeling cycle mounted under my microscope (Fig. 1). Because I now knew the segregation did not occur until the second division, I had delayed the colchicine treatment for a longer time after labeling so that I had a diploid cell with only eight chromosomes. The segrega-

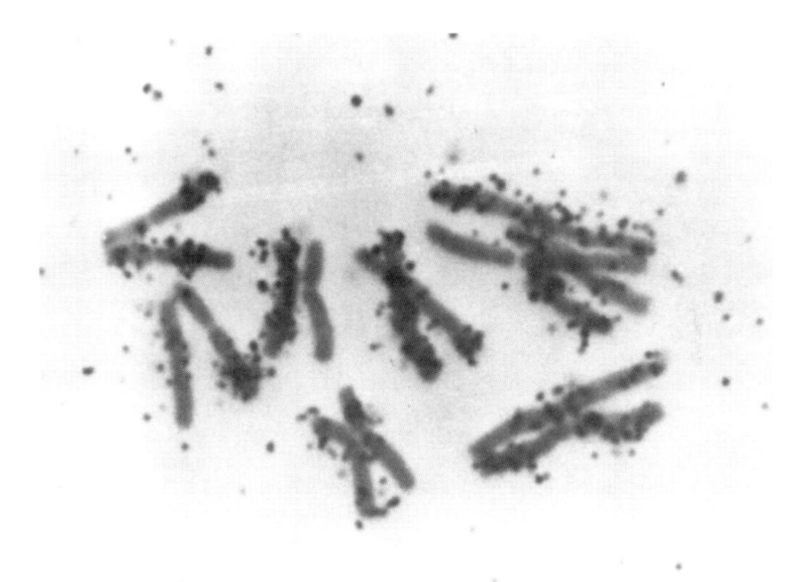

FIGURE 1. Autoradiograph of a squashed root cell of *Bellevalia* after one replication in tritium-labeled thymidine and one replication without labeled thymidine, showing semiconservative segregation of DNA subunits. (From Ref. 5.)

tion was clear and there were only a few sister chromatid exchanges. I believe my visitors went away convinced.

Unknown to me at the time, in my audience at Storrs was A.H. Sturtevant, who was very interested in the results and possibilities of its future use for studies of crossing over. I knew him only casually at that time, but his photograph along with that of T.H. Morgan hung in our seminar room at Columbia because the famous fly room, where his early, classic *Drosophila* experiments had been performed, was just down the hall.[5] Morgan and Sturtevant had moved to CalTech many years before I came to Columbia. Professor Morgan was no longer living and Sturtevant was retired, but he still came to the lab regularly as I learned later when I spent my sabbatical leave at CalTech.

Before long I had an invitation from George Beadle, now Director of the Biology Division, to come to CalTech and give a seminar on our work. I found the group very friendly and informal; George Beadle operated the slide projector for me and Linus Pauling came over from Chemistry and sat in the front row. The next morning Max Delbruck called one of his work sessions to explore the possibilities of learning more about the nature of the chromosomal subunits containing the DNA by analysing the pattern of sister chromatid exchanges. He pointed out that if the units were unlike (had different polarity, for example, like the two chains of the DNA helix) the exchanges would occur in pairs. At the second division after labeling, both chromosomes descended from an original labeled chromosome would each have an exchange at the same locus, e.g. the exchanges would occur as twins. He also admitted that his recently published hypothesis that exchanges might occur between each turn of the helix to resolve the problem of unwinding the DNA helix in replication was ruled out by my experiments. Such exchanges would randomize the label unless it was assumed that the exchanges occurred at precisely the same site at each replication, an unlikely possibility.

Back at my lab, I considered the difference between the frequency of twins expected if the subunits were unlike and the frequency if they were alike and free to rejoin at random. It turned out Delbruck had neglected the exchanges that might occur at the second replication. These exchanges would be singles whether a polarity existed or not, because only one of the two chromatids of each replicated chromosome would have a labeled subunit. A calculation showed that if one analysed the tetraploid cells at the second division, the frequency of twins for chromosomes with units having polarity would be one twin pair from the first replication and two singles from the second replication, because there would be twice as many chromosomes at the second replication. Without polarity, the ratio would be one twin for every ten singles (see Fig. 2 for an explanation). With these large chromosomes and only 16 in each tetraploid complement, I thought the difference between the two hypotheses should be easy to determine.

I soon analysed enough to be sure the ratio was far from ten singles to one twin. In fact, the twins appeared to be too frequent to fit either hypothesis unless more exchanges occurred at the first replication than at the second. The ratio was 81 twins to only 30 singles. That result was included in a paper submitted to *Genetics,*[6] but by the time it appeared or soon thereafter I had additional results that approached the predicted 2:1 ratio. That result was included in a review given in 1958 at the International Congress of Genetics, Montreal, Canada.[7]

From the beginning, the organization of the DNA in chromosomes that would allow unwinding and segregation as demonstrated was a problem that led to various models for chromosome structure.[8] With about 1 m of DNA in large chromosomes, the problem of unwinding it during replication seemed insurmountable. One early model was dubbed "the centipede," with relatively short segments of DNA attached to a central axis. This was soon abandoned for a linear duplex with hypothetical interruptions in the DNA

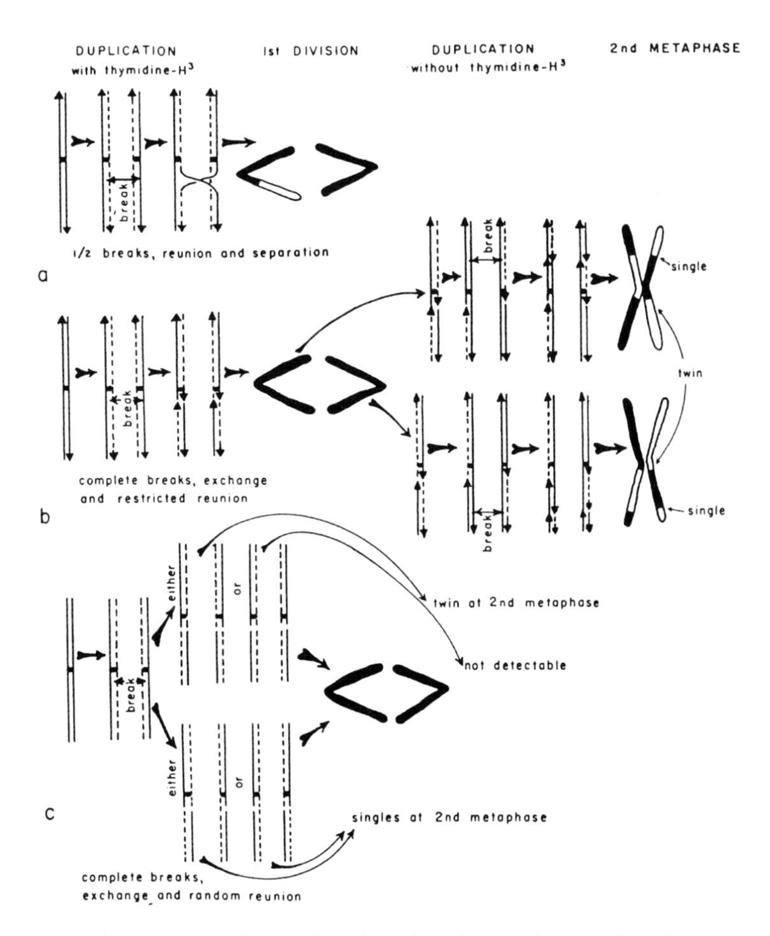

FIGURE 2. Chart showing the predicted results of sister chromatid exchanges. (a) The result of a one-half chromatid exchange with unwinding of the two subunits would be segregation at the first division after labeling. These were not observed. (b) The results of one exchange in the first replication and an equal number in each of the two daughter chromosomes at the second replication when subunits are unlike as are the two chains of a DNA helix. The predicted ratio is one twin pair for each two singles. (c) The results of similar exchanges if the subunits are free to reunite at random, i.e. with no polarity. The ratio is one twin pair for every two singles from exchanges at the first replication and with twice as many chromosomes and the same frequency at the second replication, the ratio for the total from both first and second replications would be one twin pair for every ten singles. (From Ref. 6.)

at fixed intervals that would allow each segment to unwind separately. Folding or looping of DNA segments joined by non-DNA linkers was assumed to account for the differences in size of a molecule compared to a chromo-

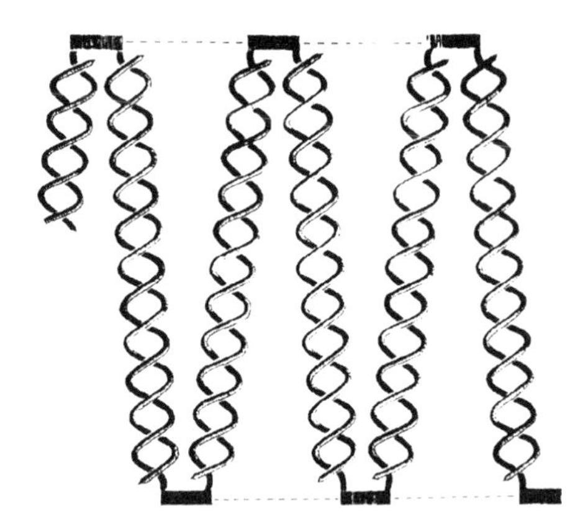

FIGURE 3. Model of chromosome structure showing discontinuous DNA molecules linked by non-DNA sections. (From Ref. 7.)

some. Figure 3 shows an early model based on this concept.[7] After Meselson and Stahl[9] demonstrated semiconservative replication of bacterial DNA at the molecular level, and later when Simon[10] demonstrated similar replication in mammalian cells, the case for a single DNA helix running the length of a chromatid became much stronger. However, it was only after topoisomerases were demonstrated and their action understood that one could better understand the possibility of folding a single 1 m-long helix into a chromosome, and yet have regular unwinding and segregation of sub-units with relatively few sister chromatid exchanges.

REFERENCES

1. Watson J.D. and Crick F.H.C. 1953. *Nature* **171:** 964–967.
2. Friedkin M., Tilson D., and Roberts D. 1956. *J. Biol. Chem.* **220:** 627–637.
3. Howard A. and Pelc S.R. 1953. *Exp. Cell Res.* **2:** 178.
4. Taylor J.H., Woods P.S. and Hughes W.L. 1957. *Proc. Natl. Acad. Sci. U.S.A.* **43:** 122–128.
5. Kohler R.E. 1994. *Lords of the Fly: Drosophila Genetics and the Experimental Life,* University of Chicago Press.
6. Taylor J.H. 1958. *Genetics* **43:** 515–529.
7. Taylor J.H. 1959. *Proc. X Int. Cong. Genet. (Montreal)* 1959, **1:** 63–78.
8. Freese E. 1959. *Cold Spring Harbor Symp. Quant. Biol.* **23:** 13–18.
9. Meselson M. and Stahl F.W. 1958. *Proc. Natl. Acad. Sci. U.S.A.* **44:** 671–682.
10. Simon E.H. 1961. *J. Mol. Biol.* **3:** 101.

DNA Replication: The 30th Anniversary of the Bacterial Model and the "Baby Machine"

STEPHEN COOPER

University of Michigan Medical School, Ann Arbor, Michigan, USA

I first met Charles E. Helmstetter ("Chick" to all who knew him back then) in July 1963, when we both arrived in Copenhagen for postdoctoral studies in Ole Maaløe's laboratory. He had come from the NIH where he had been a postdoc in Don Cummings' laboratory. There he developed a new method to synchronize bacteria—the membrane-elution method, or as it is now called, affectionately, the "baby machine."[1] Helmstetter was in Copenhagen to introduce this method into Ole's lab and to study biosynthetic patterns during the division cycle. Ole's interest in synchronization was stimulated by his association with Max Delbruck and the Phage School.[2]

I had just graduated from the Rockefeller Institute after four years with Norton Zinder studying the growth and biochemistry of f2, the male-specific RNA phage. I had decided to go to Copenhagen after reading the Schaechter-Maaløe-Kjeldgaard papers, about which I have written elsewhere.[2] I did not really grasp the elegance of those papers in 1963—it took me nearly 30 years to finally understand them.[3]

Helmstetter's earlier PhD work at the University of Chicago was on the effect of UV light on bacteria during the division cycle. At Chicago, the filter paper method was used to synchronize bacteria.[4] A pile of filter papers were placed in a holder, a bacterial culture was filtered through the papers and the initial filtrate was collected. The smaller newborn cells were assumed to trickle preferentially through the papers and enrich the initial drops. Cells in

Reprinted from *Trends in Biochemical Sciences* December 1997, 22(12): 490–494.

these drops were then resuspended in fresh medium with a synchronized culture as the assumed and expected result.

Chick continued his bacterial synchronization studies at the NIH where a new technical development—the Coulter Counter—propelled his work forward. This electronic instrument counted and sized bacteria quickly and accurately, without requiring colony formation. As it turned out, the filter paper method did not really synchronize bacteria at all. Incorporation studies with the filtration method gave an exponential pattern for thymidine synthesis during the division cycle,[4] an experimental indication that no selection or synchronization was being performed by the filter papers. As we were to eventually learn, the true pattern of synthesis during the division cycle is far from exponential.

Surprisingly, the eventual development of the baby machine method for cell synchronization was not based on the principle of small cells filtering through filter papers, but on an entirely different principle, the growth of cells bound to the filter. The idea that the cells were growing in the filter paper stack came about during a chat with a few people at a Biophysical Society meeting (including Dave Freifelder and Phil Hanawalt). Someone asked Chick how long the cells were filtered. He answered, "a few minutes." Someone replied, "then they must be growing in the filter." The conversation did not go much further that day, but that night an event took place that is reminiscent of Kekule's vision of snakes rolling about with their tails in their mouths (which was the famous inspiration for the structure of benzene). Chick lay in bed in the dark hotel room and stared at the ceiling. He began thinking about things being attached to the ceiling and growing. For unknown reasons he thought of chickens; if a hen were attached to the ceiling, the eggs would fall down. He realized this would be the case with the cells, went back to the NIH and did the experiment—long-term elution from a small stack of paper filters—and it worked.

When the cells that were attached to the filter papers divided, only newborn cells were released by division. One daughter cell remained attached to the filter and the sister cell was released. The final step in the development of the baby machine was the change to a single cellulose nitrate membrane filter to which the bacteria adhered after filtration. This gave the best synchrony, and is still the method used today.[1] Because only newborn "baby" cells are released from the membrane, the method is called "the baby machine."

One of Ole Maaløe's main interests was macromolecular synthesis during the division cycle. Ole had studied DNA synthesis ten years before, but this work was marred by a major artifact. Using temperature-shock synchronization, Ole and Karl Lark found a DNA synthesis pattern in prokaryotes similar to that found in eukaryotic cells (a gap, then DNA synthesis and then another gap).[5] Control experiments[6] (recounted in an earlier history[2]) showed that this result was not natural, but was caused by a synchronization artifact.

We North Americans (there were five of us) had a good year in Copenhagen. Every afternoon tea was set out in the library, whether or not you showed up, so we all planned our experiments around tea time. And this is where most discussions took place. On Thursdays, the lab attendant would approach a postdoc and say "Har du penge?"—Do you have the money? Someone was obliged to pay for the Wienerbrød (Danish pastry) for that day. Other discussions took place in the little cafeteria over frikadiller smørrebrød (Danish hamburger on buttered bread) and chocolate with butter sandwiches.

But all good things soon end. After a year, Chick accepted a position at Roswell Park Memorial Institute in Buffalo. I moved to Hammersmith Hospital in London, England, to William Hayes' Genetics Department. (In Copenhagen, I became interested in the genetics and biochemistry of D-methionine use,[7] and the Hayes' group was a superb bacterial genetics lab.) Chick continued to work on the membrane-elution method while at Roswell. I spent the year 1964–65 in London, then moved to Tufts Medical School in Boston, USA, to work on protein synthesis with Kivie Moldave whom I had met in Copenhagen.

Chick was not satisfied with his synchronization procedure. At Roswell he worked on every detail of the method. He obtained special cabinets with heat curtains that allowed experiments to be performed at a constant temperature without a warm room. He looked at uridine, uracil and thymidine incorporation. After two years, he came to the realization that his beloved baby machine had problems. The baby machine still perturbed cells. Even a slight change in temperature led to measurable perturbations in incorporation, and such changes were certainly occurring on the membrane. He put special thermocouples on the membrane surface to measure the surface temperature. He used dyes to analyse fluid flow across the membrane surface. But in the end, the perturbations could not be eliminated.

In frustration, and unable to sleep, Chick spent another night thinking and worrying about the method. He decided that the method had failed. As with other synchronization methods, his method did perturb cells. What could he salvage from the method he had worked on for over five years? What could he change, or modify? He gazed out the window to the snowy landscape. He went over, in his mind, all possible permutations of the experiment. The usual experiment was to put cells on the membrane, collect the cells and then label cells at different ages. Hmmm! What if the cells were labeled first, then put on the membrane? He realized he had solved the problem. The newborn cells that were eluted from the bound cells would be eluted in reverse age order at division. The newborns first off the membrane after binding would be from the *oldest* cells in the culture (those just about to divide). With time, the newborns would come from *younger and younger* cells. By measuring the radioactivity per cell in the eluate during elution, the radioactivity incorporated into the original, unperturbed cells could be determined as a function of cell age.

There was a standard "post-elution labeling" experiment planned the next morning. Though tired from a sleepless night, Chick went to the lab and told his technician to change the protocol. Label the cells first, bind the labeled cells to the membrane, and then analyse the radioactivity per cell in the eluate. Although the movie version would be better if the experiment worked that morning, the first experiment was a failure because too much radioactive thymidine was added.

The next day the experiment was repeated with less thymidine, and it worked. It worked beyond all expectations. The plateaus and dips were clear. The DNA replication pattern during the division cycle was obvious. And he had a method; the backwards method was born.[8]

Chick wrote to me at Tufts shortly after performing this experiment. He explained the results obtained with moderate- and slow-growing cells. The graphs were extremely clear. At the beginning of elution there was a clear "dip" in the radioactivity per cell. These were slow-growing cells, and the results indicated a synthetic gap in the older cells of the division cycle, a "G2-phase" in eukaryotic terminology.

When I received the letter I was filled with excitement. I sat down and wrote a 43-page, handwritten letter (dated 14 January 1965, but obviously 1966) discussing the experiment (this was before computers!). My first sentence read: "Your last letter was belated, long, exciting, and if I may be allowed—brilliant." I continued, "I think you have hit on something that may make the 'selectostat' famous and also answer some very interesting questions" (the "baby machine" name had not yet been invented and so I did not know what to call it).

Most of the letter related to the segregation problem, which at that time was one of the major problems in bacterial growth. I included drawings of different segregation possibilities, wrote about different models, and analysed various experiments with the pre-labeling method. Besides the segregation problem, I wrote about enzyme synthesis during the division cycle.[9] I also speculated about cell wall growth and drew a hypothetical picture of diaminopimelic acid (a wall-specific label) incorporation during the division cycle.[10]

The amazing part of the story comes with Robert Guthrie's entry into the picture. Robert Guthrie was the developer of the "Guthrie Test," a test for phenylketonuria (PKU) in newborn infants. Guthrie, at the University of Buffalo, had developed a simple test whereby a drop of blood from a baby is placed on an absorbent card, the card is mailed to a central laboratory, and a bacterial bioassay measures the presence of phenylketones. If a child were found to have PKU, the child was given a special diet, low in phenylalanine, and mental retardation was prevented.

In 1966, Kivie Moldave was offered a position as Chair of Biochemistry at the University of Pittsburgh. I decided not to go with him but to look for another position. I answered an advertisement in *Science,* and arranged to meet Guthrie at the annual Biochemical Society meeting. Because of my work on D-methionine metabolism (begun in Copenhagen, continued in London and finished at Tufts), Guthrie felt that I could extend the PKU test to other blood-testable inherited diseases. And so I moved to Buffalo in the summer of 1966. My office was temporarily occupied, and I had a month to kill before I could be fully at home at Children's Hospital where the PKU laboratories were located. What else was there to do but to go over to Roswell and work on DNA replication?

Chick had just finished analysing slow-growing cells. He had shown that the gap became more distinct at slower and slower growth rates. In the fastest cells he studied, glucose-minimal grown cells, there was no gap. A round of replication took 40 min. Serendipitously, this explained Ole's control experiments showing no gap in DNA synthesis.[6]

When I arrived, it was logical and obvious that we ought to look at faster growth rates. We did the thymidine labeling on glucose-casamino-acids cells and the labeling pattern was the same as in 40-minute cells. Aha! We had it. Cells growing slower than 40-minute doubling times would have a gap, because the doubling time was greater than the time for a round of replication. With faster growth rates, the DNA replication rate sped up so a round of replication was equal to the doubling time. This model proposed that in 20-minute cells the pattern of replication would look exactly like that of the 40-minute cells; the time for a round of replication would be 20 min rather than 40 min.

That night I went home and wrote a full paper describing the model. The wonderful thing was that essentially all the available data fitted the model. I was able to fit the famous Cairns picture (an autoradiograph of a single *Escherichia coli* chromosome caught in the act of replication), run-out experiments (determining the amount of residual DNA replication following inhibition of protein synthesis), and all sorts of little pieces of published data into the general model. As it turned out this model was not correct. All of this external data had no ability to discriminate between the model I wrote up that night and what was eventually the correct model.

Of course the only experimental result that did not fit was the most important one. In what I have called "The Fundamental Experiment of Bacterial Physiology,"[2,3] Schaechter, Maaløe and Kjeldgaard found a continuous variation in DNA content as growth rate was varied in different media. (I was familiar with this experiment, because it was the one that persuaded me to go to Copenhagen.) The DNA content was highest in fast-growing cells and lowest in slow-growing cells, and each growth rate had a different DNA content. The model we first considered had the same DNA content for all cells growing faster than a 40-minute doubling time.

Karl and Cynthia Lark had proposed a model to explain the different DNA contents in cells growing at different rates. They proposed that glucose-grown cells had two chromosomes replicating simultaneously, and that slower succinate-grown cells had two chromosomes replicating sequentially.[11] The Larks proposed that the glucose-grown cells were expected to have more DNA than the succinate-grown cells, thus solving the DNA content problem. As we wrestled with this problem, it became apparent that the Larks' model had identical DNA contents in both glucose- and succinate-grown cells. Two chromosomes replicating simultaneously throughout the division cycle had the same cellular DNA content as cells with two chromosomes replicating sequentially.

Over the next few days, Chick and I continued doing experiments at other growth rates. As the method was improved, and as results became sharper, we noticed a "peak" near the step in incorporation. As we now understand it, this peak was a result of multiple-fork replication. Careful analysis of the results indicated that there was a constancy in time between the termination step and division and a constancy between initiation and termination. I rewrote the paper, using the same external data, to propose a constant 40 min for a round of replication (C period) and a constant 20 min between termination and division (D period). Cells growing faster than 40-minute doubling times had periods with multiple forks for DNA replication. The patterns of DNA replication during the division cycle are illustrated in Fig. 1. Three years earlier, at a cafeteria lunch in Copenhagen, I heard Ole talk about the rumored results of Yoshikawa and Sueoka on multi-fork replication in *Bacillus subtilis*[12]; this result certainly aided in understanding the *E. coli* results.

Now the fun began. I estimate that we went through approximately 33 drafts of the two papers that were eventually published in the *Journal of Molecular Biology*.[13,14] Drafts went back and forth each day—without the benefit of Email or fax machines. Chick's attention to detail matched mine, and together we worked on every word. I redid DNA measurements to determine the molecular weight of the *E. coli* genome. As it turned out, the results were remarkably close to the current value based on the DNA sequence. This indicated that the constant C and D model was correct, that the chromosome configurations we drew (Fig. 1) were correct, and furthermore, that the membrane-elution method could be used to determine synthetic patterns during the division cycle.

I began experiments on the shift-up (the other part of the "Fundamental Experiment"), looking at DNA replication patterns as cells were shifted

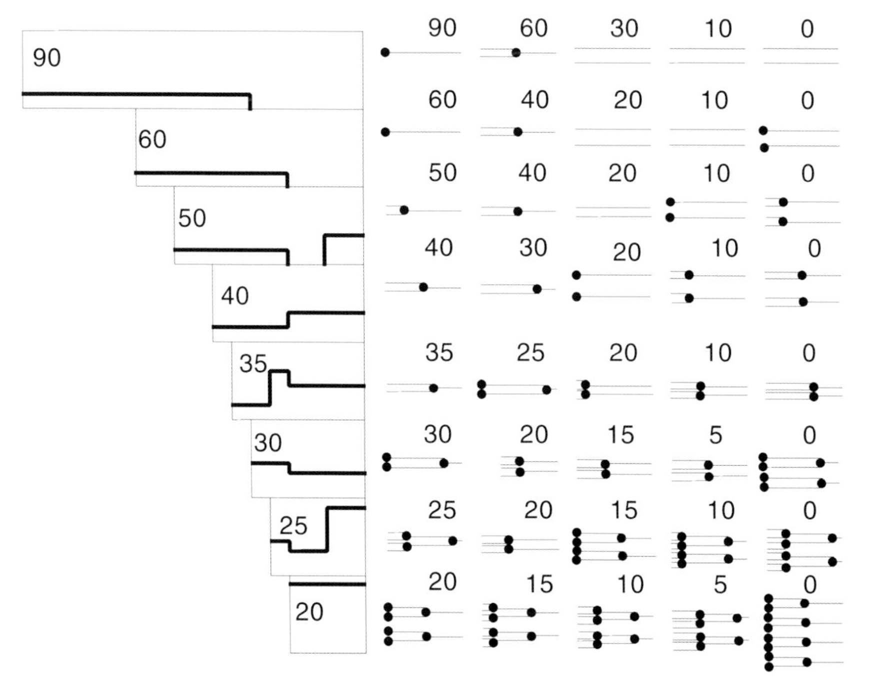

FIGURE 1. DNA replication patterns during the bacterial division cycle as originally understood in 1968. At the left are the proposed rates of thymidine incorporation for the chromosome configurations illustrated at the right. The numbers above the chromosomes indicate the number of minutes before a division. Between 20- and 60-minute interdivision times (numbers in the rate diagrams), a round of replication takes 40 min, and the time between termination and cell division is 20 min. A complete description of the derivation of these figures is given in Ref. 3.

from a slow-growth medium to a fast-growth medium. Because the Coulter Counter allowed many more experiments to be performed, the rate-maintenance phenomenon observed by Schaechter, Maaløe and Kjeldgaard (continuing the pre-shift rate of cell division for 60 min after shift-up) was confirmed for many different shift-up combinations. The complexity of the thymidine incorporation during a shift-up was difficult to fathom until I began a computer analysis with Steve Margolis, in which we simulated the shift-up conditions using the constant C and D periods and the constant initiation mass model.[15] Plotting the results revealed an amazing result. At 60 min of elution (a C + D period), there was always a drop in radioactivity per cell from the cells eluted from the baby machine. The explanation became clear. We did not need the computer any more. Rate maintenance was explained by the invariant Cs and Ds even during a shift-up. Newly inserted replication points could not determine a new cell division until at least a C + D period (60 min) had passed.[16]

The papers we published broke naturally into two parts. The first paper dealt only with the observed pattern of DNA incorporation during the division cycle; there was a minimum of interpretation. The second paper interpreted the incorporation data with chromosomal replication patterns. In this second paper, the DNA contents I had measured were used to calculate the molecular weight of the *E. coli* genome. This molecular weight calculation was done in the simplest and most straightforward way possible: the amount of DNA was determined for a known number of bacterial genomes, giving the molecular weight directly.

Our papers were published in 1968[13,14] and later that year we presented a paper at the Cold Spring Harbor Symposium.[17] In that Symposium paper, a major generalization was unveiled, as we reported the remarkable relationship of our C and D periods to the eukaryotic S and G2 periods. As growth rates varied, the C and D periods, and the S and G2 periods in eukaryotic cells, remained constant. This was the initial finding that led to the continuum model,[3,17–19] about which more next year, its 30th anniversary.

Although I left Buffalo in 1970 to move to the University of Michigan, the collaboration begun in 1966 or even in 1963 still continues. This year is the 30th anniversary of the initial description of the baby machine and the pattern of DNA replication in slow-growing cells.[8] That result, and in particular the backwards membrane-elution method, generated many other results. Some are well known and appreciated; others are not well known. The segregation problem was illuminated by the baby machine as we now know that individual DNA strands segregate non-randomly at division.[20] In addition, the pattern of cell wall synthesis during the division cycle (also a topic of that first letter over 30 years ago) was worked out and understood using the membrane-elution method.[10] The same method was used to solve the controversy over the pattern of cell growth (linear or exponential),[21,22] and to clarify the pattern of segregation of the major cellular components.[20] Most interesting, the problem that started the entire collaboration, the segregation problem, was advanced by two separate membrane-elution approaches.[23–26]

The most lasting effect of that time in Buffalo was having the feeling for a while that we were the only people in the world who knew something that no one else knew. I think it was best described by Albert Einstein who, upon gaining the insight of his General Theory of Relativity in 1915, said that "something snapped inside me." I too can say that I have felt that "snap." Even after 30 years I remember that feeling with joy and excitement. Yet best of all, 1 remember the wonderful collaboration that I had with my long-time friend and associate, Chick Helmstetter.

ACKNOWLEDGEMENTS

A. Zaritsky, W. Hlavacek, and D. Kirschner made numerous stylistic suggestions that improved this history. C. Helmstetter was helpful in musing about what happened many years ago. And my wife Alexandra, as ever, was a wonderful editor to work with on this labor of love.

REFERENCES

1. Helmstetter C.E. and Cummings D.J. 1964. *Biochim. Biophys. Acta* **82:** 608–610.
2. Cooper S. 1993. *J. Gen. Microbiol.* **139:** 1117–1124.
3. Cooper S. 1991. *Bacterial Growth and Division, Biochemistry and Regulation of Prokaryotic and Eukaryotic Division Cycles,* Academic Press.
4. Abbo F.E. and Pardee A.B. 1960. *Biochim. Biophys. Acta* **39:** 478–485.
5. Lark K.G. and Maaløe O. 1956. *Biochim. Biophys. Acta* **21:** 448–485.
6. Schaechter M., Bentzon M.W., and Maaløe O. 1959. *Nature* **183:** 1207–1208.
7. Cooper S. 1966. *J. Bacteriol.* **92:** 328–332.
8. Helmstetter C.E. 1967. *J. Mol. Biol.* **24:** 415–427.
9. Helmstetter C.E. 1968. *J. Bacteriol.* **95:** 16134–16141.
10. Cooper S. 1988. *J. Bacteriol.* **170:** 422–430.

11. Lark K.G. and Lark C.A. 1965. *J. Mol. Biol.* **13:**, 105–126.
12. Yoshikawa H. and Sueoka N. 1963. *Proc. Natl. Acad. Sci. U. S. A.* **49:** 559–566.
13. Helmstetter C.E. and Cooper S. 1968. *J. Mol. Biol.* **31:** 507–518.
14. Cooper S. and Helmstetter C.E. 1968. *J. Mol. Biol.* **31:** 519–540.
15. Margolis S.G. and Cooper S. 1971. *Comput. Biomed. Res.* **8:** 427–443.
16. Cooper S. 1969. *J. Mol. Biol.* **43:** 1–11.
17. Helmstetter C.E., Cooper S., Pierucci O., and Revelas E. 1968. *Cold Spring Harbor Symp. Quant. Biol.* **33:** 809–822.
18. Cooper S. 1979. *Nature,* **280:** 17–19.
19. Cooper S. 1989. *J. Theor. Biol.* **135:** 393–400.
20. Cooper S. 1996. in *Segregation and Cell Surface Structures* (2nd edn) (ed. F.C. Neidhardt et al.), pp. 1652–1661. ASM Press.
21. Cooper S. 1988. *J. Bacteriol.* **170:** 436–438.
22. Cooper S. 1988. *J. Bacteriol.* **170:** 5001–5005.
23. Helmstetter C.E. and Leonard A.C. 1987. *J. Mol. Biol.* **197:** 195–204.
24. Helmstetter C.E., Leonard A.C., and Grimwade J.E. 1992. *J. Theor. Biol.* **159:** 261–266.
25. Cooper S. and Weinberger M. 1977. *J. Bacteriol.* **130:** 118–127.
26. Cooper S., Schwimmer M., and Scanlon S. 1978. *J. Bacteriol.* **124:** 60–65.

DNA Strand Separation, Renaturation and Hybridization

JULIUS MARMUR

Albert Einstein College of Medicine of Yeshiva University, Bronx, New York, USA

In the textbook *Molecular Biology of the Cell,* Alberts et al.[1] list the major steps in the development of recombinant DNA technology. They state that in 1961 "Marmur and Doty discovered DNA renaturation, establishing the specificity and feasibility of nucleic acid hybridization reactions."

There were several discoveries and events that led to the experiments on DNA strand separation, renaturation and nucleic acid hybridization that have had an important impact, not only on recombinant DNA technology, but also on the analysis of nucleic acid homology, gene expression and the polymerase chain reaction.

After my doctoral training in bacterial physiology, I joined Rollin Hotchkiss' laboratory at the Rockefeller Institute (now Rockefeller University) as a postdoctoral fellow (1952–1954), where I used the quantitative *Diplococcus pneumoniae* transformation assays, pioneered by Hotchkiss, to investigate the genetic and physical linkage of two genes encoding streptomycin resistance and mannitol-1-phosphate 5-dehydrogenase.[2] My familiarity with isolating DNA that had high single and double marker-transforming activity proved to be very useful for the subsequent preparation of bacterial DNAs[3] to study the relationships between their G + C contents and either their thermal denaturation profiles or their buoyant density in caesium chloride (CsCl) gradients.

Reprinted from *Trends in Biochemical Sciences* August 1994, 19(8): 343–346.

DOTY'S LABORATORY

In the early and mid 1950s Paul Doty and his colleagues, particularly J.T. Yang, had studied the helix-to-random-coil transitions of synthetic polypeptides and proteins (such as collagen) in solution, comparing the transitions to those of a "one-dimensional crystal." They followed the disruption and reformation of helical configurations in response to critical changes in pH and temperature, by monitoring reversible changes in viscosity and optical rotation. Doty concluded: "Thus it appears that the helically configured regions of globular proteins in aqueous solution are in a state of balance from which they can increase or decrease in response to environmental changes." These ongoing studies led to hydrodynamic and light-scattering studies on the dependence of DNA denaturation on temperature, pH and ionic strength, begun in 1952 (by B.H. Bunce, S. Rice, C.A. Thomas, M.E. Reichmann and M. Litt). In 1956 Doty[4] suggested that "by establishing complete control over the denaturation process, the possibility of selectively destroying parts of the structure responsible for the transformation of particular genetic markers becomes open to exploration."

Doty would frequently meet with Rollin Hotchkiss and Muriel Roger in the latter half of the 1950s to discuss their results on the relationship between the physical and biological properties of native and denatured transforming DNA. Transformation was the only biological assay for DNA at the time and, realizing its importance from his conversations with Hotchkiss, Doty invited me to join his laboratory at Harvard. Being a member of Doty's laboratory (1956) was a new and unique experience for me (Fig. 1). Whereas in Hotchkiss' laboratory I had learned how to respect the biological properties of DNA, my experience in the Doty laboratory, where viscosity, sedimentation and light scattering were used to determine the precise size and shape of nucleic acids, added a respect for the physical properties of DNA. Not only did Doty foresee that it would be extremely useful to have a biological assay for DNA that could correlate with its secondary structure, but he also had the foresight to ask Jacques Fresco to join his laboratory in order to study the interactions of synthetic oligonucleotides. These groups interacted productively, under Doty's guidance, to plan and interpret many of the DNA denaturation and renaturation studies.

THE CsCl DENSITY GRADIENT

A very important analytical development, pioneered by Jerome Vinograd, was the introduction and use of CsCl density-gradient centrifugation, especially in

FIGURE 1. Dr P.M. Doty (left) and Dr J. Marmur (right) in Dr Doty's laboratory, circa 1959.

experiments that elegantly demonstrated that the genome of *Escherichia coli* replicated semiconservatively. These experiments were performed *in vivo* on cells in which the DNA was labeled, by growth for many generations, in a medium containing $^{15}NH_4Cl$ as the sole nitrogen source. By monitoring the buoyant density of the DNA in a CsCl gradient as a function of time after transfer to unlabeled medium, it was possible to show that after one cell division the DNA acquired a hybrid density, now interpreted as being due to the labeling of complementary strands by either ^{15}N or ^{14}N. In an additional important experiment, Meselson and Stahl[5] heated the cell extract at 100°C for 30 min and monitored the density of the *E. coli* DNA. They detected two bands, corresponding to denatured ^{15}N and ^{14}N DNA strands. While the authors were cautious in interpreting these results, it became obvious later that they had observed unambiguously, for the first time, that the DNA strands could be separated from one another *in vitro* by heating. The analytical CsCl density gradient technique was introduced to us at Harvard by Noboru Sueoka, who demonstrated a relationship between the percentage of G + C in native DNA and its buoyant density in CsCl. In a collaborative effort, we extrapolated this relationship to a large collection of bacterial DNA samples that I had prepared for thermal denaturation studies.[6,7]

Although Doty's laboratory had already developed a great deal of expertise in monitoring the molecular weight of native, sonicated and denatured DNAs by measuring viscosity, sedimentation coefficients and light scattering, attempts to measure the twofold drop in the molecular weight of calf thymus DNA, predicted by the double-stranded model proposed by Watson and Crick, had failed. The experiments were made difficult by aggregation of the large quantities of denatured DNA required for the light-scattering studies, the likelihood of single-strand breaks in the starting material and the fact that the light-scattering measurements could only be made at angles above 30° with the equipment then available. Subsequent studies with microbial DNA would prove to be highly rewarding.

DNA DENATURATION AT HIGH TEMPERATURES

The absorbancy–temperature denaturation profile to monitor hyperchromicity at the ambient temperature was developed by Jacques Fresco and Charles Epstein in the summer of 1956. Using their method we could study DNA denaturation at elevated temperatures, obviating the use of more elaborate hydrodynamic methods. The sample chamber of a Beckman spectrophotometer was bracketed with hollow jackets that allowed hot water to be circulated, to heat the DNA samples in a controlled manner. It was relatively easy to follow the denaturation of duplex DNA by monitoring the hyperchromicity at 260 nm, and thus to study the relationship between the percentage of G + C (a range of ~25–70%) in microbial DNA and its melting temperature (T_m).[6,7] A plot of the G + C content of microbial DNAs versus their T_m values indicated that in moderate salt concentrations *D. pneumoniae* DNA should have melted completely at 100°C. However, transforming DNA that was heated to 100°C and cooled rapidly retained variable levels of biological activity (1–5%). This transforming activity was even retained when the DNA was autoclaved (at 120°C). These findings were surprising, since Leonard Lerman had shown previously that genetically competent *D. pneumoniae* cells could readily bind labeled native DNA, but not denatured *E. coli* DNA.

DNA CAN RENATURE

Two important clues led us to postulate that renaturation of denatured *D. pneumoniae* DNA could occur. The first was that the residual transforming activity was variable, depending on the rate of cooling after thermal denaturation.[6,7,9] If the heated DNA was cooled slowly (annealing), by transferring it in test tubes to beakers of ever-increasing size containing water that had been heated and allowed to cool, it exhibited residual transforming activity approaching that of native DNA that had never been heated. These experiments were carried out in collaboration with Dorothy Lane, a superb senior assistant who had attained a high level of expertise in the complexities of DNA-mediated natural transformation of *D. pneumoniae* when she had worked in Hotchkiss' laboratory at the Rockefeller Institute.

The second important observation was the following: although the thermal denaturation profiles of native calf-thymus and *D. pneumoniae* DNA were very similar, exhibiting a narrow hyperchromicity of about 40% in the region of their respective T_m values, their denaturation profiles were dramatically different if they were denatured by heating then cooled rapidly in iced water. The initial portion of their denaturation profiles indicated that both DNAs behaved in a similar manner, with a steady increase in absorbancy as the random base pairs that had reformed in the initial heat-denaturation–fast-cooling step were disrupted. The profile of the calf-thymus DNA continued to exhibit a steady increase in absorbancy with increasing temperature,[10] whereas that for the bacterial DNA sample exhibited a decrease in absorbancy in a region about 25°C below the T_m for native DNA.[11] Why should there be a difference between the absorbancy profiles for denatured bacterial and mammalian DNAs, with the former exhibiting hypochromicity while being heated? After a good deal of discussion, Doty, Fresco and I realized that once the intrastrand base pairs of denatured, rapidly cooled DNA had been disrupted, complementary strands were free to renature. Since bacterial DNA is less complex in its base sequences, it renatured more rapidly than mammalian DNA.

Subsequent experiments showed that the DNA isolated from the T-even phages, with even smaller genomes, renatured more readily than did bacterial DNA. Recovery of the transforming activity of denatured DNA was optimal when the denatured DNA was maintained at 25°C below the T_m. Renatured DNA had a T_m of only 1–2°C below that of native DNA, excluding the possibility that the renatured DNA represented a complex formed by random base pairing between various single strands.

Also contributing to the interpretation of DNA renaturation were the parallel studies of Jacques Fresco and Bruce Alberts[12] on the interaction between the synthetic single-stranded ribo-oligonucleotides poly(A) and poly(AU). They showed that the "two stranded helical complexes formed always contain equal moles of homopolymer residues in the copolymer," thus maximizing the extent of base pairing. At this point, excitement about DNA denaturation/renaturation in Doty's laboratory was very high; we knew how to proceed, with most experiments yielding predictable, positive results.

How could we show that the DNA strands had separated and then reassociated when exposed to annealing conditions? There were two problems to solve: (1) preparation of enough purified denatured ^{15}N-labeled *E. coli* DNA for renaturation experiments to detect hybrid ^{15}N–^{14}N DNA by CsCl density-gradient centrifugation; and (2) elimination of single-stranded DNA that had not annealed. DNA labeled with ^{15}N was isolated in large quantity, and Larry Grossman, then at Brandeis University, supplied us with *E. coli* phosphodiesterase (the first characterized exonuclease from *E. coli,* isolated by R. Lehman) to digest single-stranded, but not double-stranded, DNA.

A CLEAR DEMONSTRATION OF STRAND SEPARATION AND ANNEALING

We realized that if we could show that denaturation and renaturation of a mixture of ^{15}N-labeled and ^{14}N DNA samples would give rise to hybrids, then it would strongly indicate that both strand separation and strand renaturation had taken place, assuming that aggregation had not occurred.[11] Although the results indicated that the re-formation of base pairs had indeed occurred, they were not altogether convincing, since the buoyant-density differences in CsCl between duplex ^{15}N-labeled, ^{14}N and hybrid DNAs were not sufficient to make a clear distinction between them.

Carl Schildkraut, a hardworking and dedicated graduate student at the time, recalls that he suggested the use of another heavy isotope, deuterium, to label nonexchangeable hydrogen atoms in DNA in order to increase the buoyant-density difference between labeled and unlabeled DNA. The problem that now arose was that *E. coli* did not readily grow in a defined medium made up in 100% D_2O. Since I had read that two scientists at the Argonne National Laboratory, H.L. Crespie and J.J. Katz, had prepared lyophilized deuterated extracts of algae, I arranged to visit their laboratory. The algae were grown in numerous illuminated, large flat containers on a

D_2O-containing medium, sterilized by an autoclave run on D_2O! I was overwhelmed with the vast number of 5 lb bottles of the lyophilized deuterated extract, and enormously pleased when I was offered all the extract that 1 could use.

When I returned to Harvard, Carl Schildkraut and I grew *E. coli* with deuterated extract in a series of flasks containing increasing ratios of D_2O to H_2O until growth took place in 100% D_2O. The buoyant-density difference between deuterated and nondeuterated DNA was about three times greater than that between the ^{15}N-labeled and ^{14}N DNAs. The hybrid DNA generated by denaturation and renaturation and exposed to single-strand-specific DNase was now clearly delineated and easily distinguished from renatured heavy and light DNAs in the critical CsCl density-gradient centrifugation experiments carried out by Schildkraut. The DNase treatment eliminated any hybrids resulting from aggregation of single-stranded termini of renatured duplexes. The proportion of heavy, hybrid and light DNAs was, as expected, 1 : 2 : 1. These experiments provided the definitive physical evidence of DNA strand separation and annealing that had been deduced from the denaturation studies. Joseph Eigner confirmed the thermal effects on denaturation and renaturation of DNA by careful viscosity and sedimentation analyses. Using these empirical hydrodynamic methods and extrapolating the time of heating back to zero to eliminate the effects of strand breaks, he even showed that the molecular weight of some *D. pneumoniae* DNA samples was reduced by one half.[11] It was not until 1972 that A.I. Krasna[13] definitively showed that denaturation of phage T7 DNA reduced its light-scattering molecular weight by one half.

My bacteriology background encouraged me to take the denaturation–renaturation experiments one step further: could hybrid DNA formation be detected when the DNAs of closely related strains were used, one (*E. coli*) labeled with deuterium, the other (*Shigella dysenteriae* or *Salmonella typhimurium*) unlabeled? From the extent of hybrid formation (and the thermal stability of the hybrid DNA) it became clear that the *E. coli–Shigella* DNAs were more closely related than the *E. coli–Salmonella* DNAs. Similar experiments were carried out with deuterium-labeled *Bacillus subtilis* DNA; these indicated that there was a relationship between the ability of DNA from heterologous strains to transform *B. subtilis* and the extent of hybrid formation with DNA from the recipient cells. Another series of experiments, carried out in collaboration with Donald Green, showed that the DNAs of the genetically related *E. coli* T-even phages (T2, T4 and T6) were also closely related in their composition and in their ability to hybridize with one another, detected by labeling one of the DNAs with a heavy isotope.

The ease of preparing purified hybrid *E. coli* DNA by growing cells in the presence of heavy isotopes followed by a single generation in unlabeled medium[5] was extremely useful in that it allowed the subsequent monitoring of the denaturation of DNA by analytical CsCl density-gradient centrifugation. Thus, it was a simple extension of this technique to expose hybrid DNA to acid, alkali or formamide to demonstrate, for instance, that these agents generated single-stranded heavy and light DNAs.

It was immediately obvious that the methods used in DNA–DNA hybridization could also be exploited to study the hybridization of RNA to denatured DNA. To this end Schildkraut,[14] using CsCl density-gradient centrifugation, showed that poly(C) would form a duplex with poly(dG), confirming earlier results obtained by Alex Rich[15] and establishing that the backbones of DNA- and RNA-like polymers were compatible in forming duplex, Watson–Crick like structures. Since messenger RNA (mRNA) was not available, experiments were also carried out to detect hybrid formation between ribosomal RNA and total *E. coli* DNA; in our first experiments to demonstrate sequence homology between naturally occurring DNA and RNA, a small degree of hybridization[15] was detected on a band of renatured DNA as a hump on the high-buoyant-density side of the band. Ben Hall has kept a letter in his files in which I asked him to send me 1 mg of T-even phage mRNA! Of course, he only had small amounts of radioactively labeled mRNA, which he and Sol Spiegelman[16] used in their classic studies on T-even phage mRNA–DNA hybridization, initially detected by preparative CsCl density-gradient centrifugation (and subsequently on nitrocellulose-membrane filters). The higher buoyant density of RNA in CsCl gradients shifted its complementary DNA into a higher-density region of the gradient.

The first public announcement of DNA denaturation, renaturation and hybridization was presented by Doty[8] at a Harvey Society Lecture on 21 January 1960 in New York City. Since our experiments had been carried out over a short period of time and the results had not yet been widely disseminated, there was a high level of excitement. When Doty showed slides of the electron micrographs (taken by Cecil Hall at MIT), demonstrating the difference between denatured DNA, seen as puddles, and renatured, rod-like DNA duplexes, I believe I detected a gasp from the audience. Appropriately, Hotchkiss was asked to thank the speaker on behalf of the Harvey Society.

Experience has shown that base pairing is the key to DNA structure and function. Base pairing also makes possible much of what we do with DNA, and each technical advance—DNA synthesis, cloning, blotting, sequencing, the polymerase chain reaction—has exploited the extraordinary specificity of nucleic acid hydrization.

REFERENCES

1. Alberts B.M. et al. 1989. *Molecular Biology of the Cell* (2nd edition), p. 181, Garland.
2. Hotchkiss R.D. and Marmur J. 1954. *Proc. Natl. Acad. Sci. USA* **40:** 55–60.
3. Marmur J. 1961. *J. Mol. Biol.* **3:** 208–218.
4. Doty P. 1956. *Proc. Natl. Acad. Sci. USA* **42:** 791–800.
5. Meselson M. and Stahl F.W. 1958. *Proc. Natl. Acad. Sci. USA* **44:** 671–682.
6. Doty P. 1961. *Harvey Lect.* **55:** 103–138.
7. Marmur J., Rownd R., and Schildkraut C. 1963. *Prog. Nucleic Acid Res.* **1:** 232–300.
8. Watson J.D. and Crick F.H.C. 1953. *Nature* **171:** 964–969.
9. Marmur J. and Lane D. 1960. *Proc. Natl. Acad. Sci. USA* **46:** 453–461.
10. Doty P.M. et al. 1959. *Proc. Natl. Acad. Sci. USA* **45:** 482–499.
11. Doty P., Marmur J., Eigner J., and Schildkraut C. 1960. *Proc. Natl. Acad. Sci. USA* **46:** 461–476.
12. Fresco J.R. and Alberts B.M. 1960. *Proc. Natl. Acad. Sci. USA* **46:** 311–321.
13. Krasna A.I. 1972. *J. Colloid Interface Sci.* **39:** 632–646.
14. Schildkraut C., Marmur J., Fresco J.R. and Doty P. 1961. *J. Biol. Chem.* **236:** PC2–PC4.
15. Rich A. 1960. *Proc. Natl. Acad. Sci. USA* **46:** 1044–1053.
16. Hall B.D. and Spiegelman S. 1961. *Proc. Natl. Acad. Sci. USA* **47:** 137–146.

Blotting at 25

ED M. SOUTHERN

Department of Biochemistry, Oxford, UK

It gives me a lot of pleasure to open a journal, to see a nice clean blot and to realize that blotting methods are still in use more than 25 years after their introduction. It is surprising that they have not been completely replaced by more powerful methods such as PCR[1] and Sanger sequencing.[2] One reason for their survival is that blotting addresses some questions not met by other methods: sequencing is not well adapted to analysing multiple copies of sequences from complex DNAs such as human genomic DNA; PCR cannot be used to analyse long stretches of DNA such as the multikilobase restriction fragments produced by some restriction enzymes. It was just such a problem that led to the development of the first blotting method.

ROOTS

At the time, I was working in Peter Walker's group at the MRC Mammalian Genome Unit in Edinburgh. The word "genome" was not widely understood in those days and there was only a handful of laboratories working in this fledgling field. Genomic DNA was known to be immensely complex and there were few methods of analysis that could address specific sequences in this complex mixture. Density gradient centrifugation had been used to purify satellite DNAs[3] and they had been shown to have a relatively simple repetitive sequence by kinetic (Cot curve) analysis of reassociation.[4] Their presence in eukaryotic DNA in such high abundance presented a great mystery, still not completely solved, and Peter recruited me in 1967 to sequence a few examples to see whether this gave any clues to their function.

Ken Murray introduced me to Fred Sanger's methods for the sequence

Reprinted from *Trends in Biochemical Sciences* December 2000, 25(12): 585–588.

analysis of RNA,[5] which I adapted to the analysis of pyrimidine tracts. Sanger's method involved the electrophoresis of RNA fragments on cellulose acetate strips, followed by "blotting through" to DEAE paper and subsequent separation in the second dimension by ionophoresis. The notion of transferring DNA fragments from one medium to another was made familiar to me by this technique. For more quantitative transfer, I replaced the cellulose acetate strip with a long polyacrylamide tube gel. The transfer step involved extruding this long worm of polyacrylamide onto the DEAE paper, where it stuck firmly during ionophoresis so that the oligonucleotides were completely eluted onto the DEAE paper.

The fingerprints (Fig. 1) showed that the sequence of the α-satellite DNA from guinea pig was remarkably simple—variants of $(CCCTAA)_n$/$(GGGATT)_n$. It was subsequently shown by Bob Moyzis and colleagues that this is the sequence of the mammalian telomere. However, the simplicity of the sequence and its high diversity suggested to us that it was unlikely to have any sequence-dependent function.[6] This was one of the pieces of evidence that led Peter Walker and others to the view that a high proportion of eukaryotic DNA was evolutionary baggage, or "junk."

The coding part of the genome seemed to be more immediately interesting and I became interested in the 5S ribosomal RNA (rRNA) genes as a result of a piece of work done with Peter Ford. He had found that the frog oocyte stores large amounts of 5S rRNA ready for the burst of translational

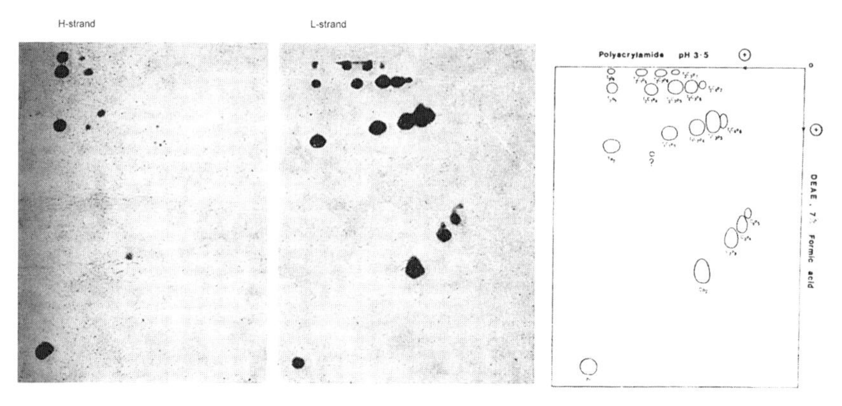

FIGURE 1. Pyrimidine-tract analysis of the H- and L-strand of guinea pig α satellite DNA indicates a simple repetitive structure with variation introduced by base substitutions. The basic sequence CCCTAA/GGGATT was later found to be the mammalian telomeric repeat.

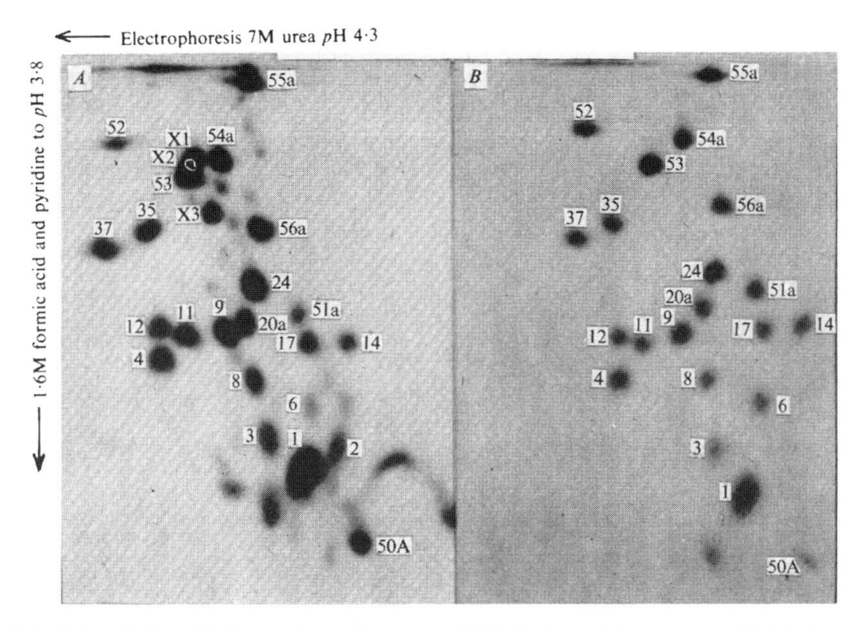

FIGURE 2. RNAse T1 fingerprints of *Xenopus* 5S RNAs from (a) oocytes and (b) kidney cells show that there are two distinct sequences that must be under separate transcriptional control. The blotting method was developed as a way of looking for the two types of gene responsible for these transcripts with the view to sequencing them to find the control elements.

activity that takes place at fertilization.[7] Using an adaptation of Sanger's methods, we carried out a detailed sequence comparison showing that these oocyte sequences differed significantly from those found in somatic cells[8] (Fig. 2). Clearly, the two types of 5S rRNA were coded by different genes that were expressed differently. This seemed to offer a good system for the study of the regulation of transcription.

THE NEED

We had the idea that, if we could isolate the two types of genes, we could sequence the upstream regions to identify sequences in the DNA that were responsible for regulating transcription. Although density gradient centrifugation had been used by Birnstiel and his colleagues to purify the ribosomal genes from *Xenopus laevis*,[9] this method did not give useful purification of the

5S rRNA genes.[10] The problem we faced was thus how to purify the 5S rRNA genes, which, although they were repeated, formed only a small fraction of the DNA. Cloning methods were developed in Stanford around this time but were not used until 1976[11] because there was a worldwide moratorium pending the outcome of a debate on their safety.

I thought that the key might be the restriction endonucleases. I had been introduced to these fascinating enzymes by Ken and Noreen Murray, who were leading figures in molecular biology in Edinburgh and internationally. In 1970, Tom Kelly and Ham Smith[12] had shown that the type-II restriction endonucleases cut at specific sites in DNA, and Kathleen Danna and Dan Nathans[13] had shown that DNA fragments could be separated by gel electrophoresis, a method that was familiar to me as I worked close to Ulrich Loening in Edinburgh. Ulrich had developed ways of separating RNA on gels some years earlier.[14]

The resolving power of gel electrophoresis is remarkable, much higher than that of density gradient centrifugation, and I wanted to exploit it as a way of purifying restriction fragments of the 5S rRNA genes. However, my colleague Gerard Roizes and I had done experiments showing that the genomic DNA of higher eukaryotes produced many thousands of different restriction fragments, which formed a smear on gel electrophoresis.[15] How could we find the fragments carrying the 5S rRNA genes? One way would have been to slice up the gel (a trick learned from Ulrich Loening), elute the DNA from each slice and detect the 5S rRNA genes by hybridization, which exploits the complementarity between the sequence of the mRNA and its cognate gene. A convenient method of hybridization to DNA trapped on nitrocellulose membranes, developed by Gillespie and Spiegelman, was in widespread use by the late 1960s.

BLOTTING

At this point, laziness, the father of invention, stepped in. The prospect of cutting lots of gels into hundreds of slices, eluting the DNA from each of them and binding this to a filter that would then be hybridized and counted, drove me to thinking that there had to be a better way of doing it. The obvious first thought was to try to hybridize the RNA to the DNA fragments in the gel, adapting the method of *in situ* hybridization that was developed in Edinburgh by Ken Jones[16] and by Mary Lou Pardue and Joe Gall[17] at Yale.

Julia Thompson joined me from Peter Ford's laboratory to explore this approach but we found that the DNA soaked out of the gel during the hybridization process, even if we dried the gel to trap the fragments. Others solved this problem and a method based on hybridization to dried gels was described by Shinnick *et al.*[18] before the publication of the blotting procedure. Julia left to return to Canada and the project was shelved until I had time to take it up a few months later.

It occurred to me that the trick was to transfer the fragments from the gel to nitrocellulose membrane, as the Gillespie and Spiegelman method of hybridizing RNA to DNA trapped on nitrocellulose was well established. Charles Thomas and colleagues[19] had described a way of dissolving agarose in concentrated $NaClO_4$, a solution with a high density, on which agarose gels floated. I thought that if I interposed a sheet of nitrocellulose paper between the gel and the surface of the $NaClO_4$ solution, the solution might permeate the membrane and dissolve the gel, leaving the DNA attached to the membrane. I see now that this was unlikely to have succeeded, but I set up an experiment to see if the gel would dissolve in this situation.

As I sat and waited for the gel to dissolve, I saw that a bead of liquid formed on top of the gel almost at once. This showed that the gel was extremely porous and immediately suggested that there was no need to dissolve the gel, that the solution to the problem was to soak the DNA fragments out of it. With the realization that the gel was porous, it was a simple matter to adapt the "blotting through" stage of the Sanger method, substituting the gel for the cellulose acetate strip and nitrocellulose membrane for the DEAE paper. It is from this that the name "blotting" derived. I have to say that I've never been keen on this name and tried for years to have the method called "gel transfer," but I concede that this does not trip lightly off the tongue and "blotting" seems to be here to stay as long as the method survives.

I should have realized that gels were porous and the idea of the transfer should have leapt into my mind much sooner, because I had seen how to do it at my primary school (Lanehead School in Burnley, Lancashire, UK). There, from the age of five, I was lucky to be under the wing of an elderly and kind teacher, Mrs Elizabeth Laycock, who used to let me help her with a number of tasks, among them the replication of examination papers (quite a useful chore to be involved in). In the days before the Xerox machine, copies were made by a process in which a master page was made by writing the text on a highly calendered, relatively impermeable sheet of paper, using a dense, violet ink. This master copy was placed, face down, on a tray filled with gelatine and left for the ink to soak in. After removing the master, copies could be taken by placing a more absorbent paper on the surface so that water from the gelatine soaked into the paper carrying an impression of the writing. The only difference between this technique and blotting is that one transfers ink, the other DNA.

INTO THE PUBLIC DOMAIN

It was not long before I had done a number of experiments to show that the transfer worked and gave a high resolution picture of the bands corresponding to genes. I did some experiments with 5S rRNA genes from *Escherichia coli* and with ribosomal genes from *Xenopus laevis* (Fig. 3).

At around this time, we had a visit from Mike Mathews, who was working in the Cold Spring Harbor Laboratory. He saw my results and was keen to take the method back with him to use in their work on viruses. I had no

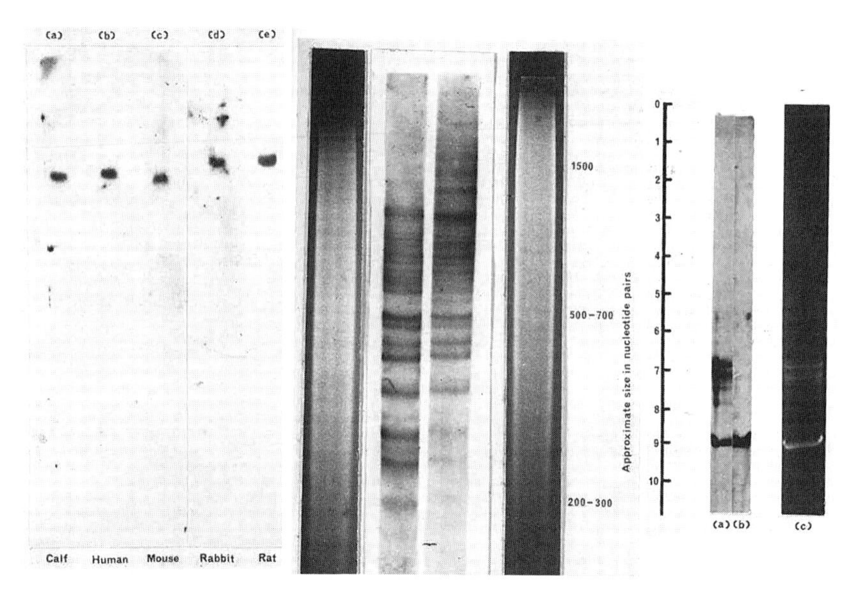

FIGURE 3. Examples of "gel transfers" from the original publication.[21] (a) HaeIII digests of *E. coli* DNA probed with 5S ribosomal RNA; (b) EcoRI digest of purified ribosomal RNA genes probed with 18S and 28S ribosomal RNAs; (c) EcoRI digests of various mammalian DNAs probed with ribosomal RNA.

written description and so I gave him something on a scrap of paper to take back to the USA. He and his colleague Mike Botchan quickly got the method to work and introduced improvements, such as the use of nick-translated DNA to produce high-specific-activity probes. This allowed the detection of single-copy sequences in total human genomic DNA.

Word spread in the research community and the people at Cold Spring Harbor asked my permission to pass the protocol to others in the field, acknowledging that it had originated from me. I was happy for them to do this and this is how my name came to be used as an adjective to describe the method. I believe that, if it had been published in the normal way, it would not have had such a firm attachment.

Later, John Alwine and George Stark developed a method for transferring RNA[20] and, as a joke, they called it "northern blotting"; naturally, when a related method for proteins appeared, it was called "western blotting." The variant that I like best is a two-dimensional analysis in which two sets of fragments are separated on different gels. One is transferred to a membrane and the other, labelled, set is placed at right angles for hybridization. This was given the charming name of "Southern Cross."

Journal publication was somewhat delayed. My first attempt at publication in the *Journal of Molecular Biology* was rejected because the editors considered it to be a "methods" paper, which the journal could only publish if it contained original and interesting results. The time it took to carry out new experiments to satisfy the editors delayed publication by many months.[21] However, during this time, the people at Cold Spring Harbor and I were sending sketches (see Fig. 4) and notes to whoever wanted to know about the method and I doubt whether many who use it have ever seen the publication. Nevertheless, I understand that it holds the record for citations for a paper published in the *Journal of Molecular Biology*.

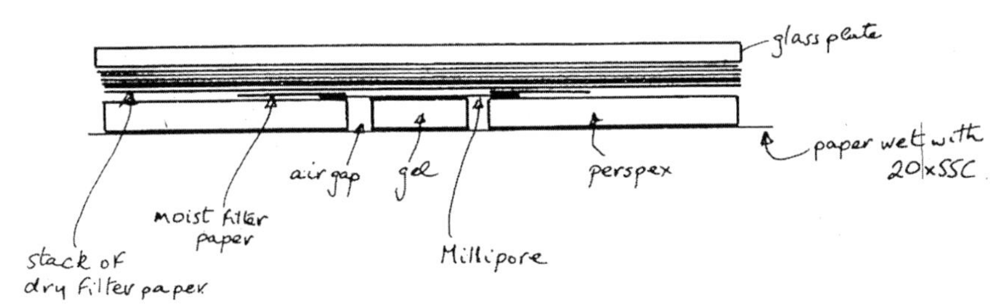

FIGURE 4. A "prepublication" of the method, sent out in advance of publication in the *Journal of Molecular Biology*.

SUBSEQUENT DEVELOPMENTS

In the 25 years since publication, the method has found a wide range of applications. Among my own favourites are those that have opened up the field of human genetics. In the summer of 1979, David Botstein visited me at Wood's Hole, where I was the visiting Lilley Fellow. David spent some time explaining his ideas for using the blotting method to analyse variation in human DNA sequences on a large scale, in order to map the human genome. This was an extension of the work of Kan and Dozy,[22] who had shown how to detect the sickle cell mutation in the β-globin gene by analysing a nearby polymorphism in a *Hpa*I site, which was in linkage disequilibrium with the sickle-cell mutation. At the time, I did not fully grasp the importance of what David was saying but when later the idea was published,[23] it became clear that here was a revolutionary approach to human genetics, and, of course, it set in train the programme which led to sequencing the human genome.

The early work I did with the method was all to do with repeated sequences—the ribosomal and histone genes, and satellite DNAs. We strove, with little success, to make the method work with single-copy genes. This was achieved by Alec Jeffreys and Dick Flavell, whose map of rabbit globin genes[24] first demonstrated the presence of intervening sequences in mammalian genomic DNA. Alec went on to develop the methods for DNA fingerprinting that have had such a large impact on identification and forensic testing. His first methods used blotting to display variation in dispersed minisatellite sequences.[25]

What happened to the original aim of purifying the 5S rRNA genes? Peter Ford and I did start a programme but it was not our central interest and we abandoned it when we heard that the formidable Don Brown had taken it up as a major project for his group. Ironically, when the transcriptional regulation of the 5S rRNA genes was sorted out, it was shown that they are transcribed by RNA polymerase III and that the promoters are internal to the transcribed part of the gene.[26] In other words, the region in the RNA sequences where Peter and I had found sequence differences between the somatic and oocyte RNAs are now known to be in the regions that interact with transcription factors.

REFERENCES

1. Saiki R.K. et al. 1985. Enzymatic amplification of β-globin genomic sequences and restriction site analysis for diagnosis of sickle cell anemia. *Science* **230:** 1350–1354.
2. Sanger F. and Coulson A.R. 1975. *J. Mol. Biol.* **94:** 441–448.
3. Flamm W.G. et al. 1967. *Proc. Natl. Acad. Sci. U. S. A.* **57:** 1729–1734.
4. Britten R.J. and Kohne D.E. 1968. *Science* **161:** 529–540.
5. Sanger F. et al. 1965. *J. Mol. Biol.* **13:** 373–398.
6. Southern E.M. 1970. *Nature* **227:** 794–798.
7. Ford P.J. 1971. *Nature* **233:** 561–564.
8. Ford P.J. and Southern E.M. 1973. *Nat. New. Biol.* **241:** 7–12.
9. Birnstiel M. et al. 1968. *Nature* **219:** 454–463.
10. Brown D.D. et al. 1971. *Proc. Natl. Acad. Sci. U. S. A.* **68:** 3175–3179.
11. Goff S.P. and Berg P. 1976. Construction of hybrid viruses containing SV40 and lambda phage DNA segments and their propagation in cultured monkey cells. *Cell* **9:** 695–705.
12. Kelly T.J. Jr and Smith H.O. 1970. *J. Mol. Biol.* **51:** 393–409.
13. Danna K. and Nathans D. 1971. *Proc. Natl. Acad. Sci. U. S. A.* **68:** 2913–2917.
14. Loening U.E. 1968. *J. Mol. Biol.* **38:** 355–365.
15. Southern E.M. and Roizes G. 1974. *Cold Spring Harbor Symp. Quant. Biol.* **38:** 429–433.
16. Jones K.W. 1970. *Nature* **225:** 912–915.
17. Pardue M.L. and Gall J.G. 1970. *Science* **168:** 1356–1358.
18. Shinnick T.M. et al. 1975. *Nucleic Acids Res.* **2:** 1911–1929.
19. Dean W.W. et al. 1973. *Anal. Biochem.* **56:** 417–427.
20. Alwine J.C. et al. 1979. Detection of specific RNAs or specific fragments of DNA by fractionation in gels and transfer to diazobenzyloxymethyl paper. *Methods Enzymol.* **68:** 220–242.
21. Southern E.M. 1975. *J. Mol. Biol.* **98:** 503–517.
22. Kan Y.W. and Dozy A.M. 1978. Polymorphism of DNA sequence adjacent to human β-globin structural gene: Relationship to sickle mutation. *Proc. Natl. Acad. Sci. U. S. A.* **75:** 5631–5635.
23. Botstein D. et al. 1980. Construction of a genetic linkage map in man using restriction fragment length polymorphisms. *Am. J. Hum. Genet.* **32:** 314–331.
24. Jeffreys A.J. and Flavell R.A. 1977. The rabbit β-globin gene contains a large insert in the coding sequence. *Cell* **12:** 1097–1108.
25. Jeffreys A.J. et al. 1985. Hypervariable "minisatellite" regions in human DNA. *Nature* **314:** 67–73.
26. Miller J. et al. 1985. Repetitive zinc-binding domains in the protein transcription factor IIIA from *Xenopus* oocytes. *EMBO J.* **4:** 1609–1614.

One Chromosome: One DNA Molecule

BRUNO H. ZIMM

University of California,
La Jolla, California, USA

By the mid 1950s, DNA had been firmly established as the carrier of genetic information. First, the 1944 experiments of Avery, MacLeod and McCarty[1] showed that bacterial transforming principle was nearly all DNA. Second, in 1952, Hershey and Chase[2] showed that DNA—not protein—was the infectious component of bacteriophage. Then, in 1953, Watson and Crick[3] established that the molecular form of DNA is a double-stranded helix.

Two questions now arose. How long are these helices? How are they packed in the chromosomes that are easily visible through a microscope?

Before 1959, the molecular weights of many samples of DNA had been measured, and all had values of 2–8 kb—whether they came from bacteriophage T2, which has ~150 kb of DNA per phage, *Escherichia coli*, which has 3000 kb per cell, or calf thymus, which has ~60 000 kb per chromosome. The values were sometimes higher and sometimes lower, depending more on the laboratory that made the measurement than on the source of the DNA or the method of preparation.

We now know what was wrong. The methods of preparation in use were violent and broke the long chains. We are now very much aware of the fragility of DNA; it is therefore shocking to read a 1950s protocol. For example, in the widely used Kay, Simmons and Dounce preparation the "crude Na–DNA is dissolved in 700 ml of distilled water . . . This process requires at least 2 hours with continuous rapid stirring using a heavy duty stirrer."[4]

Reprinted from *Trends in Biochemical Sciences* March 1999, 24(3): 121–123.

BACTERIOPHAGE

Davison[5] resolved the issue in 1959: he showed that standard laboratory operations—for example, loading an ultracentrifuge cell through a hypodermic needle, or shaking a solution with chloroform for deproteinization—broke the DNA from *E. coli* bacteriophage T2 into pieces. However, he also reported that, if reasonable care was used, the DNA could be obtained unbroken and that it had a sedimentation coefficient of about 60 S, which corresponds to a length of ~150 kb—about ten times higher than that previously found. It was then easy to show from the known chemical composition of the phage that there is one, and only one, DNA molecule per phage.[6]

BACTERIA

The phage result opened questions about larger organisms. Would it turn out that there was one molecule per cell in bacteria, or one per chromosome in eukaryotes? And, if so, could the molecule be extracted intact? The first question was answered a few years later by Cairns,[7] whose autoradiographs of *E. coli* DNA showed that a few long strands of the length expected for one cell complement were present. Then, Davern[8] showed that at least some of this DNA could be banded intact in an equilibrium CsCl gradient in an ultracentrifuge, if care was taken to minimize shearing. (D. Freifelder has compiled a collection of papers on DNA, together with interesting commentary, from this era.[9])

In our laboratory, we had specialized in hydrodynamic methods of size determination—in particular, ultracentrifugation and viscosity measurements. Once the full size of the T2 molecule was appreciated, we developed an equation that related sedimentation coefficient, intrinsic viscosity and molecular length. To do this, we had to learn how to load an ultracentrifuge sample cell without using the syringe, and we, and Crothers,[10] had to construct a new type of low-shear-rate viscometer because the T2 DNA was so sensitive to shearing. The viscometer was of the rotating cylinder (Couette) type; it had a stationary outer cylinder and a floating inner cylinder that was turned by the action of a rotating magnetic field on a piece of metal in this cylinder. The critical feature of the design was the centering of the inner cylinder by surface-tension forces. Although the instrument worked very well for clean DNA solutions, its rotor could be jammed by solid surface films from the proteins contained in crude cell lysates. In 1967, Stan Gill, a sabbatical visitor, and Doug Thompson[11] from our laboratory, modified the design. They replaced the floating rotor with a completely immersed Cartesian-diver inner rotor (see Box 1), thus eliminating the surface-film problem entirely. Using this instrument, Lynn Klotz showed[12] that he could reproducibly measure the viscosity of crude lysates of *E. coli* and *Bacillus subtilis.*

Box 1. The Cartesian Diver

A Cartesian diver is a float containing a bubble of air to supply bouyancy; the size of the bubble, and hence the tendency of the float to rise or sink, is controlled by varying the pressure on the system. The pressure is varied by adjusting a screw that pushes on the flexible stopper that seals the vessel containing the float. When well adjusted, the float stays in position in the liquid indefinitely and does not touch anything solid; it is then free to turn in response to the rotating magnetic field.

A NEW TOOL

More useful than the viscosity, however, was another property that was now (because of the Cartesian-diver rotor) observable in DNA solutions: the viscoelastic relaxation time (T), which one could obtain by measuring the elastic and viscous components of the response of the DNA solution to stress imposed by the rotor (Fig. 1). When the viscometer was converted to a viscoelastometer, a rotating magnetic field drove the rotor as before. However, in addition, the rotor's angular position was recorded as a function of time by a system using polarized light that responded to the orientation of a piece of polarizer in the rotor.[13] When the magnetic field caused the rotor to turn, the DNA molecules in the solution were stretched (Fig. 1b); when the field was turned off, the rotor recoiled backward because the stretched molecules retracted (Fig. 1c) towards the random coils characteristic of chain molecules at rest (Fig. 1a,d). The backward motion of the rotor decayed approximately exponentially with time; the longest time constant of this decay was T, which depends sensitively on the size of the DNA molecules. (In this brief description, I ignore several technical complications.[12–14])

Decades of experiments on synthetic polymers, and accompanying theory, had established the relationships between the hydrodynamic quantities, such as intrinsic viscosity, and molecular lengths of coiling chains. For a so-called homologous series—that is, a series of chain polymers that have the same molecular structure but different lengths—the log–log plot of such a hydrodynamic quantity against length is a straight line. DNA molecules form

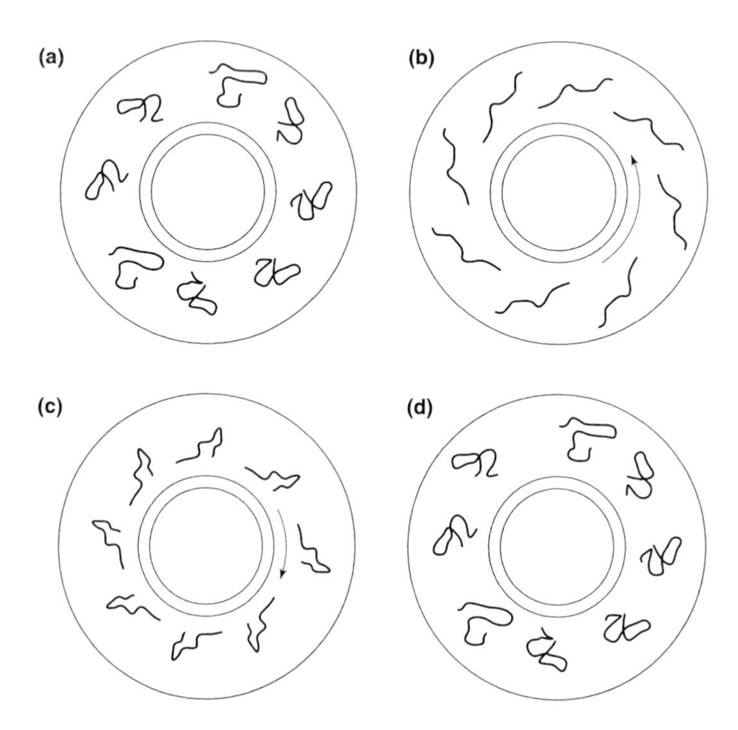

FIGURE 1. Schematic diagram of the viscoelasticity experiment in the concentric-cylinder viscoelastometer. The view is down the cylindrical axis; the outer cylinder is stationary; the inner cylinder can rotate and is driven by a magnetic field. (a) The system is at rest, and the DNA molecules form random coils. (b) The inner cylinder rotates steadily counter-clockwise, and the DNA coils are stretched out by the flowing liquid. (c) The field has been turned off; the DNA coils are recoiling back to their relaxed state. Hydrodynamic coupling through the solvent causes the rotor to rotate clockwise. (d) The DNA has returned to its random-coil state, and the rotor has (exponentially) come to rest.

such a series. In particular, T is a function of length (L):

$$\log(T) = A + b \log(L) \tag{1}$$

In the above expression, A and b are constants that depend on chain structure and solvent conditions but not on molecular length. Fortunately, all known chromosomal DNA molecules have effectively the same physical structure—aside from AT and CG variations, which have only minor effects on the chain stiffness and shape. A and b are therefore practically constant. Theory applied to the intrinsic-viscosity data on phage and smaller DNAs had established the value of b (~1.6). Thompson and Gill[15] used T2 DNA, whose size was known, to get a value for A from stopped-flow flow-birefringence measurements, and Chapman, Klotz and Thompson confirmed this, using the viscoelastometer. Klotz[12] then showed that the same constants gave very reasonable sizes (~3800 kb) for the largest DNA in crude cell lysates of both *E. coli* and *B. subtilis.*

The qualifier "largest" is important. The viscoelastic recoil is peculiar because of the way it is dominated by the largest molecules present. *T* is measured from the slope of a plot of the log of the amount of recoil against time (*t*). This plot becomes linear when *t* is high: the fast relaxations of small molecules have decayed, which leaves only the recoil of the largest molecule. The slope is then the reciprocal of *T* for the largest molecule. This value is therefore distinguished easily from those of smaller molecules, and separation of the DNA from the other components of the cell lysate is not necessary: the chromosomal DNA is, by far, the largest molecule present. Klotz estimated, from the shape of the recoil curve, that ~40% of the DNA was in the form of the largest molecule, which presumably was the intact but linearized chromosome; the smaller pieces were probably mainly degradation fragments. In a control experiment, Klotz treated the lysate with DNAase and found that the recoil then disappeared.

To get these results, Klotz devised a very gentle method of lysing the cells directly in the viscoelastometer chamber by using lysozyme, proteases and detergent after preheating the cells for 15 min at 70°C to denature endogenous DNAase. Pouring or pipetting of the shear-sensitive lysate was thus avoided.

At about the same time, Ruth Kavenoff[16] showed that she could identify *B. subtilis* DNA of the same size through sedimentation on a sucrose gradient if two conditions were satisfied: (1) she used the Klotz lysis procedure to prepare the DNA in the centrifuge tube; and (2) she extrapolated the sedimentation coefficient to zero speed and zero concentration to avoid speed-dependent and concentration-dependent artifacts. She also found that the DNA broke if the lysate was poured from one beaker to another.

DROSOPHILA MELANOGASTER

Given that large DNA could now be prepared intact, if proper precautions were taken, in 1972 Kavenoff tried the viscoelastic method on a eukaryote, *Drosophila melanogaster.* Each of the two largest *D. melanogaster* chromosomes contains ~20 times as much DNA as does *E. coli.* A fundamental biological question therefore arose: what is the relationship between the DNA and the chromosome? Does one DNA molecule run the whole length of the chromosome, or are there many separate pieces of DNA that are, for example, the same size as *E. coli* DNA? A quote from one of our papers gives a flavor of the time.

> Early evidence suggesting that a chromatid contains one piece of native DNA was provided by Taylor et al. in 1957. They found[17] that in cells labeled with tritiated thymidine for one generation both chromatids of each metaphase chromosome were uniformly labeled, and the label in each daughter chromosome was essentially conserved throughout subsequent generations. It was immediately apparent to them that this pattern of chromosome replication was analogous to the replication scheme proposed for DNA by Watson and Crick in 1953. However, they felt it "inconceivable" that there could exist DNA molecules as large as the chromosome. Their feeling was based on the few facts then known about the duplication and mechanical properties of chromosomes and DNA . . . The notion that one DNA molecule runs the length of the chromosome is consistent with the observation that unfixed chromosomes are fragmented by DNAse, but not by RNAse or protease.
>
> We have studied *Drosophila* DNA because it was a convenient eukaryotic DNA . . . The availability of a cultured cell line[18] was also important. *Drosophila* chromosomes . . . condense into metaphase chromosomes with classical morphology and fine structure. The DNA contents of some of these chromosomes had already been determined, and there are numerous well characterized karyotypic variations both within and among species.[14]

We could also have cited another important factor: the presence on campus of D.L. Lindsley, a *Drosophila* guru. Lindsley had a large collection of flies and excellent connections with the rest of the *Drosophila* world.

The metaphase chromosome arrangements (idiograms) of five *Drosophila* species and strains that Kavenoff[14] worked on are shown in Fig. 2. The *D. melanogaster* wild type possesses four pairs of homologs. The largest (represented by the paired chevrons on the right and left in Fig. 2) are designated number 2 and number 3. The vertical lines represent the pair of homologs of chromosome 1, the X-chromosome, whose DNA content is ~50% of that of chromosome 2 or 3. The pair of dots represents chromosome 4, which is ~10% of the size of chromosome 2 or 3. George Rudkin[19] had measured the DNA content of these chromosomes (by microspectrophotometry), obtaining values of ~60 × 10^3 kb for each. Remembering that the viscoelastic technique effectively measures only the largest molecule in a lysate, we thought that we might see a recoil that corresponded to the same size (as predicted by

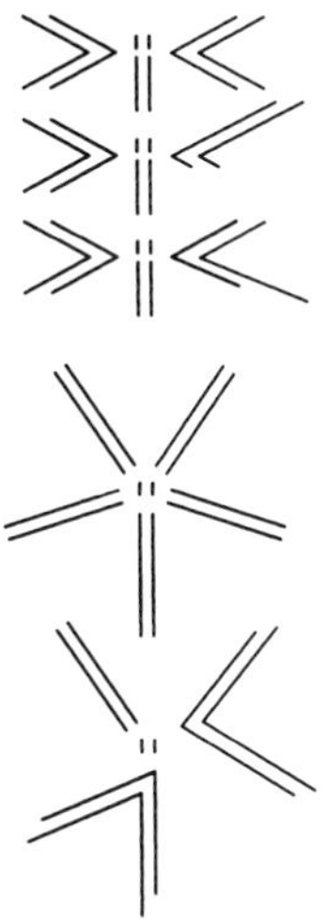

	Time (h)	Repeats	Size (kb)
Drosophila melanogaster			
Wild type	1.67	44	61×10^3
Inversion	1.77	9	63×10^3
Translocation	2.98	10	87×10^3
Drosophila virilis	2.12	5	71×10^3
Drosophila americana	5.0	3	119×10^3

FIGURE 2. *Drosophila* species and mutant chromosomes at metaphase. Each line represents one chromosome; the kink indicates the position of the centromere (if the centromere is not at a chromosome end). The relaxation time (Time) in hours, the number of independent measurements (Repeats), and the size corresponding to each relaxation time (calculated by using Klotz's formula) are shown.

Klotz's T versus molecular-length function): that is indeed what we found. However, using Klotz's function, we had to depend on a very long extrapolation (of more than two decades!) from T2-phage DNA.

More convincing was comparison of strains—a comparison that did not depend on the long extrapolation. The third, translocation, strain shown in Fig. 2 had a large piece split off one of the X chromosomes and attached to one of the homologs of chromosome 2. This increased the size of the largest molecule in the lysate by ~41%. Gratifyingly, T increased by 78% (in comparison with the wild type)—as predicted by equation 1. The inversion strain gave a value identical to that of the wild type, which showed that the DNA ran continuously through the centromere.

The relationship between the two closely related species, *D. virilis* and *D. americana*, was especially interesting. The V-shaped chromosomes in *D. americana* presumbly arose by fusion of the rod-shaped chromosomes of *D. virilis* or a common ancestor—as is shown by the pairing of individual arms of *D. americana* with the rod-shaped chromosomes of *D. virilis* in hybrids of the two species. Our values for the molecular lengths agreed with this picture: the value for *D. americana* was about twice that for *D. virilis.* Thus, the rule that DNA molecules run unbroken from one end of the chromosome to the other also held for eukaryotes.

ACKNOWLEDGEMENT

I thank Patricia Jennings for advice on the arrangement of this manuscript.

REFERENCES

1. Avery O., MacLeod C. and McCarty M. 1944. *J. Exp. Med.* **79:** 137–157.
2. Hershey A.D. and Chase M. 1952. *J. Gen. Physiol.* **36:** 39–56.
3. Watson J.D. and Crick F.H.C. 1953. *Nature* **171:** 737–738.
4. Kay E.R., Simmons N.S., and Dounce A.L. 1952. *J. Am. Chem. Soc.* **74:** 1724–1726.
5. Davison P.F. 1959. *Proc. Natl. Acad. Sci. U.S.A.* **45:** 1560–1568.
6. Davison P.F., Freifelder D., Hede R., and Levinthal C. 1961. *Proc. Natl. Acad. Sci. U. S. A.* **47:** 1123–1129.
7. Cairns J. 1962. *J. Mol. Biol.* **4:** 407–409.
8. Davern C.I. 1966. *Proc. Natl. Acad. Sci. U. S. A.* **47:** 1123–1129.
9. Freifelder D. 1978. *The DNA Molecule: Structure and Properties,* W.H. Freeman.
10. Zimm B.H. and Crothers D.M. 1962. *Proc. Natl. Acad. Sci. U. S. A.* **48:** 905–911.
11. Gill S.J. and Thompson D.S. 1967. *Proc. Natl. Acad. Sci. U. S. A.* **57:** 562–566.
12. Klotz L.C. and Zimm B.H. 1972. *J. Mol. Biol.* **72:** 779–800.
13. Klotz L.C. and Zimm B.H. 1972. *Macromolecules* **5:** 471–481.
14. Kavenoff R. and Zimm B.H. 1973. *Chromosoma* **41:** 1–27.
15. Thompson D.S. and Gill S.J. 1967. *J. Chem. Phys.* **47:** 5008–5017.
16. Kavenoff R. 1972. *J. Mol. Biol.* **72:** 801–806.
17. Taylor J.H., Woods P.S., and Hughes W.T. 1957. *Proc. Natl. Acad. Sci. U. S. A.* **43:** 122–128.
18. Schneider I. 1972. *J. Embryol. Exp. Morphol.* **27:** 353–365.
19. Rudkin G.T. 1965. *In Vitro* **1:** 12–20.

Through the Looking Glass: The Discovery of Supercoiled DNA

JACOB LEBOWITZ

University of Alabama at Birmingham, Alabama, USA

The discovery of supercoiled DNA was first reported 25 years ago in a paper entitled, "The Twisted Circular Form of Polyoma Viral DNA" by Vinograd, Lebowitz, Radloff, Watson and Laipis.[1] This personal reflection describes the different experimental and conceptual processes that eventually led to the discovery as they actually occurred.

The concept that chromosomes may be circular emerged from genetic studies of *E. coli* by Jacob and Wollman[2] and was confirmed by Cairn's autoradiograms.[3] The demonstration in 1962 by Fiers and Sinsheimer[4] that the single-stranded DNA from the small coliphage ϕX174 is in the form of a ring opened up the possibility that the genomes of other phages and animal viruses would contain circular forms of DNA. The phenomenon of DNA supercoiling was first considered because a new conceptual model was needed to explain all the disparate observations that had arisen in the analysis of the circular polyoma DNA.

The tumorigenicity of polyoma virus[5–7] made the characterization of the viral genome of obvious importance, and Roger Weil initiated studies on the DNA in Renato Dulbecco's laboratory at the California Institute of Technology in 1962–1963. Weil reported[8] that the DNA extracted from polyoma virus exhibited the following unusual properties: (1) the DNA renatured monomolecularly; and (2) heating the DNA at 100°C for 10–20 min followed by rapid cooling did not reduce the infective titer, suggesting that the presumed loss of helical configuration did not affect biological activity. Con-

Reprinted from *Trends in Biochemical Sciences* May 1990, 15(5): 202–207.

currently, Lionel Crawford[9] reported that there were two sedimenting components of polyoma virus DNA, a fast (F) 21S form and a slow (S) 14S form. In order to extend the characterization of the structure of polyoma DNA, Roger Weil moved from Dulbecco's laboratory, down the corridor of the basement of the Church Laboratory of Chemical Biology at Caltech to join Jerome Vinograd. Simultaneously (and completely independently), Renato Dulbecco and Marguerite Vogt set out to characterize the structure of polyoma DNA. Consequently, two laboratories at the same institution, on the same floor, were independently pursuing the same research objective. I was quite startled by this situation, having recently arrived as a post-doctoral fellow to pursue hemoglobin research with Vinograd.

During 1963 I observed the progress of the polyoma research while I tried, without success, to understand the mechanism of hemoglobin chain dissociation and reassociation. Within a relatively short time Dulbecco and Vogt, and Weil and Vinograd performed sets of sedimentation velocity and buoyant density experiments which laid the foundation for a model for polyoma DNA. Figure 1 summarizes key data from both laboratories.[10,11] The only difference in the results obtained was the identification of an additional slow component of 14S by analytical ultracentrifugation, not identified by preparative ultracentrifugation analysis. However, the focus of all the studies centered on the 20S and 16S components. The F and S forms at neutral pH have identical molecular weights and so their different $s_{w,20}$ values were probably due to conformational differences (i.e. the F form's reduced frictional coefficient must indicate a more compact structure). Also, the strands of the F form cannot separate under alkaline conditions whereas the S form strands can.

Each laboratory then pursued a different approach. Dulbecco and Vogt[10] analysed the effects of DNase I on the F and S components and found that the F form was converted to the S form at pH 7, and when the conversion was followed by alkaline band sedimentation, the 53S was converted to a 16S component (see Fig. 1). Under both neutral and alkaline conditions, no intermediate products were formed, and the conversion rates of the 20S to 16S and 53S to 16S showed identical first-order kinetics; one hit appeared to be a converting event. In contrast, the DNase I cleavage of the neutral or alkaline 16S forms followed higher-order kinetics.

It was evident from the DNase I results that a change in conformation of the DNA could be introduced by a single cleavage event. Fiers and Sinsheimer[4] had shown that DNase I converts the fast component of ϕX174 sin-

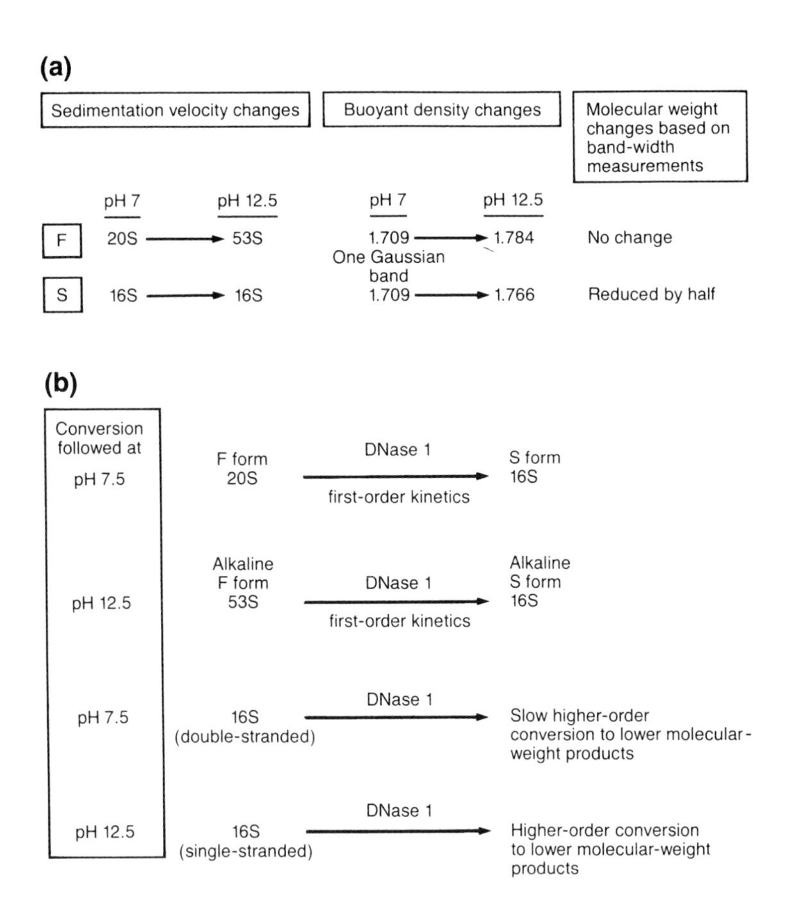

FIGURE 1. Summary of key results obtained independently by Dulbecco and Vogt[10] and Weil and Vinograd[11] to characterize the structure of polyoma viral DNA. For their sedimentation studies, Dulbecco and Vogt performed zone or band sedimentation velocity analysis and density gradient equilibrium centrifugation in a preparative ultracentrifuge, collecting radioactive fractions to determine band positions. Weil and Vinograd performed all their studies in the analytical centrifuge. (a) Summary of the sedimentation velocity and buoyant density changes for the fast (F) and slow (S) sedimentating forms of polyoma DNA when these forms were subjected to alkaline denaturation conditions at pH 12.5. The buoyant density band widths provided the indicated estimates of the molecular weight changes caused by alkaline denaturation. (b) Summary of the results obtained by Dulbecco and Vogt for the DNase I conversion of the F to the S form of polyoma DNA, and the S form to lower molecular-weight products examined at either pH 7 or pH 12.5. The respective order of the kinetics found for the respective DNase I conversions is also shown.

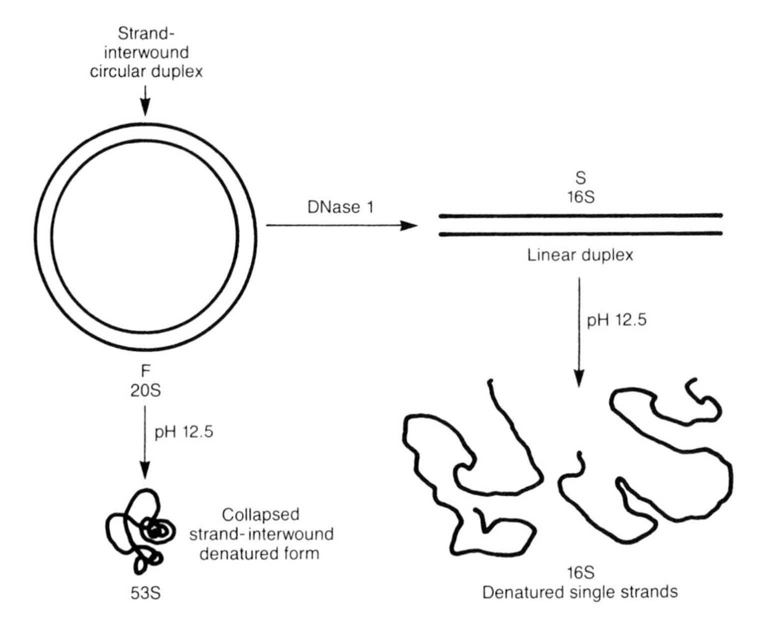

FIGURE 2. The 1963 model proposed both by Dulbecco and Vogt, and by Weil and Vinograd for polyoma DNA from the data shown in Fig. 1 and the evidence from electron micrographs of the DNA obtained by Walther Stoeckenius.

gle-stranded DNA to a slow component by first-order kinetics and concluded that single-stranded rings were converted to single-stranded linear molecules by DNase I. The similarity of the DNase I conversion kinetics for ϕX174 and polyoma DNAs suggested common conformational changes. However, for polyoma it would be a conversion of a strand-interwound 20S circular duplex into a 16S linear duplex by the action of DNase I (see Fig. 2).

THE DNASE I PARADOX

A paradox existed for Dulbecco and Vogt, however: how could DNase I convert a circular duplex into a linear double-stranded molecule by first-order kinetics if the endonuclease introduces only single-stranded breaks? Multiple DNase I scissions would be required before phosphodiester-bond breaks in both strands coincided, or were very close, and double-strand breakage could occur. Consequently, there appeared to be a profound difference in the effect of DNase I on the F vs. the S components. Conversion of the slow forms into lower molecular-weight products was consistent with single-strand breaks and higher-order kinetics. The first-order conversion of the neutral or alkaline fast forms of polyoma, however, was consistent with a single-strand scission. This paradox first suggested the possibility that the duplex ring molecule might contain a single-strand segment that was cut by DNase I and this cleavage was responsible for the F to S conversion. To explore this possibility, the F form was subjected to several converting hits with DNase I, and then exonuclease I was added to digest any single-stranded DNA with exposed 3′ OH ends. No measurable hydrolysis products could be found where 1% hydrolysis could have been detected. By ruling out the involvement of a single-strand segment to account for the conversion kinetics, it appeared that only different DNase I cleavage mechanisms could account for the conversion of a circular duplex to a linear duplex. Dulbecco and Vogt[10] proposed that DNase I generated a double-strand cleavage on the ring duplex and they speculated that this unusual behavior was due to the lability of the ring molecules. "Perhaps the ring form imposes statistically higher stress on the phosphodiester bonds; hydrolysis of one bond may increase considerably the stress on the complementary bond, making it highly susceptible to hydrolysis by the same enzyme molecule, or to breakage." This stress would not exist in the linear duplex, and DNase I would now only be able to introduce single-strand breaks. The conversion of a circular duplex to a linear duplex appeared to account for all the observed data. At pH 12.5, the interwound strands of the circular duplex could not separate and would denature into some form of condensed structure. The strands of the linear form, however, would separate and apparently migrate with an $s_{w,20}$ value close to sedimentation coefficient of the linear duplex form at pH 7 (Fig. 2). The lability of the circular form's phosphodiester bonds would account for the appearance of the S form when the DNA was isolated from the virus. All the results could be accounted for by the model proposed in Fig. 2 if the circular duplex could be converted to a linear duplex as proposed above.

THE RISE AND FALL OF THE 1963 MODEL FOR POLYOMA DNA

To pursue their analysis of polyoma DNA further, Weil and Vinograd collaborated with Walther Stoeckenius, then of the Rockefeller Institute, to obtain electron micrographs of polyoma DNA.[11] Three configurations of polyoma DNA were visible by electron microscopy: coiled molecules without free ends, fully extended cyclic molecules and linear molecules. The latter two forms could readily be measured and were essentially identical in size. The validity of the model for polyoma DNA appeared to be confirmed. A striking feature of the electron micrographs was the high proportion of tightly coiled molecules.[11] It was proposed that these molecules arose by intramolecular crosslinking with protein used in spreading the DNA and, by inference, must be part of the population of circular duplexes.

Although precise dates are difficult to recall, Roger Weil returned to Switzerland to establish his own laboratory approximately six months after the publication of the Weil and Vinograd study.[11] Around this time Bill Studier presented a seminar on his post-doctoral work at Stanford,[12] which involved a comprehensive study of the changes in $s_{w,20}$ vs. pH for linear bacteriophage DNAs (some of his results are presented in Fig. 3a). It was of considerable interest to Jerry Vinograd to perform a similar alkaline titration of the respective forms of polyoma DNA and Bob Watson (his research technician), with my assistance, initiated the study. I subsequently took over this project completely, being delighted to drop hemoglobin research and focus exclusively on polyoma DNA.

In a short time we obtained the pH-band sedimentation velocity titrations of the 20S and 16S forms of polyoma DNA shown in Fig. 3b. The alkaline titration curve of the strand-separable 16S form, which we accepted to be a linear duplex, was quite similar to the titration curve obtained for phage DNAs (Fig. 3a), but the titration curve for the 20S circular form was very unusual; alkaline-induced melting first generated a decrease in $s_{w,20}$; a minimum was reached; then the sedimentation velocity increased continuously until a plateau is reached at 53S. The generation of a denatured 53S collapsed structure, in which no strand separation could be detected, was fully expected, but we could not explain the initial dip in $s_{w,20}$ for the circular form which reached a minimum at the sedimentation velocity of the putative linear form.

In addition we discovered that we could convert the 20S to the 16S form using a variety of reducing agents,[1] suggesting that a special linker might exist that created the duplex ring structure (a hypothesis previously proposed by Fiers and Sinsheimer[4] for ϕXDNA).

A PERPLEXING OBSERVATION

Vinograd was anxious to obtain electron micrographs to assist the polyoma DNA structural characterization, and an undergraduate, Phil Laipis, used the

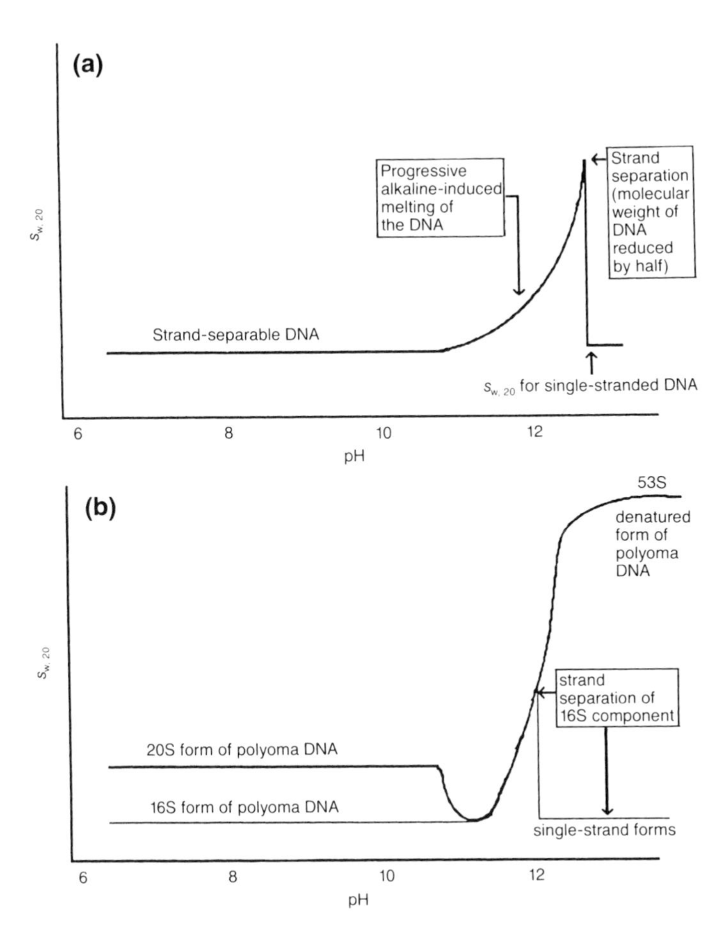

FIGURE 3. (a) Changes in sedimentation velocity ($s_{w,20}$) for strand-separable DNA as a function of increasing pH. The $s_{w,20}$ value for the DNA increases at pH 11 because the ionization of imino sites of thymines and guanines leads to duplex unwinding throughout the DNA and increased flexibility at the melted sites. These hydrodynamic changes, induced by alkaline titration, lead to a continual decrease in the frictional coefficient which increases $s_{w,20}$ until strand separation occurs. Strand separation causes the molecular weight to be reduced by half, leading to an instantaneous drop in $s_{w,20}$. The $s_{w,20}$ value of the single strands at pH 12.5 is very close to that of the linear duplex in the pH range 7–11. (b) Changes in $s_{w,20}$ of the components of polyoma DNA as a function of increasing pH. The 16S component shows the behavior expected of strand-separable DNA [described in (a)]. In contrast, the 20S form first undergoes a decrease in $s_{w,20}$ until it reaches the value of the 16S form. Both forms comigrate from pH 11 to 12 where the slow component strand separates while the $s_{w,20}$ of the fast component continues to increase until it reaches a plateau value of 53S.

Kleinschmidt and Zahn[13] technique for spreading DNA for visualization by the electron microscope. Roger Radloff (a graduate student) separated the 20S and 16S forms by preparative zone centrifugation. When Phil obtained electron micrographs of the 20S and 16S forms he was totally perplexed: there were no linear molecules to be seen, only circular forms were present in the respective electron micrographs of the 20S and 16S fractions. When Phil announced this result, I was convinced that he had mixed up the 20S and 16S fractions and examined the 20S circular DNA fraction twice. I watched while he re-examined the fractions but there was no mix up. Linear molecules were not present in either of the fractions, only circular molecules, with the 20S fraction containing more overlapping or twisted circular molecules.

Since reducing agents could convert the 20S to the 16S form, might it not be possible that linear molecules were recyclizing under the conditions used to spread DNA? That is, the presumed special linker could re-form the circular duplex from the linear duplex. We initiated a kinetic study to follow the $FeCl_2$-induced conversion of the 20S to the 16S form by electron microscopy. DNA aliquots were removed from the reaction mixture at different times, and prepared for electron microscopy as rapidly as possible: once again only circular forms could be seen. Although doubts about the previous polyoma DNA model were growing, there was still the possibility of very rapid reversibility under the electron-microscopy spreading conditions, but this reversibility did not happen in solution. However, this behavior would surely not be the case for endonuclease conversion; phosphodiester bonds would not reform in solution or in a protein film used to trap the DNA for electron microscopy. Consequently, when the 16S form generated by DNase I was shown to be circular by electron microscopy, double-strand cleavage events were completely ruled out and we concluded that single-strand scissions were responsible for the observed conformational conversion.

How many single phosphodiester scissions were required for this conversion? The results of Dulbecco and Vogt[10] indicated that one scission was sufficient. We decided to repeat the DNase I conversion kinetics, following the conformational changes by electron microscopy and by sedimentation velocity under both neutral and alkaline conditions. Since the 53S form arises from the denaturation of the 20S form, both these forms should decrease at the same rate if only one single-strand break was necessary for conformational conversion of the 20S to 16S form. However, if more than one scission was necessary for the 20S to 16S conversion then we should observe a faster rate of conversion for the 53S form to 16S single strands in alkaline conditions since the first scission would allow for strand separation. The results were unequivocal, the 20S to 16S, and 53S to 16S conversions occurred with identical first-order kinetics.[1] Consequently, the 16S form was most likely a nicked circular duplex and zone sedimentation under denaturing conditions should separate single-strand rings from linear strands. In all our previous alkaline sedimentation studies we had stopped the run when the 16S component was approximately half way down the analytical ultracentrifuge cell. Perhaps we had simply not allowed adequate time to resolve alkaline single-stranded forms. We extended the sedimentation time and found that the 16S alkaline form resolved into two sedimenting components, the presumed 18S single-stranded circles and 16S single-stranded linear molecules.

"YOU HAVE THE STRUCTURE!"

From the above data it was clear that the 16S form had to be a nicked circular duplex (form II) but the structure of 20S (form I) remained a mystery. Bob Sinsheimer suggested that we focus on why the 20S form appeared to be partially supertwisted in the electron micrographs. After their discussion, Vinograd obtained coiled telephone cables in order to model possible DNA structures and I found him engrossed in his office wrapping a black cable around a white cable to create a strand-interwound duplex. He held the cable ends to form a circle and started wrapping the double-stranded cables around each other to generate supertwisting or supercoiling of the interwound cables. As I was looking at this telephone cable model, the alkaline pH titration results entered my mind and I realized that the twisted structure had to be right; the sedimentation velocity dip was the key. At that moment, I believe I shouted to Jerry Vinograd, "You have the structure!" I explained that if the strand-interwound telephone cables were pulled apart at any point, the supertwists should come out. This was equivalent to DNA strand-separation—the initial unwinding of the duplex would produce a coupled unwinding of duplex turns and superhelical turns, provided that the latter had the right sense. This process would explain the alkaline titration dip since the superhelical structure would first lose superhelical turns and the $s_{w,20}$ would decrease. However, further DNA unwinding would reverse the process and $s_{w,20}$ would increase as strand unwinding caused superhelical turns of the opposite sense. I believe we tried to pull apart the interwound telephone cables to model the coupled removal of duplex turns and superhelical turns, but the built-in loops of each cable caused knotting and the result was not a convincing demonstration. Nevertheless, we both realized that the superhelical circular model could explain both the titration data and the single-hit conversion of the 20S form to the 16S form. A single-strand scission in the supercoiled circular duplex would be a point of free rotation that would release the superhelical turns or supercoils generating a nicked duplex. How the supercoiled model accounted for all the results is summarized in Fig. 4.

PROOF FOR THE SUPERCOILED MODEL

We set out to explain the supercoiled model to the other members of Vinograd's lab. Roger Radloff appeared skeptical, correctly pointing out that we had not proven unequivocally that the 16S form at pH 7 was a simple nicked circular duplex even though we had resolved two sedimenting structures (putative single-strand ring and single-strand linear molecules) in alkali. We invited Roger to provide the proof, which he did. He heat denatured a population containing 60% 20S and 40% 16S forms and subsequently added *E. coli* exonuclease I. Single-strand molecules with free 3′ OH ends would be digested by this enzyme whereas both the denatured strand-interwound circular molecules and the single-strand rings would be completely resistant. Alkaline sedimentation should reveal the loss of the 16S single-stranded linear molecules but no loss of either the 18S single-stranded rings or the 53S denatured form. The anticipated results were found: only the alkaline 16S component had free 3′ OH ends and could be digested with exonuclease I. Consequently, the 16S form must be a nicked circular duplex at neutral pH, the one-hit (a single phosphodiester bond cleavage) conversion product of the supercoiled 20S form.

One last concern remained: how could we account for the linear molecules that Walther Stoeckenius obtained in his electron micrographs? Weil and Vinograd[11] had identified a third component of polyoma virus, with an $s_{w,20}$ of 14.5S, which had been shown to be a linear duplex. We simply had not paid any attention to the slowest sedimenting component. *E. coli* endonuclease I introduces double-stranded breaks in DNA and therefore the 20S form should be converted to a linear duplex under one-hit conditions. This result was obtained and the linear molecules generated by double-strand cleavage also had an $s_{w,20}$ of 14.5S. Consequently, the linear molecules observed by Stoeckenius must have been the originally identified third 14.5S component (form III).

The characterization of polyoma DNA was complete as shown in Fig. 4 and published in Ref. 1. It is evident from Fig. 4 that DNA supercoiling arose

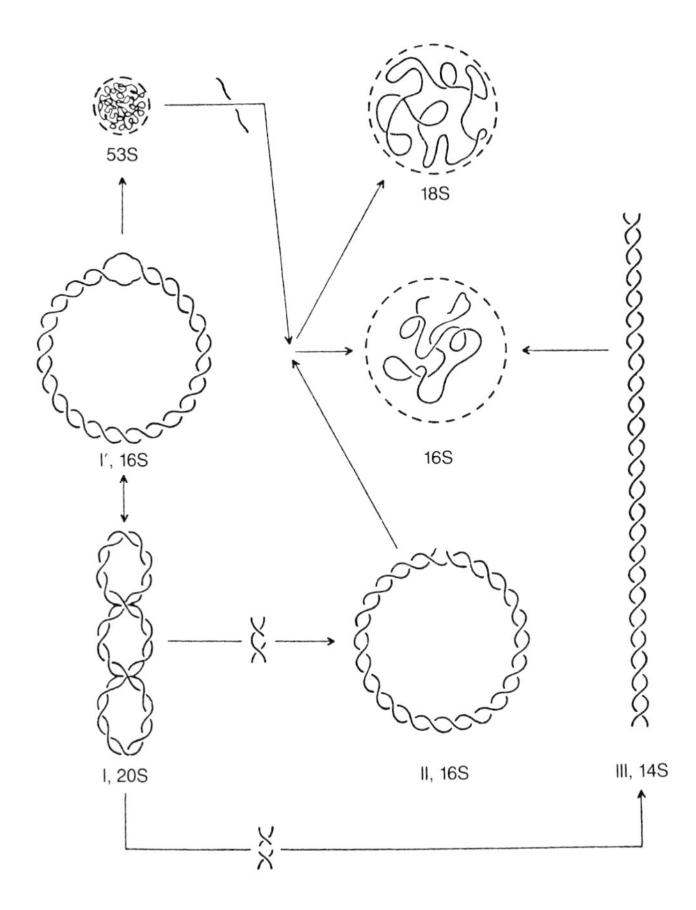

FIGURE 4. Diagrammatic representation of the several forms of polyoma DNA. The dashed circles around the denatured forms indicate the relative hydrodynamic diameters. The superhelical turns for form I should be right-handed. Alkaline titration first leads to the I′ structure in which DNA unwinding has removed the right-handed superhelical turns. Further alkaline-induced unwinding eventually leads to the collapsed 53S denatured form. A single phosphodiester cleavage (nick site) generates a point of free rotation for the removal of the supercoils and the generation of a nicked relaxed molecule, form II. Double-strand breakage leads to the 14S linear duplex, form III. Alkaline denaturation of form II or nicking of the 53S form under denaturing conditions leads to the single-strand circles and linear molecules whereas only the latter would occur for denaturation of form III.

from a deficiency of duplex turns. The rewinding of this twist deficiency requires a coupled introduction of one superhelical turn for every rewound turn. We did make one mistake; the proposal of left handedness for the interwound supercoils.[1] This mistake arose because we started model building with ropes fixed at either end and observed that a deficiency of right-hand duplex turns was compensated for by the ropes forming left-handed toroidal supercoils (equivalent to winding the rope left-handedly around a cylinder). We did not realize that if we had allowed the left-handed toroidal supercoils to convert to interwound supercoils (Fig. 4), the interwound superhelical turns would wind right handed, since left-handed toroidal turns convert into right-handed interwound superhelical turns. Loren MacHattie of Harvard pointed out this possible mistake at a Biophysical Society meeting, and when I returned to Caltech, Ted Young (a graduate student in Sinsheimer's lab), illustrated the point. He simply took an elastic rubber tube and attached two parallel unwound smaller diameter rubber tubes to the right and left ends of the original tube to complete a circle. This easily simulates a circular DNA molecule with a melted open region (see Fig. 4). The introduction of right-hand duplex turns into the two-strand region by rotation immediately introduces left-handed toroidal torsion which interconverts into right-handed interwound superhelical turns. I quickly built my own rubber model and once Jerry Vinograd had examined it he was clearly convinced of our error. We decided to rectify matters in a symposium presentation that Jerry was preparing which would be published. The symposium also offered the opportunity for an additional test of the model. Any denaturation should cause a coupled unwinding of duplex turns and supercoils. Subsequently the $s_{w,20}$ dip seen for the alkaline titration should also occur in an acid titration. In conjunction with Susan Mickel (then a research technician), we performed the acid titration and verified our prediction. An acid induced dip, completely comparable to the alkaline behavior was observed.[14] Our complete titration results were added to the symposium presentation along with the correction that the interwound superhelical turns were right handed.[14]

ONE CAN'T BELIEVE IMPOSSIBLE THINGS

In the course of the discovery of supercoiled DNA there were many instances when one group member or another would say, "Why that's impossible!," so I placed on the door of the laboratory the following quote from *Alice in Wonderland* (*Through the Looking-Glass* by Lewis Carroll):

> "I can't believe that!" said Alice.
>
> "Can't you?" the Queen said in a pitying tone. "Try again: draw a long breath, and shut your eyes."

> Alice laughed. "There is no use trying," she said: "One can't believe impossible things."
>
> "I daresay you haven't had much practice," said the Queen. "When I was your age, I always did it for half-an-hour a day. Why, sometimes I've believed as many as six impossible things before breakfast . . ."

ACKNOWLEDGEMENTS

It is with the greatest pleasure that I thank Roger Radloff and Phil Laipis for their contributions to the discovery of supercoiled DNA and for reviewing this manuscript to ensure that my memory is not flawed. One cannot write about the discovery of supercoiled DNA without expressing gratitude to the late Jerome Vinograd as a teacher and pioneer in the characterization of DNA and, most importantly, for providing a laboratory environment that made scientific discovery a marvellous experience. I am grateful to *TIBS* for this second trip through the "Wonderland of Science."

REFERENCES

1. Vinograd J., Lebowitz J., Radloff R., Watson R., and Laipis P. 1965. *Proc. Natl. Acad. Sci. USA* **53:** 1104–1111.
2. Jacob F. and Wollman E.L. 1958. *E. coli. Symp. Soc. Exp. Biol.* **12:** 75–92.
3. Cairns J. 1963. *J. Mol. Biol.* **6:** 208–213.
4. Fiers W. and Sinsheimer R.L. 1962. *J. Mol. Biol.* **5:** 424–434.
5. Stewart S.E., Eddy B.E., and Borgese B.E. 1958. *J. Natl. Cancer Inst.* **20:** 1223–1243.
6. Stewart S.E., Eddy B.E., and Stanton M.F. 1959. *Proc. Can. Cancer Conf.* **3:** 287–305.
7. Vogt M. and Dulbecco M. 1960. *Proc. Natl. Acad. Sci. USA* **46:** 365–370.
8. Weil R. 1963. *Proc. Natl. Acad. Sci. USA* **49:** 480–487.
9. Crawford L.V. 1963. *Virology* **19:** 279–282.
10. Dulbecco R. and Vogt M. 1963. *Proc. Natl. Acad. Sci. USA* **50:** 236–243.
11. Weil R. and Vinograd J. 1963. *Proc. Natl. Acad Sci. USA* **50:** 730–738.
12. Studier F.Wm. 1965. *J. Mol. Biol.* **11:** 373–390.
13. Kleinschmidt A.K. and Zahn R.K. 1959. *Z. Naturforsch.* **14b:** 770–779.
14. Vinograd J. and Lebowitz J. 1966. *J. Gen. Physiol.* **49:** 103–125.

Ernest F. Gale and Protein Synthesis: The Difficulties of Analysing a Complex System

HANS-JÖRG RHEINBERGER

Max Planck Institute for the History of Science, Berlin, Germany

The relationship between biochemistry and the emerging discipline of molecular biology has a complicated history and has taken many twists—twists that were dependent on research traditions, institutional structures, disciplinary affiliations, experimental or philosophical commitments, or local contingency. Here, I focus on an experimental endeavor that came to occupy a controversial position during the clarification of the mechanism of protein synthesis in the 1950s and finally was marginalized. This endeavor had its roots in studies on the assimilation of amino acids by the bacterium *Staphylococcus aureus*, and it was elaborated by Ernest F. Gale and his co-workers at the University of Cambridge in England.[1,2]

ANTIBIOTICS AND THE CHOICE OF *S. AUREUS*

Ernest Gale began his research career in 1936 at the Biochemical Laboratory in Cambridge, where he worked with Marjory Stephenson on the adaptation of sugar-metabolizing enzymes in *Escherichia coli* and on factors that influence the deamination of amino acids. During World War II, he investigated the production of amines by bacteria and the amino acid decarboxylases that are responsible for this metabolism.[3] After the War, he engaged in a comprehensive project on the assimilation of amino acids by bacteria. The metabo-

Reprinted from *Trends in Biochemical Sciences* September 1998, 23(9): 362–364.

lism of amino acids was at the heart of this investigative enterprise. The driving force for the choice of the organism, the Gram-positive bacterium *S. aureus*, was the decision to look into the action of penicillin. Gale's interest in this antibiotic reflected his personal experience of the war, matched medical needs and opened up prospects for funding. He recalls an incident that contributed to his decision to work in this area below.

> In the middle of the War, I had received from ICI a small sample of the newly produced penicillin (activity 60 units per milligram) but, before we could use it for any experimental work, I received, in the middle of one night, a cry for help from the Professor of Clinical Medicine in the hospital that one of his nurses was dying from a staphylococcal infection. [My] supply ran out in the early morning—by that time the nurse was conscious and her temperature had fallen almost to normal. [This] had a profound effect on me, and I decided that we should try to find out what was happening. [Penicillin] and many of the later antibiotics were mainly effective against Gram-positive organisms, so we should work with them. The infections most dramatically cleared by penicillin were staphylococcal ones (E.F. Gale, pers. commun.).

Gale started to build up his system by using conventional microbiology and microbial-nutrition chemistry. Many strains of *S. aureus* have lost the ability to synthesize several naturally occurring amino acids and thus can be manipulated readily by starving them and by supplementing the medium with amino acids, other biochemical control substances and antibiotics. By 1950, Gale had standardized his *in vivo* bacterial system with respect to incubation conditions, selection of particular strains and specific amino acids, and protein-product analysis.[4]

In his early investigations of the mode of action of penicillin, Gale had already considered a possible effect of the antibiotic on ribonucleic acid metabolism. On the occasion of his Commonwealth Travelling Fellowship in 1951, he took his *S. aureus* and some penicillin on a visit to the RNA expert Torbjörn Caspersson in Stockholm. The levels of cellular RNA varied considerably during the growth cycle and, in the presence of penicillin, they increased markedly. In 1952, back in Cambridge, Gale shifted his research agenda. In collaboration with his skilled technician Joan P. Folkes, he began to focus on the relationship between nucleic acids and protein synthesis.[5] In a series of experiments that followed the routine protocol for measuring the metabolism of amino acids, the two researchers found evidence for a positive correlation between the nucleic acid content of bacterial cells and their rate of protein synthesis. Because of this change of approach, Gale—as early as 1952—was among the first to gain direct experimental access to the connection between nucleic acids and proteins in bacteria.

RADIOACTIVE AMINO ACIDS: A DIFFERENCE BETWEEN SYNTHESIS AND INCORPORATION

Only a year later, Gale and Folkes enhanced the sensitivity of their system by an order of magnitude by introducing radioactive [^{14}C]glutamic acid, which had become commercially available from the Amersham Radiochemical Centre. The two workers observed two major effects. With only the radioactive amino acid in the incubation mixture (condition 1), no net synthesis could be measured, but glutamic acid was nevertheless incorporated into the existing protein. The two Cambridge researchers therefore conceptualized this incorporation as an exchange reaction.[6] When a full complement of amino acids, together with the radioactive amino acid, was present in the incubation mixture (condition 2), labeled glutamic acid became part of the protein, and net protein synthesis took place.

Because there were substantial differences in the action of antibiotics, with respect to both incorporation (condition 1) and protein synthesis (condition 2), Gale, unlike others at the time, decided not to consider the incorporation reaction as a simplified model of polypeptide formation.[7] He wrote, "Whatever may be the final interpretation of these findings, it is clear that it is dangerous to use incorporation studies as measures of protein synthesis in living cells."[6] In the years to come, this argument against incorporation studies became one of Gale's leitmotivs. In June 1951, he had paid a visit to Paul Zamecnik at the Massachusetts General Hospital in Boston. This was about the time that Philip Siekevitz and Zamecnik obtained their first amino acid incorporation results using a fractionated, cell-free rat-liver homogenate. Obviously, Gale was skeptical, but at the same time he was attracted by the possibility of exploiting the radioactive-tracer technique in order to establish a cell-free bacterial system.

TOWARD A CELL-FREE BACTERIAL SYSTEM: NUCLEIC ACIDS RECONSIDERED

In 1953, Gale announced his first step towards a bacterial in vitro protein-synthesis assay.[8] *S. aureus* was not an organism that could be disrupted easily, but Gale and Folkes finally managed to disintegrate the cells by means of supersonic vibration in buffered glucose. By subsequent low-speed centrifugation, they recovered a fraction of disrupted cells that behaved in a manner comparable to that of intact cells—under condition 1 and under condition

2. Most importantly, chase experiments using nonradioactive glutamic acid indicated that exchange of this amino acid occurred under incorporation conditions but not under synthesis conditions. Gale's skepticism towards in vitro incorporation assays grew. With respect to the role of nucleic acids, the results were even more intriguing. Removal of RNA abolished the amino-acid-incorporation reaction; re-addition of RNA restored it. The degree of incorporation, however, varied greatly, depending on the particular radioactive amino acid used.[9–11]

Gale was well aware of the speculations on a template function for microsomal ribonucleic acid, which, as he put it in 1955, "have been bandied about between the initiated during recent years."[10] He envisaged that two phases were involved in such a reaction. The first was a combination of—presumably activated—amino acids with nucleotide residues. The second was a combination of the amino acids through peptide bonds. Gale's reasoning was that, in the first phase, the "amino-acid residues on the nucleic acid chain could be expected to exchange with corresponding residues in the medium."[10]

At this point, Gale and Folkes were very near to reformulating the template model of protein synthesis in terms of amino-acid–ribonucleic-acid intermediates. However, they appear to have been reluctant to extrapolate from their findings. With respect to the differential effect of RNA on different amino acids in the incorporation reaction, Gale instead pondered over the possibility that the method was not in fact suitable for the study of protein synthesis at all. The effects of nucleic acids on protein synthesis could have been related to specific amino acids rather than to specific proteins. He concluded, "We have already had many indications that the former is true," and continued, "Consequently we have abandoned the use of isotopes and incorporation studies as means of studying protein synthesis per se, and have turned our attention instead to [the measurement of] changes in protein-nitrogen and in specific enzymes in the disrupted cell preparation."[10]

DISRUPTED CELLS AND β-GALACTOSIDASE

Faithful to this conclusion, Gale turned his attention to specific enzymes whose synthesis could be assessed in the disrupted-cell system. Here, I restrict myself to discussion of β-galactosidase, an enzyme that he had already studied in *Escherichia coli* as early as 1937.[12] What he called the development of this enzyme depended on the inducing sugar lactose. Gale and Folkes found that the production of the enzyme was inhibited when they depleted the disrupted *S. aureus* cell of nucleic acids and that the addition of endogenous RNA restored the enzyme's development. However, RNA preparations from previously induced cells did not show this restoring effect. Obviously, the RNA involved in the process was not a stable molecule. Under conditions of severe depletion of nucleic acids, the whole process even became dependent on the presence of freshly added *S. aureus* DNA. It therefore seemed clear to Gale that DNA was responsible for the organization of RNA or protein–RNA residues, which were, in turn, responsible for the organization of amino acid residues before they formed peptide bonds. Moreover, it seemed "clear that, in inducible systems at any rate, protein synthesis is accompanied by, if not dependent upon, RNA synthesis."[13] This was Gale's version of what Francis Crick, three years later, termed the Central Dogma.

Jacques Monod had visited Gale's laboratory on several occasions. However, he did not seem to appreciate Gale's observations on the Gram-positive bacterium *S. aureus.* Monod is said to have coined the dictum "What is true for *E. coli,* is also true for E. lephant." Ironically, Monod also "believed that the Gram-positive system [could] be different from [the] Gram-negative one and [that] we had been misguided to use the *S. aureus* as [a] source" (E.F. Gale, pers. commun.). Instead of immediately drawing on Gale's finding, the group at the Pasteur Institute took five more years to identify an RNA intermediate in their *E. coli* galactosidase system.

Clearly, in 1955, RNA was proving to have surprising as well as confusing characteristics. Remember that at this point no distinction between ribosomal RNA, messenger RNA and transfer RNA had yet been made; Gale's disrupted-cell system had a single RNA-containing fraction. There was no way of operationally distinguishing three (or even two) different RNAs—the first a component of the microsomal particle, the second a player in the incorporation reaction and the third a prerequisite for the synthesis of a specific enzyme (such as β-galactosidase). With the benefit of hindsight, we can distinguish the different signals in Gale's system as messenger RNA and transfer RNA. However, from the viewpoint of an experimental process that was driven, as Thomas Kuhn would say, "from behind,"[14] we are left with the impression of a labyrinth in which Gale and Folkes were groping around but not finding the way out. In contrast to what we might expect today, not even Monod knew how to follow up these results.

OPTIONS: INCORPORATION FACTORS

The system that Gale and Folkes had in their hands was "pregnant with many possibilities"—as Sol Spiegelman remarked in a discussion in 1955.[15] There was the incorporation reaction, its RNA dependence and its amino acid specificity; there was the β-galactosidase-synthesis reaction and its coupling to RNA-synthesis; there were the differential effects of antibiotics and of other inhibitors on both reactions; and there was a strange observation that, to the neglect of the others, Gale and Folkes decided to pursue in the years that followed.

This intriguing observation was the fact that RNase-digested ribonucleic acid stimulated both the amino-acid-incorporation reaction (condition 1) and the protein-synthesis reaction (condition 2). By fractionating the digest, Gale hoped to find specific small oligonucleotides that might cause the stimulatory effects. At the beginning, it looked as if several di- and tri-nucleotides were promoting the incorporation of particular amino acids: ACC appeared to stimulate the incorporation of aspartic acid; GUU appeared to stimulate that of glutamic acid; and AUU appeared to stimulate that of leucine (see Fig. 1). "It would follow," Gale speculated, "that an experimental basis for something approaching a nucleic acid-template of the nature postulated by Dounce and others is beginning to appear."[16] However, particularly when he tried to prepare highly purified oligonucleotide fractions, the experimental effects began to shift and became associated with substances that could be removed from the nucleotide fractions. Judging from their physicochemical properties, Gale soon came to view these substances as non-nucleotide compounds—hence the rather unspecific term incorporation factors.[17,18]

Gale reported these findings at a Symposium of the Biochemical Society in February 1956. Francis Crick was also present and on this occasion gave a short explanation of his adaptor hypothesis. In his statement on adaptor molecules, he referred to Gale's incorporation factors as being of possible experimental relevance within the context of his hypothesis.[19] Although Crick was a neighbor of Gale in Cambridge, he visited the Unit for Chemical Microbiology rarely and, as Gale recalls, "seemed to think that our work was outside his interests" (E.F. Gale, pers. commun.). According to Crick, however, it was Gale who was not interested in collaboration (S. De Chadarevian, pers. commun.).

The attempts to isolate and chemically characterize his incorporation factors "side-tracked" Gale "into a long and tedious investigation."[20] Gale and Folkes failed to produce sufficient amounts of material to characterize the factors chemically and, resignedly, presented their efforts in 1958, "in

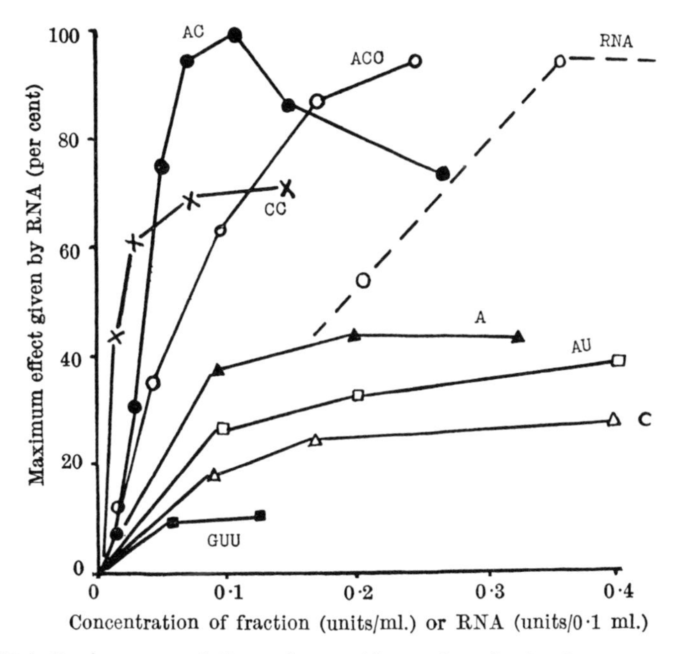

FIGURE 1. Replacement of ribonucleic acid by nucleotides for the activation of aspartic acid incorporation in disrupted *staphylococci*.[16] (Figure reproduced, with the permission, from Ref. 16.)

order that other workers may be able to make and study the factor."[21] The discussion continued until the factor was eventually shown to be a glycolipid fraction.[22,23]

At that same time, the controversy surrounding the reversibility of *in vitro* amino acid incorporation was resolved, surprisingly. After Zamecnik's group and several other laboratories had isolated soluble RNA (with its amino acid attachment capacity), the reversible uptake of amino acids into the protein fraction of disrupted *S. aureus* cells appeared in a new light: it could be envisaged as a bonding of activated amino acids to an RNA component of this very fraction.[24]

Between 1957 and 1958, Mahlon Hoagland (a colleague of Zamecnik) spent a year with Francis Crick in Cambridge, and took the opportunity to settle the long-standing debate between Gale and Folkes, on one side, and Zamecnik and Hoagland, on the other. At the Fourth International Congress of Biochemistry in Vienna, in September 1958, Gale reported that "a preparation of the 'soluble prn [polyribonucleotide]' from liver, provided by Dr

Hoagland, is as effective as staphylococcal nucleic acid in promoting glycine incorporation in the disrupted staphylococcal cell."[25] In the ensuing discussion, Hoagland[26] considered this to be a "heavy blow" to the intermolecular-exchange theory that had tended to cast doubt on the Boston *in vitro* system as a model system for protein synthesis.

Gale's work on *S. aureus* had set him on a long, meandering path. It led him from biochemical studies of amino acid metabolism, through the action of penicillin, to the ribonucleic-acid–protein connection and, finally, to one of the central topics of molecular biology—the problem of coding. Towards the end of the 1950s, however, he felt that the difficulties he had encountered in his analysis were insurmountable. As he put it, "the system we had chosen to study had shown itself to be complex and difficult to analyse" (E.F. Gale, pers. commun.). Consequently, Gale decided to stick to classical microbial biochemistry, to change his research agenda and to investigate, among other things, the action of fungal antibiotics in the test organism *Candida*. His textbook on the molecular action of antibiotics became a standard reference in the field.[27] Gale continued to work on antibiotics until his official retirement in 1981.

CONCLUDING REMARKS

Is there a lesson that can be learned from Gale's development of his experimental system? One could say that it attests to a lack of ingenuity on his part. After all, the major breakthroughs in protein-synthesis research became associated with other names. This is certainly too simple a view. The trajectory of an experimental system can be considered to consist of a progressive reproductive refinement that is punctuated by a series of bifurcation points. A characteristic of such bifurcation points is that, at the time when a decision (as to which of the possible lines of inquiry to follow) must be made, the information that would be necessary in order to assess all the possible consequences of such a decision is usually not available. This is an intrinsic, epistemic characteristic of the research process, and it can be reduced neither to an individual's judgement nor to the particular circumstances of a local research environment. Therefore, the future of an experimental system is never defined by the past alone; it is unpredictable.

Together with the use of *Staphylococci* as a model system, Gale's accomplishments were largely ignored by the mainstream of molecular biology and, consequently, have escaped the attention of historians almost entirely. In this essay, I have tried to remedy this injustice and pay tribute to a bold endeavor.

REFERENCES

1. Rheinberger H.-J. 1996. *J. Hist. Biol.* **29:** 387–416.
2. Rheinberger H.-J. 1997. *Toward a History of Epistemic Things. Synthesizing Proteins in the Test Tube*, Stanford University Press.
3. Gale E.F. 1946. *Adv. Enzymol.* **6:** 1–32.
4. Gale E.F. 1953. *Adv. Protein Chem.* **8:** 285–391.
5. Gale E.F. and Folkes J.P. 1953. *Biochem. J.* **53:** 483–492.
6. Gale E.F. and Folkes J.P. 1953. *Biochem. J.* **55:** 721–729.
7. Gale E.F. and Folkes J.P. 1953. *Biochem. J.* **53:** 731–735.
8. Gale E.F. 1953. In *6th International Congress of Microbiology, Symposium on Bacterial Metabolism*, pp. 109–125, Fondazione Emanuele Paterno, Roma.
9. Gale E.F. and Folkes J.P. 1954. *Nature* **173:** 1223–1227.
10. Gale E.F. 1955. In *A Symposium on Amino Acid Metabolism* (ed. W.D. McElroy and H.B. Glass), pp. 171–192, Johns Hopkins Press.
11. Gale E.F. and Folkes J.P. 1955. *Biochem. J.* **59:** 661–675.
12. Stephenson M. and Gale E.F. 1937. *Biochem. J.* **31:** 1311–1315.
13. Gale E.F. and Folkes J.P. 1955. *Biochem. J.* **59:** 675–684.
14. Kuhn T.S. 1992. *The Trouble With the Historical Philosophy of Science*, Dept of the History of Science of Harvard University.
15. Spiegelman S. 1956. In *Enzymes: Units of Biological Structure and Function* (ed. O.H. Gaebler), pp. 67–89, Academic Press.
16. Gale E.F. and Folkes J.P. 1955. *Nature* **175:** 592–593.
17. Gale E.F. 1956. In *CIBA Foundation Symposium on Ionizing Radiations and Cell Metabolism* (ed. G.E.W. Wolstenholme and C.M. O'Connor), pp. 174–184, Little, Brown and Co.
18. Gale E.F. 1957. In *The Structure of Nucleic Acids and Their Role in Protein Synthesis* (ed. E.M. Crook), pp. 47–59, Cambridge University Press.
19. Crick F.H.C. 1957. In *The Structure of Nucleic Acids and Their Role in Protein Synthesis* (ed. E.M. Crook), pp. 25–25, Cambridge University Press.
20. Portugal F.H. and Cohen J.S. 1977. *A Century of DNA*, MIT Press.
21. Gale E.F. and Folkes J.P. 1958. *Biochem. J.* **69:** 611–619.
22. Demain A.L. et al. 1962. *Biochem. J.* **83:** 424–429.
23. Gale E.F. and Folkes J.P. 1962. *Biochem. J.* **83:** 430–438.
24. Gale E.F. 1959. In *Recent Progress in Microbiology* (ed. G. Tunevall), pp. 104–114, Charles C. Thomas.
25. Gale E.F. 1959. *Proceedings of the Fourth International Congress of Biochemistry, Vienna 1958*, (Vol. 6), pp. 156–165, Pergamon Press.
26. Hoagland M.B. 1959. *Proceedings of the Fourth International Congress of Biochemistry, Vienna 1958*, (Vol. 6), pp. 166–170, Pergamon Press.
27. Gale E.F. et al. 1972. *The Molecular Basis of Antibiotic Action*, John Wiley.

Reminiscences about Nucleic Acid Cytochemistry and Biochemistry

JEAN BRACHET

Université Libre de Bruxelles,
Rhode-St-Genese, Belgique

There was sharp opposition between "anatomists" and "physiologists" when I was a medical student in the University of Brussels, some 60 years ago. This split was exemplified by the presence of two separate buildings, called respectively Institutes of Anatomy and of Physiology, in the newly erected Medical School. The first housed embryologists, histologists and pathologists, the second physiologists, biochemists, bacteriologists and pharmacologists. Biochemistry was a recent outgrowth of the older and larger physiology laboratory; the young professor, E.J. Bigwood, was at that time mainly interested in redox potentials. There was no inner communication between the two buildings except a long dark underground corridor; we called it the "tunnel." Students used it, but in general senior "anatomists" and "physiologists" were not much interested in meeting each other.

I became an "anatomist" in 1927, although I had a much greater interest in organic chemistry than in human bones. We had been told by our professor of histology, Pol Gérard, that in merotomy experiments (bisection of an egg or unicellular organism) anucleate cytoplasmic fragments survive and even display normal activities for some time. This fascinated me and I decided to study the interactions between nucleus and cytoplasm in intact cells (I am still working on them today).

This choice led me to the embryology laboratory headed by my father, who very wisely advised me to work under his young colleague, Albert Dalcq.

Reprinted from *Trends in Biochemical Sciences* June 1987, 12(6): 244–246.

Dalcq had been among the very first to demonstrate that calcium ions are of paramount importance for the maturation and fertilization of starfish eggs; he was then analysing the respective roles of the sperm and egg nuclei in frog development by X-irradiation and local treatment with trypaflavine. His experiments showed that non-nucleated fertilized eggs can undergo a few irregular cleavages, but never gastrulation. I was lucky to work with Dalcq because, in those days, he displayed a real interest in biochemistry. He had even spent a couple of months in David Keilin's laboratory in Cambridge where he had learned a few biochemical techniques, with the hope of following cytochrome synthesis during development. But he soon realized that he was and would always remain a morphologist; he was fond of cytochemistry, enjoying his microscopic investigations of the localization of phosphatases in egg and sperm; he would never have crushed an egg for the analysis of biochemical parameters (even for phosphatase activity measurements).

As soon as I had learned the classical histological techniques of fixation, embedding, sectioning and staining, Dalcq proposed a research subject for me: a study of the localization of "thymonucleic acid" in growing oocytes with the recent cytochemical method of Feulgen and Rossenbeck.[1]

According to the biochemistry textbooks, then as now, there are two main classes of nucleic acids: one of them, now known as DNA, had a queer sugar residue which was identified only in 1930 as deoxyribose by Levene, Mikeska and Mori.[2] This category of nucleic acids was believed to be localized in the nuclei of only animal cells; the prototype of these "animal nucleic acids" was thymonucleic acid from calf thymus. The other type of nucleic acid (our RNA), known to contain a pentose residue that was later identified as D-ribose, was thought to be specific to plant cells. Yeast zymonucleic acid was the best-known of these "plant nucleic acids" (also called phytonucleic acids). The role played by the two kinds of nucleic acids in the nuclei was mysterious: their small size (they were believed to be tetranucleotides of about 1300 Da) precluded any genetic function; it was suggested that they might act as intracellular buffers[3] or as colloids giving a high viscosity to the nuclei.[4] This was all I could find about nucleic acids in biochemistry textbooks around 1930.

R. Feulgen was a distinguished biochemist who had tried for many years to identify the mysterious sugar present in thymonucleic acid (DNA): he discovered that this sugar gives aldehyde reactions and thought that it was glucal, an aldehyde derivative of glucose. Feulgen also found that DNA reacts with fuchsin sulfurous acid (the classical Schiff aldehyde reaction) to give a violet compound after removal of the purines by mild acid hydrolysis. Finally, he applied this aldehyde colour reaction to tissue sections after fixation of the cells with a rather harsh fixative (a mixture of saturated sublimate and acetic acid). Feulgen's main important result was that *all* cell nuclei, vegetal as well as animal, stained positively with his procedure. However, this very important finding (DNA is present in all cell nuclei) was not taken seriously by many biochemists who believed in colour reactions obtained in test tubes, but not on tissue sections. Their scepticism increased when Feulgen showed that, under certain conditions, the cytoplasm also gave a Schiff reaction due to a class of lipids, the plasmalogens. He made a sharp distinction (which remains true today) between the "nucleal" reaction for DNA and the "plasmal" reaction for plasmalogens.

The now classical Feulgen "nucleal" reaction was described for the first time in a well-known German biochemical journal,[1] but no morphologist had the curiosity to read Feulgen and Rossenbeck's original paper. Dalcq had heard of the Feulgen reaction by reading a French journal of histology in which a cytochemist, Jean Verne, had summarized Feulgen's results (two years after the publication of Feulgen's paper). It was quite an event when I went through the "tunnel" to the biochemistry library to read Feulgen's original paper: nearly an act of treason to my friends of the Anatomy Institute!

My own observations and those of others on oocytes of a large number of animal species led to the conclusion that, if the oocytes were adequately fixed, their slender lampbrush chromosomes stained positively with the Feulgen reaction at all stages of oogenesis.[5,6] This implied that, contrary to earlier reports, DNA is a constant constituent of these chromosomes: that this nucleic acid might play a genetic role was contrary to the then current belief that genes were made of proteins. I found later[7] that a Feulgen-positive core becomes visible under the microscope when the nucleoli disintegrate during meiotic maturation in amphibian oocytes. This was the first indication that the nucleolar organizers contain DNA; it took many years before molecular biologists discovered that this DNA is ribosomal DNA (rDNA) and that the nucleoli direct the synthesis of the cytoplasmic ribosomal RNAs (rRNAs).

However, the most important question for chemical embryologists around 1930 was: is DNA synthesized when the fertilized egg divides quickly and repeatedly in smaller and smaller cells during cleavage? Two opposing theories attempted to answer this question. Jacques Loeb proposed that there would be a total, *de novo* nucleic acid synthesis, at the expense of small precursors, during embryonic development. Emil Godlewski[9] believed that during oogenesis eggs accumulated all the materials (including DNA) which are required for the multiplication of the nuclei during cleavage: there would be only a *migration* of pre-existing cytoplasmic nucleic acids into the nuclei and no net nucleic acid synthesis.

With the Feulgen reaction (under correct technical conditions) I could not find any evidence for the existence of a large DNA reserve in the cytoplasm of oocytes and unfertilized eggs from different animal species. During sea urchin egg cleavage, the intensity of Feulgen staining increased in parallel with the increase in the number of the nuclei. The cytochemical evidence was thus in favour of Loeb's net synthesis theory. But we have seen that biochemists did not think much of the Feulgen reaction. They believed that it was not specific since it is a mere aldehyde reaction, contamination with plasmalogen was always possible, there was no evidence that the Feulgen test is quantitative. These doubts were strongly expressed in a book which was almost Holy Gospel for me in 1931, Joseph Needham's *Chemical Embryology.*[10] Characteristically its section on "nuclein and nitrogenous extractives" (creatine and creatinine were handled in the same section as nucleic acids!) amounted to only 16 pages out of 1724. This is not surprising since, in those days, everybody was interested in energy production, intermediary metabolism, mechanisms of cellular oxidations, and very few people cared about nucleic acids. In fact, Needham was an exception, having himself worked on nucleic acid synthesis during embryonic development of aquatic eggs.

On the subject of histochemical methods, Needham wrote: "Histochemical methods are much more uncertain than purely chemical ones."[10] This scepticism was still present in a later book published in 1942 by J. Needham[11]: "Great though the pioneer value of histochemical work may be it is particularly vulnerable to technical criticism" and "The Feulgen test, in the absence of proper precautions, is given by aldehydic phosphatides (plasmal); if possible, it should never be used *in vitro*." The last criticism was justified: several people had tried to estimate the DNA content of crushed unfertilized sea urchin eggs with the Feulgen reaction; they could not remove completely the plasmalogens and had concluded incorrectly that the eggs contain very large amounts of DNA. Needham, who was Reader in Biochemistry in the world-famous Cambridge Laboratory of Biochemistry (headed by the Nobel Prize winner Sir Frederick Gowland Hopkins), expressed very well the negative position held by a majority of biochemists toward cytochemistry.

Going back to nucleic acid synthesis during egg cleavage, my findings with the cytochemical Feulgen reaction were in complete contradiction with the existing biochemical evidence which entirely supported the migration theory: Masing[12] had found, long ago, that the total purine content of sea urchin eggs does not increase markedly during development. More recently, J. and D. Needham[13] had reported that "nucleoprotein phosphorus" also remains almost constant during the early development of several marine invertebrate eggs, including those of the sea urchins.

The discrepancy between the cytochemical and the biochemical data was thus incontrovertible. I knew that biochemical methods would have to be used if biochemists were ever to be convinced. Luckily, Z. Dische[14] had just published his diphenylamine colorimetric method for the estimation of deoxyribose (and thus of DNA) in animal tissues; at my request, he kindly sent me a sample of thymonucleic acid, a brownish, poorly soluble powder. This allowed me, in Roscoff in 1931, to make quantitative estimations of the DNA content of developing sea urchin eggs. These fully substantiated my earlier cytochemical findings: unfertilized sea urchin eggs contained very little DNA and this nucleic acid was synthesized during cleavage. However, Masing and the Needhams were also right! I measured the purine content of developing sea urchin eggs (with a very complicated and lengthy chemical method—there were no UV-spectrophotometers in those days) and entirely confirmed Masing's old results: unfertilized sea urchin eggs contain large amounts of nucleic acid purines and there is little purine synthesis during development.[15]

I could see only one way out of the contradiction: to assume that, contrary to what was printed in all biochemistry textbooks, sea urchin eggs contain large amounts of a *plant* nucleic acid, a RNA. This unorthodox proposal was of course not easily accepted by the scientific community. But I was greatly encouraged when I asked J. Needham's advice: he had found the matter important enough to discuss it with Hopkins, who had given him advice that I never forgot: "Tell this young man that he should not believe everything that is written in textbooks, but make experiments."

I went once more to Roscoff and measured the pentose content of sea urchin eggs (with a method devised for estimating pentosans in straw that was looked down upon with irony by my French friends Monod, Lwoff and Ephrussi). These eggs indeed contained large amounts of pentoses associated with the nucleoprotein fraction. Unfertilized eggs of several species of marine invertebrates also had a high RNA content.

The biochemists were now satisfied, but not the morphologists. Histochemistry was a very important and lively topic for discussion in our Anatomy Institute because one of its members, Lucien Lison, had written a thoughtful and critical book on the subject.[16] When I related my results to my teacher Albert Dalcq, he merely said: "I shall never believe your story until you show me your RNA under my microscope." This negative opinion was of course shared by all my friends in the Anatomy Institute. I endeavoured to satisfy them (and myself!). After several unsuccessful attempts I dug out from a textbook on histological techniques the so-called Unna stain (a mixture of two basic dyes, methyl green and pyronine). Unna believed that methyl green stained oxidizing sites (chromatin) and pyronine reducing sites (nucleoli and cytoplasm). I immediately suspected that methyl green would stain DNA and pyronine RNA; the only way to prove this hypothesis was to remove the two nucleic acids from the histological sections by specific nuclease (DNAse and RNAse) digestion. The use of nucleases for cytochemical purposes had been sharply and correctly criticized by Lison[16] on the grounds that enzymes are never pure. However, when Kunitz[17] crystallized ribonuclease, the long-needed tool for the cytochemical analysis of nucleic acids became available. It immediately turned out that pyronine indeed stains RNA and methyl green double-stranded DNA.[18] The so-called Unna-Brachet method for RNA cytochemical detection has been very widely used by embryologists and histologists; it is still taught to students in French-speaking Colleges and Universities.

I made a very puzzling finding when I applied my cytochemical method for nucleic acid detection to a variety of animal tissues. I found that there was a close and unexpected correlation between the RNA content of a cell and its ability to synthesize proteins: for instance, the exocrine part of the pancreas, which synthesizes large amounts of enzymes, stained much more strongly with pyronine than the Langerhans islets which produce only small quantities of hormones. In order to convince the biochemists, I estimated with chemical methods the pentose content of various tissues from different origins[19]: these quantitative estimations confirmed my cytochemical findings and lent support to the hypothesis that RNA must be involved in protein synthesis.

At that time, T. Caspersson had constructed in Stockholm a delicate and very sensitive UV-cytophotometer which allowed him to localize in the cells and to measure quantitatively the UV-absorbing nucleic acids.[20] He found independently that RNA is localized in the nucleoli and the cytoplasm and that there is a correlation between RNA content and protein synthesis. Simultaneously we reached the same conclusion: RNA somehow directs protein synthesis[21–22]. This conclusion was not easily accepted by the many biochemists who believed that protein synthesis results from the reversal of proteolysis. It took many years before molecular biologists found correct explanations for the mysterious part played by the various RNAs in protein synthesis. It should be added that cytochemical studies by Schultz and Caspersson[23] on *Drosophila* salivary gland cells went one step further: these led them to the conclusion that RNA is synthesized under the control of the neighbouring DNA sequences. We now know that the various RNAs are indeed transcribed on specific DNA segments.

As one can see, cytochemistry had brought us to the very heart of what much later became molecular biology: we knew that DNA is synthesized when cells divide, that it controls RNA synthesis and that RNA directs protein synthesis. However, nobody understood the mechanisms of replication, transcription and translation until biochemists working on enzymes, biophysicists elucidating the structure of macromolecules, geneticists analysing bacteria and phage genetics provided the answers and changed our vague hypotheses into hard facts.

Already in 1940, I had learned a lesson that I shall never forget: one should always try to combine the biochemical and cytochemical approaches if both biochemists and morphologists (as well as yourself) are to be convinced. I tried to persuade fellow scientists of this truth in two books.[24,25] The title of the second (*Biochemical Cytology*) led to sharp adverse reactions from a few anatomy professors as late as 1960; I was openly accused by one of them to have produced a lethal, unviable hybrid between cytology and biochemistry. Today, thanks to the development of new and powerful methods (electron microscopy and differential centrifugation of homogenates first developed by Albert Claude,[26] autoradiography, immunocytochemistry, *in situ* hybridization of specific DNA sequences), the old battle has been won. Far from being lethal, the hybrid has been exceedingly fertile. There are few papers published today in the leading journals of cell biology and developmental biology where electrophoresis gels are not found next to micrographs depicting the intracellular localization of the substance of interest. Cytochemistry and biochemistry are no longer enemies: they help each other on the long, arduous way of scientific discovery.

REFERENCES

1. Feulgen R. and Rossenbeck H. 1924. *Z. Physiol. Chem.* **135:** 203–248.
2. Levene P.A., Mikeska A., and Mori T. 1930. *J. Biol. Chem.* 85: 785–787.
3. Levene P.A. and Bass L.W. 1931. *Nucleic Acids,* Chemical Catalog Co.

4. Mathews A.P. 1927. *Physiological Chemistry* (4th edn), p. 188, William Woods and Co.
5. Brachet J. 1929. *Arch. Biol.* **39:** 677–697.
6. Brachet J. 1937. *Arch. Biol.* **48:** 529–548.
7. Brachet J. 1940. *Arch. Biol.* **51:** 150–165.
8. Loeb J. 1913. *Artificial Parthenogenesis and Fertilization,* Chicago University Press.
9. Godlewski E. 1918. *Wilhelm Roux Arch. Entwicklungsmech. Org.* **44:** 499–529.
10. Needham J. 1931. *Chemical Embryology* (Vol. 3), Cambridge University Press.
11. Needham J. 1942. *Biochemistry and Morphogenesis,* Cambridge University Press.
12. Masing E. 1914. *Wilhelm Roux Arch. Entwicklungsmech. Org.* **40:** 666–667.
13. Needham J. and Needham D.M. 1930. *J. Exp. Biol.* **7:** 317–348.
14. Dische Z. 1930. *Mikrochemie* **2:** 4–32.
15. Brachet J. 1933. *Arch. Biol.* **44:** 519–576.
16. Lison L. 1936. *L'histochimie animale,* Gauthier-Villars.
17. Kunitz M. 1939. *Science* **90:** 111–113.
18. Brachet J. 1940. *C.R. Séances Soc. Biol. Paris* **133:** 88–89.
19. Brachet J. 1941. *Enzymologia* **10:** 87–96.
20. Caspersson T. 1936. *Skand. Arch. Physiol.* (suppl. 8).
21. Caspersson T. 1941. *Z. Naturwiss.* **29:** 33–43.
22. Brachet J. 1941. *Arch. Biol.* **53:** 205–257.
23. Schultz J. and Caspersson T. 1940. *Proc. Natl Acad. Sci. USA* **26:** 507–515.
24. Brachet J. 1944. *Embryologie Chimique,* Desoer Liège and Masson.
25. Brachet J. 1957. *Biochemical Cytology,* Academic Press.
26. Claude A. 1947. *Harvey Lect.* **43:** 121–164.

Jean Brachet's Alternative Scheme for Protein Synthesis

DENIS THIEFFRY* AND RICHARD M. BURIAN†

*Universidad National Autónoma de México, Cuernavaca, Morelos, México; †Virginia Polytechnic Institute and State University, Blacksburg, Virginia, USA

One of the central areas of investigation in 20th-century biology has concerned the mechanisms of protein synthesis. As is typical in the history of science, most of the textbooks and histories that focus on this area discuss a limited set of contributions, which are then considered to be "seminal" or "founding" contributions. In the case of protein synthesis, the most common references are: the one gene one enzyme hypothesis of Beadle and Tatum; Watson and Crick's double helical model of the structure of DNA; Crick's 1958 account of protein synthesis; and the operon model of Jacob and Monod (see Box 1).

Box 1. Bibliography on the history of the development of molecular biology

1. Olby R. 1974. *The Path to the Double Helix,* Macmillan.
2. Judson H.F. 1979. *The Eighth Day of Creation,* Simon and Schuster.
3. Debru C. 1987. *Philosophie Moléculaire,* Vrin.
4. Brock T.H. 1990. *The Emergence of Bacterial Genetics,* Cold Spring Harbor Laboratory Press.
5. Kay L.E. 1993. *The Molecular Vision of Life, Caltech, the Rockefeller Foundation and the Rise of the New Biology,* Oxford University Press.
6. Morange M. 1994. *Histoire de la Biologie Moléculaire,* La Découverte.

Reprinted from *Trends in Biochemical Sciences* March 1996, 21(3): 115–117.

Here, we describe and discuss a set of contributions that has been neglected in most historical accounts of the development of our understanding of protein synthesis. These are the contributions of Brachet, Jeener, Chantrenne and their collaborators from the end of the 1940s to the beginning of the 1970s, working for most of that time in a set of little rustic houses, located in the experimental garden *Jean Massart,* close to the Abbey of Rouge-Cloître, south of Brussels, Belgium (see Box 2).

As we shall see, Brachet and his co-workers achieved a consistent description of the mechanism of protein synthesis. Their work was based on a wide

Box 2. Selected bibliography by or about the Rouge-Cloître group

By Brachet

1. Brachet J. 1965. In *Biochemistry of Animal Development* (Vol. 1) (ed. R. Weber), pp. 1–9, Academic Press.
2. Brachet J. 1975. *Am. Zool.* **15:** 485–491.
3. Brachet J. 1980. *Comp. Biochem. Physiol.* **67B:** 367–372.
4. Brachet J. 1980. In *Florilege des Sciences en Belgique,* pp. 289–314. Acadèmie Royale de Belgique, Classe des Sciences, Bruxelles.
5. Brachet J. 1986. In *A History of Embryology,* pp. 245–259, Cambridge University Press.
6. Brachet J. 1987. *Trends Biochem. Sci.* **12:** 244–246.
7. Brachet J. 1989. *Biochim. Biophys. Acta* **1000:** 1–5.

By Brachet's collaborators

8. Chantrenne H. 1990. In *Selected Topics in the History of Biochemistry: Personal Recollections, III* (Comprehensive Biochemistry Vol. 37) (ed. G. Semenza and R. Jaenicke), pp. 201–213, Elsevier Science Publishers BV.
9. Thomas R. 1992. *Genetics* **131:** 515–518.
10. Alexandre H. 1992. *Int. J. Dev. Biol.* **36:** 29–41.

By historians of molecular biology

11. Fantini B. 1978. "Histoire de l'Embryologie Chimique," Ph.D. thesis, University of Paris.
12. Burian R.M. 1994. In *Les Sciences Biologiques et Médicales en France 1920–1950,* pp. 205–220, CNRS Editions.
13. Burian R.M. 1996. In *New Perspectives on the History and Philosophy of Molecular Biology,* Kluwer.

The most extended bibliography can be found in the "Notice sur Jean Brachet," published in the *Annuaire 1990* of the Académie Royale de Belgique. For other useful information, see the papers presented by R. Burian, H.-J. Rheinberger, J. Sapp, D. Thieffry and R. Thomas at the 1996 Meeting of the International Society for History, Philosophy and Social Studies of Biology, Leuven, Belgium.

set of experiments involving various techniques and model organisms. It was, in some ways, parallel to the famous work on phage and bacteria, but developed out of an independent program of research. The description of protein synthesis that resulted, considerably more complex and subtle than those provided by Crick or Jacob and Monod, has been forgotten by most current molecular biologists and even by many historians of molecular biology. In spite of this loss of historical memory, the contributions of Brachet and his colleagues played an important role in the formation of our contemporary understanding of protein synthesis.

PENTOSENUCLEIC ACIDS AS PERMANENT CONSTITUENTS OF ALL CELLS

At a time when it was still generally thought that *thymonucleic acid* (DNA) was present only in animals and *zymonucleic acid* (RNA) in plants, Brachet showed how to distinguish between them unambiguously, using both cytochemical staining and biochemical assays; he also developed methods to titrate the two types of nucleic acids. Moreover, he localized both nucleic acids precisely in sea urchin eggs, in a number of vertebrate and invertebrate tissues and in a few protists. This led him to conclude that both nucleic acids are permanent constituents of all cells. Moreover, in parallel with Caspersson,[1] Brachet clearly pointed out a correlation between the concentration of zymonucleic acid (RNA) and the level of protein synthesis activity.[2]

RNA AND PROTEIN SYNTHESIS

In the early 1940s, Albert Claude while trying to isolate the Rous sarcoma virus from extracts of chicken embryos, noticed the presence of small nucleoproteic particles, in both infected and healthy animals.[3] Using an air-driven ultracentrifuge Brachet and Jeener showed that these particles could also be isolated from other animal tissues, as well as from plant leaves and yeast extracts, and that they contain most of the RNA present in the extracts.[4]

Unhappily, this work was interrupted in July 1942, because of the closing of the Université Libre de Bruxelles by the German invaders. Both Brachet and Jeener continued to work clandestinely, until Brachet was arrested by the occupier. It was only at the end of 1944, after the liberation, that Brachet and Jeener were able to resume their experimental work.

During the early 1940s, while Brachet conducted cytochemical experiments on the location of nucleic acids, Caspersson conducted parallel ultramicroscopic cytological studies in Sweden. These led him to propose a theory designating the nucleus as the main center of protein synthesis. Brachet constantly referred to Caspersson's theory after first quoting him in an addendum to a major review of his own results in 1942.[5] However, since this first reference, Brachet developed a series of critical analyses of Caspersson's methods, results and hypotheses and, on several occasions, published a scheme summarizing Caspersson's theory[6] (Fig. 1).

Brachet's most important criticism of Caspersson's theory dealt with the idea that the nucleus is the principal site of protein synthesis. Indeed, many of the experimental results that Brachet and his collaborators obtained indicated that protein synthesis occurs mainly in the cytoplasm, and that the nuclear control of protein synthesis has to be indirect. In this respect, probably the most clear-cut results were those obtained with enucleated *Acetabularia,* some of which are discussed briefly below.

By the end of the 1940s, it was known that RNA is present in several cell constituents, but very little was known about possible specific characteristics of these differently located RNAs. Pioneering radioactive-labeling experiments by Jeener and Szafarz clearly indicated the existence of differences in RNA metabolism in different regions of the cell.[7] Quantifying the radioactivity of centrifuged RNA-containing fractions of homogenized rat liver, Jeener and Szafarz found a correlation between localization and turnover rate. Moreover, they showed that changes in physiological conditions affect-

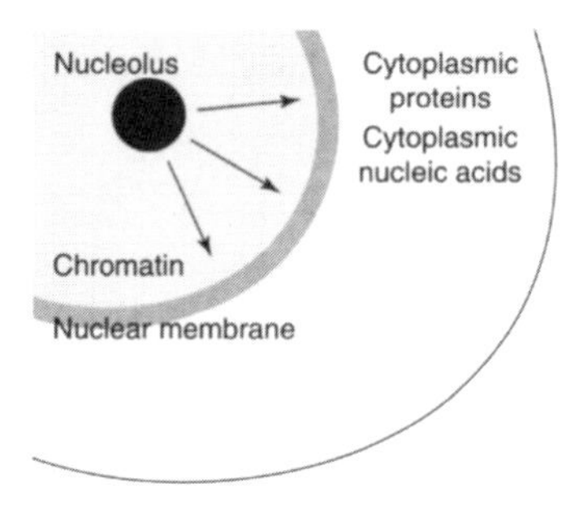

FIGURE 1. Brachet's summary of Caspersson's 1940s theory of protein synthesis. In this scheme, the nucleus is designated as the center of protein synthesis. The figure has been redrawn based on Brachet's 1944 original.[6]

ed the metabolism of these fractions differently. Finally, in the conclusion of their 1950 paper, Jeener and Szafarz made the prophetic suggestion that RNA would be transferred from the nucleus to the cytoplasm. These studies were further developed by Ficq, who showed that differences in RNA metabolism exist even within subcellular structures.[8]

Among other notable early results achieved by the little group, let us mention those obtained by Jeener from around 1953, in parallel with the work of Fraenkel-Conrat, Markham, Gierer, Schramm and others, supporting an hereditary role of RNA in tobacco mosaic virus (TMV). Indeed, using base analogs and ribonuclease, he showed that RNA integrity is more essential for virus multiplication than integrity of the protein sheath.[9]

During the early 1950s, Brachet and Chantrenne started an important series of experiments that dealt with the quantification of RNA concentration and protein synthesis in enucleated amoebae and algae. With these experiments, they showed that the RNA content remains more or less constant after enucleation and that protein synthesis continues for several weeks. Moreover, starting from some remarkable earlier results obtained by Hämmerling,[10] they showed that enucleated *Acetabularia mediterranea* could regenerate cellular structures like the cap, even though these structures were known to be under the control of nuclear genes.[11]

Brachet's cytological staining experiments distinguishing the locations and roles of DNA and various RNAs, Jeener's experiments on the role of RNA in TMV multiplication, and Brachet and Chantrenne's experiments on regeneration with enucleated giant protists yielded a major proportion of the earliest experimental results, which demonstrated decisively that RNA is essential for protein synthesis. This work, together with a whole series of metabolic studies of RNA and attempts to develop causal hypotheses to account for differentiation and morphogenesis played a major role in establishing the background for Jacob and Monod's development of the hypothesis of mRNA. In the following years, these and other results were continuously used by the Rouge-Cloître researchers to support the idea of a fundamental role of RNA in protein synthesis.

These initial studies were followed by many others, presented in dozens of publications during the following decade, many of them specifically addressing the roles of RNA in protein synthesis. The contributions were based on experimental studies involving a great variety of organisms, methods and techniques. During the 1950s, Brachet, Jeener and Chantrenne, in particular, developed an ever clearer picture of the involvement of RNAs in protein synthesis. We cannot summarize all of the results and interpretations of the Rouge-Cloître group here. Rather, we will focus on a few schemes proposed by Brachet for protein synthesis. It would be of particular interest, however, to compare this work with that of such groups as those of Chargaff, Cohen and Zamecnik (the last of who has recently been studied at length by Rheinberger[12,13]) with special emphasis on their attempts to achieve protein synthesis *in vitro* during this decade. Such comparisons, we believe, would highlight important aspects of the convergence and divergence of results achieved by very different techniques, and the interactions between those studying protein synthesis and those studying the genetics of phage and bacteria or pursuing the genetic code.

SCHEMES FOR PROTEIN SYNTHESIS

In 1952, Brachet published a short book containing a wide variety of experimental results about the role of nucleic acids in both the cell and the developing embryo. Specific chapters were devoted to the role and localization of nucleic acids in the cell, the role and localization of the nucleic acids in the developing embryo, the role of the nucleus and the cytoplasm in development and differentiation, and to the role of the nucleus and the cytoplasm in the metabolism of unicellular organisms.[14]

Although dealing with various different biological systems and experimental techniques, Brachet reached a coherent global view of the relationships between nucleic acids and protein synthesis, summarizing it in the only figure of his little book, presented on the last page (see Fig. 2).

Clearly, Brachet's view included several hypotheses, which were dismissed during the following decades, including the ideas that microsomes contained a variety of self-duplicating particles ("plasmagenes") and that microsomes could serve as precursors of mitochondria.[15] However, Brachet clearly distinguished such speculative hypotheses from those that he considered well grounded, e.g. the hypotheses that the cytoplasm is the main site of protein synthesis and that the nucleus plays an indirect role in the synthesis of most proteins.

During the following years, Brachet and his collaborators continued to pursue these lines of research, publishing numerous papers and several books dealing with protein synthesis. Probably one of the sharpest articles is Brachet's 1960 paper in *Nature,* in which he summarized his data and views on protein synthesis as follows:

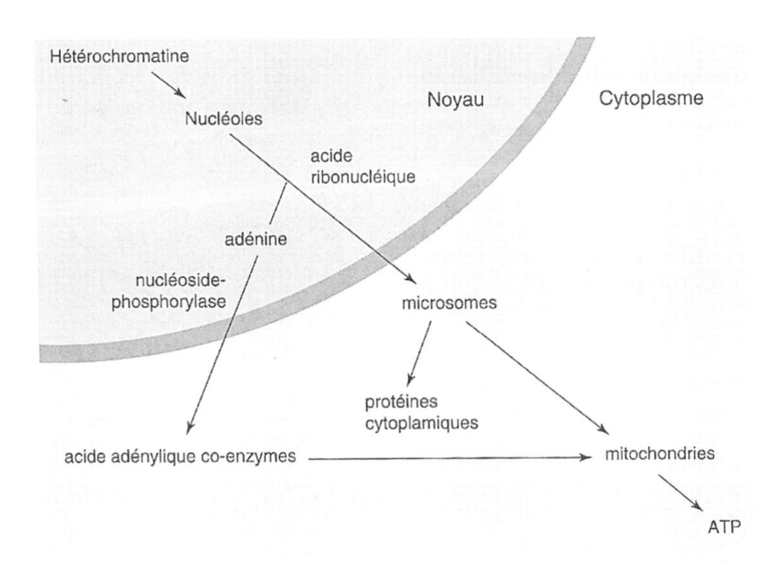

FIGURE 2. A global view of the relationship between nucleic acids and protein synthesis, and the role of nucleus and cytoplasm in this process. Several hypotheses about protein synthesis are summarized, some of which have been dismissed since, for example, the idea that microsomes contained numerous self-replicating particles or "plasmagenes" and were precursors of mitochondria. However, such speculative hypotheses were distinguished from those that had convincing experimental data supporting them, for example, that the cytoplasm was the main site of protein synthesis and that the nucleus plays an indirect role in the synthesis of most proteins. Figure redrawn based on Brachet's original.[14]

If we compare biological systems so different from each other as Amoebae, algae, eggs and reticulocytes, we find big quantitative differences; but they probably are just the result of a secondary factor, namely, the level of energy production in the enucleate cytoplasm. (. . .)

Granting these easily understandable quantitative differences, there is no doubt that all the experiments made so far point toward the same general conclusion: the cell nucleus is the main site of synthesis of nucleic acids (deoxy- and ribonucleic acids). Its role in protein synthesis is a less direct one: it apparently controls the synthesis of cytoplasmic protein through the production of nuclear ribonucleic acid. The latter must, ultimately, receive the genetic information from deoxyribonucleic acid; but there is no evidence so far that "deoxyribonucleic acid makes ribonucleic acid." Nor is there any experimental evidence for the view that deoxyribonucleic acid could directly "make" proteins, even chromosomal proteins. Finally, it should not be forgotten that, in the case of *Acetabularia*, independent synthesis of protein and ribonucleic acid (in the chloroplasts at least) is possible in the complete absence of the nucleus and, therefore, of genes and deoxyribonucleic acid.

For these reasons, the common "slogan": DNA makes RNA, and RNA makes protein, is, in my opinion, an over-simplification which may prove dangerous. The following scheme, although still certainly very simple, is probably somewhat closer to the truth[16] (Fig. 3).

It will be a task for the future to remove the question marks and to increase the number of arrows present in this still largely hypothetical scheme.

Thus, shortly before Jacob and Monod's famous paper, Brachet had reached a global and coherent view of the role of nucleic acids in the synthesis of proteins. Note that, at least until the early 1960s, the Rouge-Cloître group, unlike such groups as that of Paul Zamecnik at the Massachusetts General Hospital, did not concentrate on the biochemical mechanisms of which an RNA molecule served as template to yield a protein. Nonetheless, by localizing and titrating RNAs in different cellular and developmental contexts, Brachet and his colleagues contributed key parts of our understanding of the functions of distinct RNAs, doing so in ways that could not have been accomplished by straightforward pursuit of the *in vitro* synthesis of protein from properly activated and processed RNA signals. The comparative issues suggested here are in considerable need of continuing historical study.

One of the most important contributions of the Brachet group to our understanding of protein synthesis was probably the demonstration of the clear involvement of RNA in protein synthesis. Another equally significant observation was the importance of understanding the spatio-temporal features of RNA metabolism as a means of contributing to analysis of protein synthesis. Indeed, the characterization of the various roles of RNAs and their

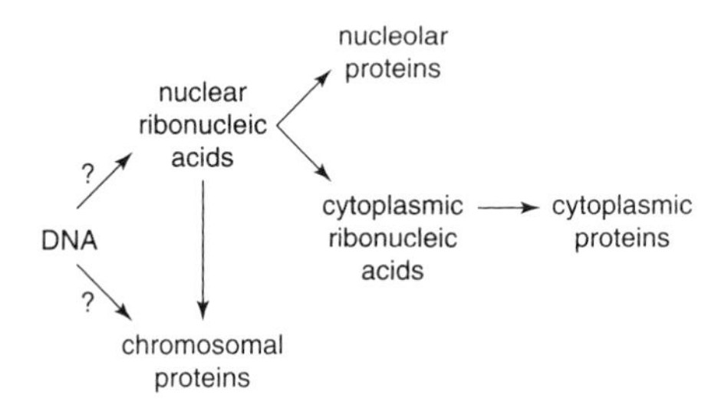

FIGURE 3. Brachet considered the slogan "DNA makes RNA, and RNA makes protein" to be an oversimplification and proposed this hypothetical scheme of protein synthesis in his 1960 paper.[16] Figure redrawn based on Brachet's original.[16]

relations to DNA and protein contributed a number of central features to our contemporary account of the role of nucleic acids in protein synthesis.

A clear account of the state of thought reached by the Rouge-Cloître group by the beginning of the 1960s, can be found in Chantrenne's book *The Biosynthesis of Proteins.*[17] This book was written just before the publication of Jacob and Monod's famous papers and contains a very nice account of the existing experimental data and ideas on protein synthesis. It includes chapters devoted specifically to the genetic control of protein synthesis, the sites of protein formation within the living cell, nucleic acids and protein synthesis, the chemical pathways of protein biosynthesis and the regulation of protein synthesis. But, perhaps its chief virtue is that it constitutes a major advance on any previous work in providing an extremely broad context, using a great variety of biological systems, for understanding the factors affecting protein synthesis and the effects of the differences among biological systems on that process. Far more than any other work of that era, except perhaps, Waddington's, it forged the first steps toward linking our knowledge of protein synthesis in prokaryotes to that of protein synthesis in eukaryotes.

The contributions of the Rouge-Cloître group to the elucidation of the mechanisms of protein synthesis continued well beyond this point. In the following years, a delicate set of experiments was performed by Chantrenne and his collaborators, ultimately leading to the isolation of the first true mRNA,[18] followed by its expression in a heterologous system.[19] This probably should be counted as the last step in the proof of the role of RNA in protein synthesis. Surprisingly, these contributions were not accepted for publication in renowned journals and are almost never cited in historical studies devoted to molecular biology.

Experimenting with various organisms, including bacteria, protists and metazoa, Brachet and his collaborators pictured the problem of protein synthesis in a very subtle fashion. The situation in eukaryotes was surely far from simple: one knew that most of the DNA was located in the nucleus, but some DNA was also found in the cytoplasm (associated with chloroplasts and mitochondria); moreover, there was conflicting evidence regarding possible synthesis of proteins in the nucleus, possibly directly from nuclear DNA. Brachet was well aware of these difficulties and translated them into an explicit scheme for the synthesis of proteins, which was well suited to handling difficulties regarding the extrapolation from prokaryotes to eukaryotes.

In retrospect, when compared with the almost contemporaneous schemes by Crick and Jacob and Monod, Brachet's scheme might seem somewhat complex and confusing. However, it laid bare the necessity of following the biochemistry of nucleic acid metabolism and arrived at essentially the same views as those of Crick, Jacob and Monod, and Zamecnik on a number of key issues regarding protein synthesis. Furthermore, Brachet's criticisms of the simplest models—DNA makes RNA, and RNA makes protein—also anticipated the difficulties that this model encountered during the 1970s.

ACKNOWLEDGEMENTS

D.T. wishes to express thanks to R. Jeener, H. Chantrenne and R. Thomas for spending much time discussing various aspects of the early work of the Rouge-Cloître group. R.B. wishes to express thanks to J. Gayon, J.-P. Gaudillière, S. Gilbert, H.-J. Rheinberger and J. Sapp for numerous helpful discussions bearing on the topic of this paper. We also wish to thank H. Chantrenne, R. Thomas, J. van Helden and E. Fox Keller for their critical reading of the manuscript.

REFERENCES

1. Caspersson T. 1941. *Naturwiss.* **29:** 33–43.
2. Brachet J. 1942. *Arch. Biol.* **53:** 207–256.
3. Claude A. 1941. *Cold Spring Harbor Symp. Quant. Biol.* **9:** 262–271.
4. Brachet J. and Jeener R. 1944. *Enzymologia* **11:** 196–212.
5. Brachet J. 1941. *Arch. Biol.* **53:** 207–257.
6. Brachet J. 1944. *Embryologie Chimique,* Editions Desoer.
7. Jeener R. and Szafarz D. 1950. *Arch. Biochem.* **26:** 54–67.
8. Ficq A., Pavan C., and Brachet J. 1959. *J. Exp. Cell Res.* (suppl.) **6:** 105–114.
9. Jeener R. 1956. *Adv. Enzymol.* **17:** 477–498.
10. Hämmerling J. 1934. *Arch. Entwicklungsmech.* **131:** 1–31.
11. Brachet J. and Chantrenne H. 1956. *Cold Spring Harbor Symp. Quant. Biol.* **21:** 329–337.
12. Rheinberger H.-J. 1992. *Stud. Hist. Phil. Sci.* **23:** 305–331.
13. Rheinberger H.-J. 1992. *Stud. Hist. Phil. Sci.* **23:** 389–422.
14. Brachet J. 1952. *Le Rôle des Acides Nucléiques dans la Vie de la Cellule et de l'Embryon,* Editions Desoer—Massons and Cie.
15. Sapp J. 1987. *Beyond the Gene. Cytoplasmic Inheritance and the Struggle for Authority in Genetics,* Oxford University Press.
16. Brachet J. 1960. *Nature* **186:** 194–199.
17. Chantrenne H. 1961. *The Biosynthesis of Proteins,* Pergamon Press.
18. Burny A. and Marbaix G. 1965. *Biochim. Biophys. Acta* **103:** 409–417.
19. Chantrenne H. and Marbaix G. 1972. *Biochimie* **54:** 1–5.

Deciphering the Genetic Code—A Personal Account

MARSHALL NIRENBERG

National Institutes of Health, Bethesda, Maryland, USA

This is an autobiographical description of the events that led to the breaking of the genetic code and the subsequent race to decipher the code. The code was deciphered in two stages over a five-year period between 1961 and 1966. During the first stage, the base compositions of codons were deciphered by the directing cell-free protein synthesis with randomly ordered RNA preparations. During the second phase, the nucleotide sequences of RNA codons were deciphered by determining the species of aminoacyl-tRNA that bound to ribosomes in response to trinucleotides of known sequence. Views on general topics such as how to pick a research problem and competition versus collaboration also are discussed.

I would like to tell you how the genetic code was deciphered from a personal point of view. I came to the National Institutes of Health (NIH) in 1957 as a post-doctoral fellow with Dewitt Stetten, Jr, a wise, highly articulate scientist and administrator, immediately after obtaining a PhD in biochemistry from the University of Michigan in Ann Arbor. The next year, I started work with William Jakoby and, by enrichment culture, I isolated a *Pseudomonad* that grew on γ-butyrolactone and purified three enzymes involved in the catabolism of γ-hydroxybutyric acid.[1]

There was a weekly seminar in Stetten's laboratory in which Gordon Tomkins, who worked in a different laboratory, participated. Gordon was

Reprinted from *Trends in Biochemical Sciences* January 2004, 29(1): 46–54.

brilliant, with a wonderful associative memory and a magnificent sense of humor. His seminars were superb, especially his description of the step-by-step developments in the problem that he intended to discuss. Towards the end of my post-doctoral fellowship, Gordon replaced Herman Kalckar as head of the Section of Metabolic Enzymes and offered me a position as an independent investigator in his laboratory. The other independent investigators in the laboratory were Elizabeth Maxwell and Victor Ginsberg, who were carbohydrate biochemists, and Todd Miles, a nucleic-acid biochemist. It was a wonderful opportunity and I decided then that if I was going to work this hard I might as well have the fun of exploring an important problem.

In my opinion, the most exciting work in molecular biology in 1959 were the genetic experiments of Monod and Jacob on the regulation of the gene that encodes β-galactosidase in *Escherichia coli* and that the mechanism of protein synthesis was one of the most exciting areas in biochemistry. Some of the best biochemists in the world were working on cell-free protein synthesis, and I had no experience with either gene regulation or protein synthesis, having previously worked on sugar transport, glycogen metabolism and enzyme purification. After thinking about this for a considerable time, I finally decided to switch fields. My immediate objective was to investigate the existence of mRNA by determining whether cell-free protein synthesis in *E. coli* extracts was stimulated by an RNA fraction or by DNA. In the longer term, my objective was to achieve the cell-free synthesis of penicillinase, a small inducible enzyme that lacks cysteine so that I could explore mechanisms of gene regulation. I thought that in the absence of cysteine the synthesis of penicillinase might proceed, whereas synthesis of most other proteins might be reduced.

In England, Pollock[2] had shown that penicillinase is inducible in *Bacillus cerus* and had isolated mutants that differed in the regulation of the penicillinase gene. In 1959, tRNA was recently discovered but mRNA was unknown. At that time, the only clues that RNA might function as a template for protein synthesis were a report by Hershey et al.,[3] showing that a fraction of RNA is synthesized and degraded rapidly in *E. coli* infected with T2 bacteriophage, and a paper by Volkin and Astrachan,[4] which showed that infection of *E. coli* by T2 bacteriophage resulted in the rapid turnover of a fraction of RNA that had the base composition of bacteriophage rather than the DNA of *E. coli.* If mRNA did exist, I thought that it might be contained in ribosomes because amino acids were known to be incorporated into protein on these organelles. I estimated it would take me two years to set up a cell-free system to determine whether RNA or DNA stimulated protein synthesis, which was a pretty accurate estimate.

I knew this was a risky problem to work on because starting out as an independent investigator you are supposed to hit the deck running and prove that you are an effective, productive investigator. One evening I saw Bruce Ames working in his laboratory. Because I thought he was one of the best young scientists at the NIH I described my research plan and asked for his evaluation. He just looked at me and said "It is suicidal." Although we both agreed that it was a dangerous project to work on, I thought suicidal was a little extreme. On the one hand I wanted to explore an important problem, on the other I was afraid of failure, but the wish to explore was much greater than the fear of failure.

As soon as I moved to Gordon's laboratory I started to make cell-free extracts that incorporated amino acids into protein, and to prepare DNA and RNA from ribosomes of penicillinase inducible and constitutive strains of *B. cerus.* I devised a sensitive assay for penicillinase and starting with conditions that had been devised by Lamborg and Zamecnik and his colleagues,[5] I tried to obtain the *de novo* synthesis of penicillinase following addition of either RNA or DNA fractions from either *B. cerus* or *E. coli.* Systematically, I explored the optimum conditions for cell-free synthesis and showed that RNA prepared from ribosomes of *B. cerus* that expressed penicillinase constitutively stimulated penicillinase synthesis by 10–15%, but RNA from either uninduced ribosomes or DNA had no effect. However, the stimulation of penicillinase synthesis was small and it was clear that I needed a more sensitive assay.

Usually around noon, Gordon Tomkins would come into my laboratory with a sandwich and we would go into the hall and talk about my work, his work, and various exciting results that had been published. I always stopped to talk to him, even though the extract that I was preparing was slowly dying in an ice bucket, because these were wonderful conversations. Gordon encouraged me and created an exciting atmosphere for young investigators.

After working on this for about a year and a half, Heinrich Matthei came to my laboratory as a postdoctoral fellow. Heinrich was a plant physiologist from Germany who was a post-doctoral fellow at Cornell who wanted to work on protein synthesis. He came under the impression that, because the NIH is such a big institution, many people would be working on protein synthesis. He stopped in Roy Vagelos's laboratory and Roy sent him to me because I was the only person at the NIH who was studying cell-free protein synthesis. We needed a more sensitive assay, so I suggested that Heinrich use the cell-free amino-acid-incorporating system that I had optimized to measure the incorporation of radioactive amino acids into protein. Heinrich insisted on preparing 20 ^{14}C-labeled amino acids by growing algae in the presence of ^{14}C-bicarbonate, hydrolyzing the protein and purifying each of the ^{14}C-labeled amino acids, because this is what he had done previously.

Using this more sensitive assay it was immediately apparent that RNA from ribosomes, but not DNA, stimulated incorporation of radioactive amino acids into protein.[6,7] I jumped for joy because this was the first definitive demonstration *in vitro* that mRNA existed and was required for protein synthesis. We fractionated RNA from ribosomes and found, as expected, that only a small portion stimulated amino acid incorporation into protein.[8]

We made three trivial technical advances that had a tremendous effect on our work. First, I established conditions that enabled us to freeze and thaw *E. coli* extracts with little or no loss in the ability to incorporate amino acids into protein. Second, as shown in Figure 1, basal, endogenous incorporation of ^{14}C-labeled valine into proteins in the absence of mRNA was high, hence, the increase in amino acid incorporation caused by mRNA was relatively small. We confirmed the reports of Kameyama and Novelli[9] and Tissieres *et al.*[10] that DNase I inhibited the incorporation of amino acids into protein in cell-free *E. coli* extracts. Therefore, we incubated *E. coli* extracts in the presence of DNase I but without a radioactive amino acid for 40 min until endogenous amino acid incorporation had almost stopped.[16,34,49] Then we divided the extracts into small portions and froze them for use later. Because the endogenous incorporation of radioactive amino acids into protein was low in these extracts, it was stimulated markedly by the addition of mRNA. Third, the standard method of washing radioactive protein precipitates in trichloracetic acid to remove radioactive amino acids involved repeated centrifugation and resuspension of protein pellets, which was very laborious and time consuming. One evening I compared this standard method with washing protein precipitates by filtration through Millipore filters. The results were identical. By using frozen–thawed, preincubated *E. coli* extracts and washing radioactive precipitates on Millipore filters we could do as much in one day as had previously taken us 8–10 days.

I then obtained yeast rRNA and tobacco mosaic virus (TMV) RNA and we found that both were as active as mRNA. However, RNA from TMV was 30–50 times more active than ribosomal RNA at stimulating amino acid incorporation into protein. I called Heinz Frankel-Conrat in Berkeley, a world

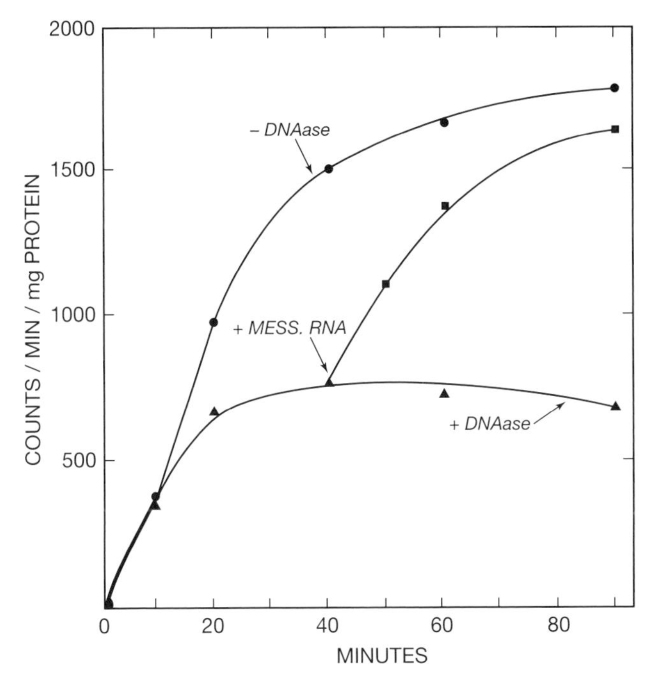

FIGURE 1. Incorporation of ^{14}C–labeled valine into protein in *Escherichia coli* extracts. Endogenous incorporation of radioactive amino acids into protein in *E. coli* extracts was high. However, amino acid incorporation ceased after incubation for ~40 min with DNase I. I found that I could freeze *E. coli* extracts and thaw them without loss of activity, so I incubated *E. coli* extracts in the absence of radioactive amino acids for 40 min, divided the extracts into small aliquots and froze them for use later in different experiments. Endogenous incorporation of radioactive amino acids was greatly reduced in such extracts, and addition of mRNA preparations from ribosomes clearly stimulated amino acid incorporation into protein.[16,34,49] Reproduced from Ref. 16.

expert on TMV who had a mutant with an amino acid replacement in the viral coat protein, to tell him our results. He invited me to come to his laboratory to synthesize radioactive protein directed by RNA from wild-type and mutant TMV, with the intention that he and a colleague would purify and characterize the products to determine whether the radioactive protein synthesized was TMV coat protein. I felt like Marco Polo exploring a new area.

Before going to Frankel-Conrat's laboratory I obtained some poly(U) and instructed Heinrich to make 20 different solutions, each with 19 cold amino acids and one radioactive amino acid, to detect poly(U)-dependent incorpo-

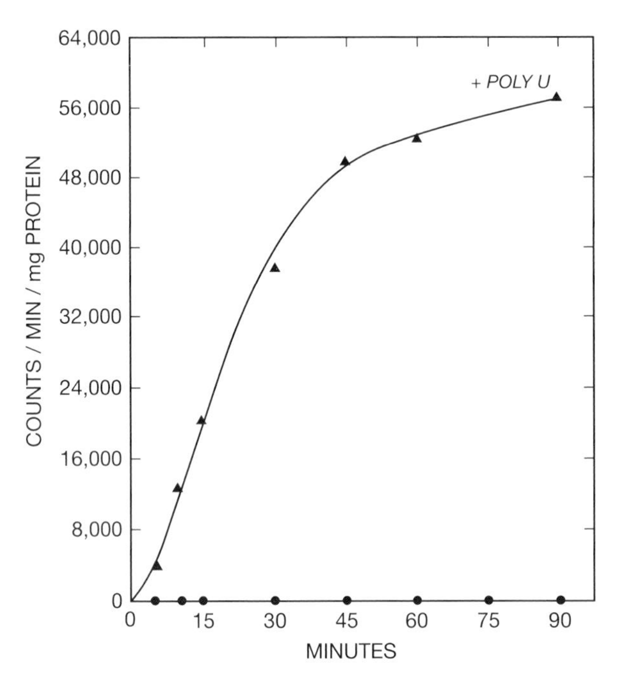

FIGURE 2. Poly(U) greatly stimulates the incorporation of radioactive phenylalanine into poly-phenylalanine.[8]

ration of a single radioactive amino acid into protein. After working in Frankel-Conrat's laboratory for about a month, Heinrich called me very excitedly to tell me that poly(U) was extraordinarily active in stimulating the incorporation of only phenylalanine into protein[8] (Figure 2). I immediately returned to Bethesda. We also showed that single-stranded poly(U) functions as mRNA, but double-stranded or triple-stranded poly(U)–poly(A) helices do not[8] (Figure 3). This was the first RNA antisense experiment. In addition, we showed that poly(C) directs the incorporation only of proline into protein.[8]

I thought the poly(U) result wouldn't be believed unless we characterized the radioactive polyphenylalanine product of the reaction very carefully. Hydrolysis of the ^{14}C-labeled polyphenylalanine by HCl recovered stoichiometric amounts of ^{14}C-labeled phenylalanine. I also thought we should show that the solubility of the ^{14}C-polyphenylalanine was the same as that of authentic polyphenylalanine, but because I knew nothing about this I went to Chris Anfinson's laboratory, which was directly under mine, to ask for names of investigators who might have characterized polyphenylalanine. Michael Sela

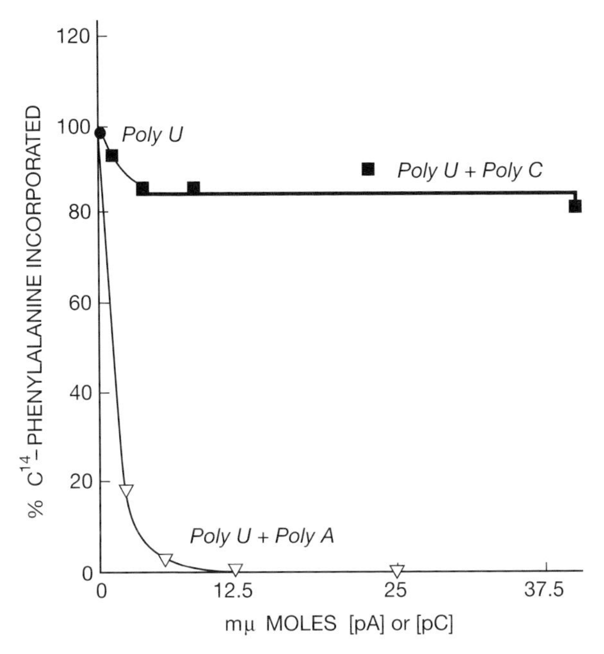

FIGURE 3. Addition of poly(A) completely inhibits the mRNA activity of poly(U) by the formation of double-stranded and triple-stranded helices. By contrast, addition of poly(C) has little effect on the mRNA activity of poly(U). This experiment, done in 1961, was the first anti-sense RNA experiment.[8]

was the only person in the laboratory at the time; I knew that he worked with synthetic polypeptides so I asked if he knew anything about the solubility of polyphenylalanine. He said "I do not know much, but I can tell you two things: one, polyphenylalanine is insoluble in most solvents; and second, it does dissolve in 15% hydrobromic acid dissolved in concentrated acetic acid." I looked at him in delight as well as astonishment because I had never heard of such a solvent. Fifteen years later I learned that Michael Sela was the only person in the world who knew that polyphenylalanine dissolved in this esoteric solution because it is used to characterize C termini of proteins and he had mistakenly added it to polyphenylalanine, which, to his surprise, dissolved.

I was scheduled to give a talk in 1961 at the Vth International Congress of Biochemistry in Moscow. Just before leaving for Russia, I married Perola Zaltzman, a biochemist from Rio de Janeiro who worked with Sidney Udenfriend at the NIH, and we planned to meet for a leisurely, two week vacation after the meeting. I gave my talk in Moscow to an audience of ~35 people.[11]

However, Francis Crick invited me talk again in a large symposium that he was chairing on nucleic acids, which I did to an extraordinarily enthusiastic audience. After returning to Bethesda, Fritz Lipmann generously gave me a partially purified transfer enzyme and we showed that phenylalanine-tRNA is an intermediate in the synthesis of polyphenylalanine directed by poly(U).[12] A picture of Heinrich Matthei and me taken in 1962 is shown in Figure 4.

Soon afterwards I gave a talk at the Massachusetts Institute of Technology. A few years earlier Severo Ochoa from New York University had been awarded the Nobel Prize for his discovery, with Marianne Grunberg-Manago, of polynucleotide phosphorylase, which catalyzes the synthesis of randomly ordered polynucleotides. While I was answering questions from the audience, Peter Lengyel came up to the podium and told the audience that he and others in Ochoa's laboratory had used randomly ordered synthetic polynucleotides that contained several different nucleotide residues to direct

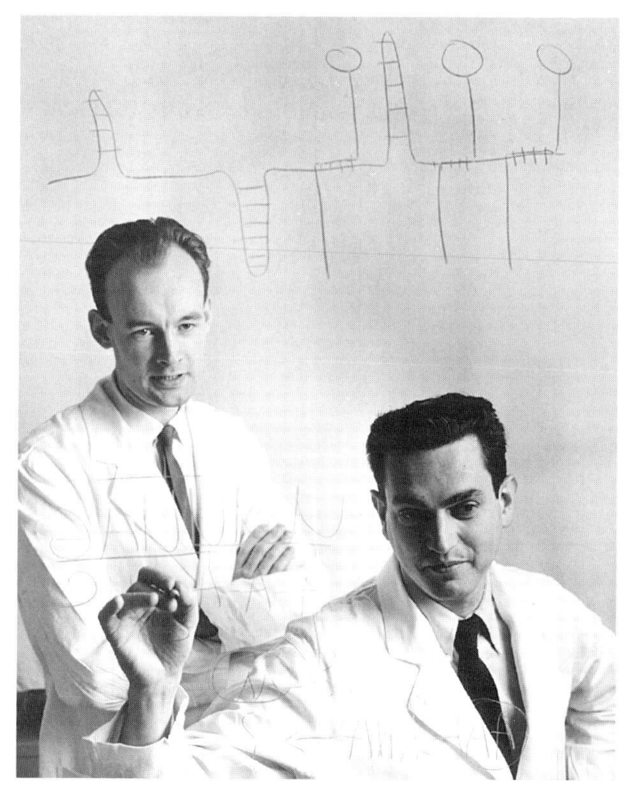

FIGURE 4. A picture of Heinrich Matthaei (left) and myself in 1962.

the incorporation of other amino acids into protein. I flew back to Washington feeling very depressed because, although I had taken only two weeks to show that aminoacyl-tRNA is an intermediate in protein synthesis, I should have spent the time focusing on the more important problem of deciphering the genetic code. Clearly, I had to either compete with the Ochoa laboratory or stop working on the problem.

The next morning (Saturday) I went to the library in Leon Heppel's corridor to look up methods of synthesizing, purifying and characterizing randomly ordered polynucleotides whose synthesis was catalyzed by polynucleotide phosphorylase. Robert Martin (whose wife writes the *Miss Manners* newspaper column) was in the library and, when I told him what had happened, he suggested we synthesize randomly ordered polynucleotides that weekend. And that is exactly what we did. Bob, who is a superb, energetic investigator, stopped his own work and during the next few months synthesized and characterized many randomly ordered polynucleotides. Bob Martin played a major role in deciphering the genetic code.

Between 1961 and 1964 Bill Jones, Bob Martin and I determined the base compositions of RNA codons by directing amino acid incorporation into protein using many randomly ordered polynucleotide preparations that contained different combinations and proportions of bases.[13–19] In Figure 5 are the minimum species of bases required for mRNA codons. Although

Polynucleotides	**Amino acids**			
U	PHE			
C	PRO			
A	LYS			
G	—			
UC	LEU	SER		
UA	LEU	TYR	ILE	ASN
UG	LEU	VAL	CYS	TRP
CA	HIS	THR	GLN	ASN
CG	ARG	ALA		
AG	ARG	GLU		
UAG	ASP	MET		
CAG	ASP	SER		

FIGURE 5. The specificity of randomly ordered polynucleotide templates in stimulating amino acid incorporation into protein in *Escherichia coli* extracts. Only the minimum kinds of bases necessary for template activity are shown, so many amino acids that respond to randomly ordered polynucleotides composed of two or more kinds of bases are omitted. The base compositions of RNA codons were derived from these experiments.[49] Reproduced from Ref. 49.

Severo Ochoa was a fierce competitor during this time,[20–28] we had never met. So, in 1961 I called him when I was in New York to arrange a meeting. I thought it would be more civilized to cooperate, or perhaps split the problem in some way, rather than compete with one another. Ochoa was very gracious, he invited me to his laboratory and introduced me to his post-doctoral fellows, and we had tea in the library. However, there was no way we could collaborate. Later, to my horror, I found that I enjoyed competing. I focused on solving the problems that we were investigating and on working more effectively, rather than on winning or losing. The competition stimulated me to become more focused and I accomplished far more than I would have in its absence. From the beginning I vowed never to cut corners or reduce the rigor with which experiments were done to win the competition. Therefore, the quality of our work remained high throughout the deciphering process.

Years after the genetic code was deciphered, Mirko Beljansky told me that during a year's sabbatical in Ochoa's laboratory, Ochoa suggested that he see whether synthetic polynucleotides could direct cell-free protein synthesis in *E. coli* extracts. Beljansky spent a year trying to direct cell-free protein synthesis with poly(A), but was unsuccessful because the polylysine product of the reaction is a basic protein and is not precipitated by trichloracetic acid. Similarly, in Jim Watson's laboratory, Tissieres also tried to direct cell-free protein synthesis with poly(A), but did not detect the polylysine product for the same reason. Therefore, the idea of using synthetic polynucleotides to direct cell-free protein synthesis originated independently in at least three laboratories.

We showed that the code is formed of triplets by the amounts of radioactive histidine, threonine, asparagine, glutamine, lysine and proline that were incorporated into protein by five poly(A-C) preparations that contained different ratios of A and C.[18,19] The observed codon frequencies were compared with the theoretical frequencies of triplets or doublets in each poly(A-C) preparation calculated from the base composition determined for each polynucleotide preparation. The data showed that the codon for histidine is a triplet that contains one A and two Cs and the codons for asparagine and glutamine are triplets that contain two As and one C. The data also showed that two triplets or a doublet correspond to threonine.

A group picture of the people in my laboratory and their spouses taken in early 1964 is shown in Figure 6. There were ~20 people in my laboratory in all who deciphered the genetic code and about half of them (and their spouses) are in this picture. The occasion was a party for Brian Clark (middle of the first row in a suit), who was returning to Cambridge, England after several years in my laboratory. The post-doctoral fellows who came to my laboratory were superb and our work on deciphering the genetic code was very much a group project.

Although we and Ochoa's group had deciphered the nucleotide compositions of RNA codons, the nucleotide sequences were unknown. We tried several ways to solve this problem, but were unsuccessful. However, Kaji and Kaji[29] had shown that poly(U) stimulates the binding of radioactive polyphenylalanine-tRNA to ribosomes and Regina Cukier in my laboratory was using randomly ordered polynucleotides to direct the binding of amino-acyl-tRNA to ribosomes. I wondered how small a message one could use that

FIGURE 6. Going away party for Brian Clark who was returning to Cambridge, England after several years in the laboratory. The photograph was taken early in 1964. This is the only group photograph I have of the people in my laboratory while we were deciphering the genetic code. From left to right: Sid and Joan Pestka, Mrs. Marshall and Dick Marshall, Tom Caskey (partially hidden), Ty Jaouni, Mrs. Rottman, Norma Heaton (the marvelous technician with whom I worked for 39 years), Fritz Rottman, Brian Clark, Phil Leder, Shirley Shapiro (the secretary for the laboratory), Joel Trupin, Mrs. Trupin, myself, Bill Groves, Perola Nirenberg, Mrs. O'Neill and Charlie O'Neill.

would remain functional in directing aminoacyl-tRNA binding to ribosomes. Leon Heppel gave me a doublet, a triplet and a hexanucleotide that consisted only of U residues. I thought that ternary complexes, such as a ^{3}H-phenylalanine-tRNA–UUU trinucleotide–ribosome complex, would be retained by Millipore filters, whereas ^{3}H-phenylalanine-tRNA would be washed through the filter. The first experiment worked beautifully; the trinucleotide, UUU, stimulated binding of ^{3}H-phenylalanine-tRNA to ribosomes, but the doublet, UU, was without effect.[30] A similar experiment with oligo(A) preparations[49] is shown in Figure 7. It was clear that we could use this technique to determine the nucleotide sequences of RNA codons. However, most of the 64 trinucleotides had not been synthesized previously or isolated. Our major problem was to devise methods to synthesize trinucleotides of known sequence. Phil Leder saw an advertisement offering 0.5 g of each of the 16 doublets for $1500 per doublet. Although expensive (a few years earlier my salary for 6 months was $1500), we brought the lot. However, we received only 15 doublets because the US Customs Service confiscated one and used it all to test for drugs.

Marianne Grunberg-Manago was visiting Maxine Singer at that time, and both were experts on the use of polynucleotide phosphorylase. Phil Leder and Richard Brimacombe from my laboratory joined them in optimizing conditions for the polynucleotide phosphorylase catalyzed synthesis

ADDITION	**[^{14}C]-Lys-tRNA BOUND TO RIBOSOMES**
	μ **μmoles**
ApA	**0.01**
ApApA	**1.92**
ApApApA	**1.92**
ApApApApA	**1.92**
ApApApApApA	**2.71**

FIGURE 7. Nucleotide sequences of RNA codons were determined by stimulating the binding to ribosomes of an appropriate species of radioactive aminoacyl-tRNA which recognized the trinucleotide. The aminoacyl-tRNA–trinucleotide–ribosome ternary complex was adsorbed on a Millipore filter and unbound aminoacyl-tRNA was removed by washing.[30,49]

of oligoribonucleotides from doublet primers.[31] Trinucleotides then were separated from oligonucleotides of different chain lengths by electrophoresis.

In addition, in 1964, Leon Heppel (who was one of the best nucleic acid biochemists in the world at the time) suggested an esoteric method for the synthesis of trinucleotides and higher homologues, reported in a single sentence in one of his papers.[32] He showed that, in the presence of a high concentration of methanol, pancreatic RNase A catalyzes the synthesis of trinucleotides and higher homologues from oligoribonucleotide primers and pyrimidine 2′-3′-cyclic phosphates. We used both methods to synthesize triplets, as shown in Figure 8. Polynucleotide phosphorylase catalyzes the addition of nucleotide residues to the 3′-ends of doublets and other oligonucleotide primers, whereas pancreatic RNase A catalyzes the addition of nucleotides to the 5′ termini of oligonucleotide primers. Maxine Singer, Marianne Grunberg-Manago, and Leon Heppel, as well as Phil Leder, Mert Bernfield and Richard Brimacombe made major contributions to deciphering the genetic code. Bob Martin, Leon Heppel, and Maxine Singer are shown in Figure 9, and pictures of Phil Leder and Tom Caskey are shown in Figure 10.

$$\text{1) ApG + UDP} \xrightleftharpoons[]{\text{POLYNUCLEOTIDE PHOSPHORYLASE } Mg^{++}} \text{ApGpU + ApG(pU)}_n\text{+ P}$$

$$\text{2) ApG + URIDINE-2',3'-CYCLIC PHOSPHATE} \xrightleftharpoons[]{\text{RNase A}} \text{UpApG + (Up)}_n\text{ApG}$$

FIGURE 8. In 1964, most of the 64 possible trinucleotides had been neither synthesized nor isolated. The major problem was to devise methods for the synthesis of trinucleotides. The two enzymatic methods that we used are shown. (1) Maxine Singer (an expert on polynucleotide phosphorylase), Marianne Grunberg-Manago (who had discovered polynucleotide phosphorylase while in Severo Ochoa's laboratory), Philip Leder and Richard Brimacombe discovered conditions for the addition of one or a few nucleotide residues to the 3′-end of doublet primers, catalyzed by polynucleotide phosphorylase.[31] (2) Leon Heppel suggested an esoteric method for triplet synthesis that he, Whitfield and Markham had discovered in which pyrimidine-2′,3′-cyclic phosphates are added to the 5′-ends of doublet primers catalyzed by pancreatic RNase A in the presence of a high concentration of methanol.[32] Merton Bernfield synthesized half of the trinucleotides in our laboratory using this method.

BOB MARTIN

LEON HEPPEL

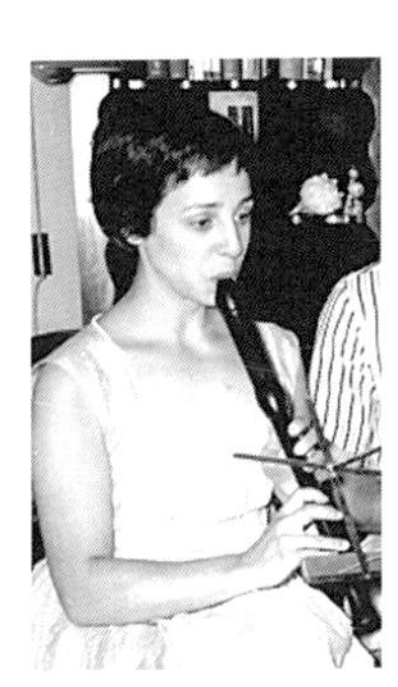

MAXINE SINGER

FIGURE 9. Three investigators at the NIH played major roles in deciphering the genetic code. Bob Martin synthesized and characterized many randomly ordered polynucleotides and had a very important role in deciphering the base compositions of RNA codons. Leon Heppel was one of the few nucleic acid biochemists in the world at that time. He gave me compounds and advice when I needed it, and suggested the use of pancreatic RNase A to catalyze trinucleotide and higher homologue synthesis, a method that he had discovered earlier.[32] Maxine Singer came to the NIH as a post-doctoral fellow working with Leon Heppel. She was an expert on polynucleotide phosphorylase and helped devise conditions for the synthesis of trinucleotides catalyzed by polynucleotide phosphorylase.[31]

PHIL LEDER

TOM CASKEY

FIGURE 10. Phil Leder came to my laboratory as a post-doctoral fellow and played a major role in deciphering the genetic code. He was the first to decipher the nucleotide sequence of a codon.[33] Tom Caskey was another post-doctoral fellow in my laboratory. Together with Dick Marshall, he compared the genetic code of *Escherichia coli* with that of *Xenopus* and hamsters and showed that the code is universal.[41] Later, Tom Caskey and his colleagues worked on the mechanism of termination of protein synthesis.

It took my colleagues, Philip Leder, Merton Bernfield, Joel Trupin, Sid Pestka, Fritz Rottman, Richard Brimacombe, Charles O'Neal, French Anderson, Don Kellogg, Ty Jaouni and myself a year to synthesize the 64 trinucleotides and test each against 20 different radioactive aminoacyl-tRNA preparations to decipher the nucleotide sequences of RNA codons.[30,33,35–40] Philip Leder deciphered the first nucleotide sequence of a valine RNA codon using trinucleotide-dependent binding of aminoacyl-tRNA to ribosomes.[33] We soon found that the third bases of synonym RNA codons varied systematically[33,35–40] and, using purified aminoacyl-tRNA fractions,[39–41] we found four patterns of degeneracy, as shown in Figure 11. Francis Crick coined the term "wobble" to describe the third-base degeneracy of RNA codons.[42] Generously, Robert Holley gave us a highly purified fraction of yeast alanine-tRNA that he and his colleagues had sequenced in what was the first determination of the nucleotide sequence of a nucleic acid.[43] Philip Leder and I showed that this species of alanine-tRNA recognizes the codons GCU, GCC and GCA, and that the nucleotide residue in the tRNA anti-codon that recognizes U, C or A in the third position is hypoxanthine.[40] After we had published the nucleotide sequences of 54 of the 64 RNA codons[30,33,35–40] the great

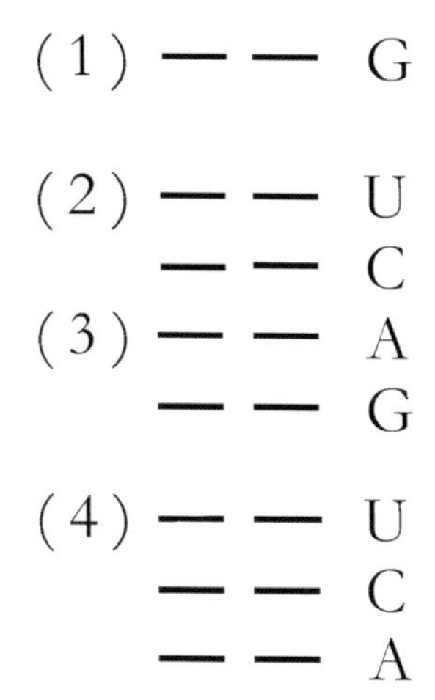

FIGURE 11. Four kinds of degeneracy for synonym codons were found using purified aminoacyl-tRNA preparations. Robert Holley gave us a highly purified preparation of yeast Ala-tRNA, which was the first nucleic acid to be sequenced,[43] and we showed that it recognized the codons GCU, GCC and GCA.[40] Hypoxanthine in the t-RNA anticodon recognizes U, C or A in the third position of RNA codons.

nucleic acid chemist, Gobind Khorana and his colleagues reported the nucleotide sequences of RNA codons determined using chemically synthesized trinucleotides.[44] The genetic code is shown in Figure 12. Clark and Marcker[45] showed that *N*-formyl-methionine-tRNA initiates protein synthesis by recognizing the codon AUG, whereas AUG at internal positions corresponds to methionine residues. UAA, UAG and UGA were reported to correspond to the termination of protein synthesis by Sidney Brenner and Alan Garen and their colleagues.[47–48] The hydrophobic amino acids Phe, Leu, Ile, Met and Val correspond to chemically similar codons that have U as the second base. By contrast, the hydrophilic amino acids Tyr, His, Gln, Asn, Lys, Asp and Glu correspond to codons with A as the second base. In addition, amino acids with chemically similar side chains, such as Asp and Glu, and Asn and Gln, have chemically similar codons. Clearly, the arrangement of codons and amino acids is not random.

After deciphering the genetic code we asked the question "is the code universal?" Thomas Caskey and Richard Marshall purified tRNA fractions from *Xenopus* embryos and hamster liver. Comparing the genetic code of *E. coli* to that of *Xenopus* and hamster, we found that the code is essentially universal[41] These results had a profound philosophical impact on me because

THE GENETIC CODE

UUU UUC PHE UUA UUG LEU	UCU UCC UCA UCG SER	UAU UAC TYR UAA TERM UAG TERM	UGU UGC CYS UGA TERM UGG TRP
CUU CUC CUA CUG LEU	CCU CCC CCA CCG PRO	CAU CAC HIS CAA CAG GLN	CGU CGC CGA CGG ARG
AUU AUC AUA ILE AUG MET	ACU ACC ACA ACG THR	AAU AAC ASN AAA AAG LYS	AGU AGC SER AGA AGG ARG
GUU GUC GUA GUG VAL	GCU GCC GCA GCG ALA	GAU GAC ASP GAA GAG GLU	GGU GGC GGA GGG GLY

FIGURE 12. It took us about a year to synthesize the 64 trinucleotides and test each against 20 radioactive aminoacyl-tRNA preparations to determine the nucleotide sequences of RNA codons.[30,33,35–40] Gobind Khorana and his colleagues synthesized the 64 trinucleotides chemically and also determined nucleotide sequences of some RNA codons.[44]

they indicate that all forms of life on this planet use essentially the same language. Some dialects have been reported subsequently in some organisms, but all are modifications of the same genetic code.

We deciphered the genetic code over a period of about five years, from 1961 to 1966. This was a group project and the post-doctoral fellows in my laboratory during this period (Table 1) contributed in many important ways. In addition, Robert Martin, Leon Heppel, Maxine Singer and Marianne Grunberg-Manago played major roles in deciphering the genetic code, and Ochoa and Khorana and their colleagues also contributed to the deciphering of the genetic code. Deciphering the genetic code was the first project that I worked on as an independent investigator, and it was an extraordinarily exciting, fun-filled project to explore and solve. Although many excellent problems related to the code and protein synthesis remained after the code was deciphered, I decided to switch to the more challenging field of neurobiology, a field I am still exploring.

TABLE 1. *Post-doctoral fellows in my laboratory 1961–1966. Deciphering the genetic code was very much a group affair and many of these individuals contributed greatly to this work. There were nine or fewer fellows at any one time, but not all worked on the code.*

Heinrich Matthaei	Thomas Caskey
Oliver W. Jones	Joseph Goldstein
Samuel Barondes	Edward Scolnick
Philip Leder	Richard Tomkins
Sidney Pestka	Arthur Beaudet
Fritz Rottman	French Anderson
Merton Bernfield	Charles O'Neal
Brian Clark	William Groves
William Sly	Raymond Byrne
Judith Levin	Regina Cukier
Joel Trupin	Donald Kellogg
Richard Brimacombe	Michael Wilcox
Richard Marshall	Frank Portugal
Dolph Hatfield	

ACKNOWLEDGEMENTS

I thank Michael Spencer for his help in digitizing and formatting the figures and tables.

REFERENCES

1. Nirenberg M.W. and Jakoby W.B. 1960. Enzymatic utilization of γ-hydroxybutyric acid. *J. Biol. Chem.* **235:** 954–960.
2. Pollock M.R. 1962. Exoenzymes. In *Bacteria* (4) (ed. I.G. Grinsalus and R.Y. Stanier), pp. 121–178.
3. Hershey A.D. et al. 1953. Nucleic acid economy in bacteria infected with bacteriophage T2. I. Purine and pyrimidine composition. *J. Gen. Physiol.* **36:** 777–789.
4. Volkin E. and Astrachan L. 1956. Phosphorus incorporation in *Escherichia coli* ribonucleic acid after infection with bacteriophage T2. *Virology* **2:** 149–161.
5. Lamborg M.R. and Zamecnik P.C. 1960. Amino acid incorporation into protein by extracts of *E. coli. Biochim. Biophys. Acta* **42:** 206–211.
6. Matthaei J.H. and Nirenberg M.W. 1961. The dependence of cell-free protein synthesis in *E. coli* upon RNA prepared from ribosomes. *Biochem. Biophys. Res. Commun.* **4:** 404–408.
7. Matthaei J.H. and Nirenberg M.W. 1961. Characteristics and stabilization of DNase sensitive protein synthesis in *E. coli* extracts. *Proc. Natl. Acad. Sci. U. S. A.* **47:** 1580–1588.
8. Nirenberg M.W. and Matthaei J.H. 1961. The dependence of cell-free protein synthesis in *E. coli* upon naturally occurring or synthetic polyribonucleotides. *Proc. Natl. Acad. Sci. U. S. A.* **47:** 1588–1602.
9. Kameyama T. and Novelli G.D. 1960. The cell-free synthesis of β-galactosidase by *Escherichia coli. Biochem. Biophys. Res. Commun.* **2:** 393–396.
10. Tissieres A. et al. 1960. Amino acid incorporation into proteins by *Escherichia coli* ribosomes. *Proc. Natl. Acad. Sci. U. S. A.* **46:** 1450–1463.
11. Nirenberg M.W. and Matthaei J.H. 1961. The dependence of cell-free protein synthesis in *E. coli* upon naturally occurring or synthetic template RNA. In *Biological Structure and Function at the Molecular Level* (ed. V.A. Engelhardt), pp. 184–189, The MacMillan Co.
12. Nirenberg M.W. et al. 1962. An intermediate in the biosynthesis of polyphenylalanine directed by synthetic template RNA. *Proc. Natl. Acad. Sci. U. S. A.* **48:** 104–109.
13. Martin R.G. et al. 1961. Ribonucleotide composition of the genetic code. *Biochem. Biophys. Res. Commun.* **6:** 410–414.
14. Matthaei J.H. et al. 1962. Characteristics and composition of RNA coding units. *Proc. Natl. Acad. Sci. U. S. A.* **48:** 666–677.
15. Jones O.W. and Nirenberg M.W. 1966. Degeneracy in the amino acid code. *Biochim. Biophys. Acta* **119:** 400–406.

16. Nirenberg M.W. et al. 1963. Approximation of genetic code via cell-free protein synthesis directed by template RNA. *Fed. Proc.* **22:** 55–61.
17. Nirenberg M.W. and Jones O.W., Jr. 1963. The current status of the RNA code. In *Symposium on Informational Macromolecules* (ed. H. Vogel), pp. 451–465, Academic Press.
18. Nirenberg M.W. et al. 1963. On the coding of genetic information. *Cold Spring Harb. Symp. Quant. Biol.* **28:** 549–557.
19. Jones O.W. and Nirenberg M.W. 1966. Degeneracy in the amino acid code. *Biochim. Biophys. Acta* **119:** 400–406.
20. Lengyel P. et al. 1961. Synthetic polynucleotides and the amino acid code. *Proc. Natl. Acad. Sci. U. S. A.* **47:** 1936–1942.
21. Speyer J.F. et al. 1962. Synthetic polynucleotides and the amino acid code, II. *Proc. Natl. Acad. Sci. U. S. A.* **48:** 63–68.
22. Lengyel P. et al. 1962. Synthetic polynucleotides and the amino acid code, III. *Proc. Natl. Acad. Sci. U. S. A.* **48:** 282–284.
23. Speyer J.F. et al. 1962. Synthetic polynucleotides and the amino acid code, IV. *Proc. Natl. Acad. Sci. U. S. A.* **48:** 441–448.
24. Basilio C. et al. 1962. Synthetic polynucleotides and the amino acid code, V. *Proc. Natl. Acad. Sci. U. S. A.* **48:** 613–616.
25. Wahba A.J. et al. 1962. Synthetic polynucleotides and the amino acid code, VI. *Proc. Natl. Acad. Sci. U. S. A.* **48:** 1683–1686.
26. Gardner R.S. et al. 1962. Synthetic polynucleotides and the amino acid code, VII. *Proc. Natl. Acad. Sci. U. S. A.* **48:** 2087–2094.
27. Wahba A.J. et al. 1963. Synthetic polynucleotides and the amino acid code, VIII. *Proc. Natl. Acad. Sci. U. S. A.* **49:** 116–122.
28. Speyer J.F. et al. 1963. Synthetic polynucleotides and the amino acid code. *Cold Spring Harb. Symp. Quant. Biol.* **28:** 559–567.
29. Kaji A. and Kaji H. 1963. Specific interaction of soluble RNA with polyribonucleic acid induced polysomes. *Biochem. Biophys. Res. Commun.* **13:** 186–192.
30. Nirenberg M. and Leder P. 1964. RNA codewords and protein synthesis. I. The effect of trinucleotides upon the binding of sRNA to ribosomes. *Science* **145:** 1399–1407.
31. Leder P. et al. 1965. Synthesis of trinucleoside diphosphate with polynucleotide phosphorylase. *Biochemistry* **4:** 1561–1567.
32. Heppel L.A. et al. 1955. Nucleotide exchange reactions catalyzed by ribonuclease and spleen phosphodiesterase. 2. Synthesis of polynucleotides. *Biochem. J.* **60:** 8–15.
33. Leder P. and Nirenberg M.W. 1964. RNA codewords and protein synthesis. II. Nucleotide sequence of a valine RNA codeword. *Proc. Natl. Acad. Sci. U. S. A.* **52:** 420–427.
34. Nirenberg M.W. 1964. Protein synthesis and the RNA code. *Harvey Lectures Series 59*, pp. 155–185.
35. Bernfield M.R. and Nirenberg M.W. 1965. RNA codewords and protein synthesis. IV. The nucleotide sequences of multiple codewords for phenylalanine, serine, leucine and proline. *Science* **147:** 479–484.

36. Trupin J.S. et al. 1965. RNA codewords and protein synthesis. VI. On the nucleotide sequences of degenerate codeword sets for isoleucine, tyrosine, asparagine, and lysine. *Proc. Natl. Acad. Sci. U. S. A.* **53:** 807–811.
37. Nirenberg M.W. et al. 1965. RNA codewords and protein synthesis. VII. On the general nature of the RNA code. *Proc. Natl. Acad. Sci. U. S. A.* **53:** 1161–1168.
38. Brimacombe R. et al. 1965. RNA codewords and protein synthesis. VIII. Nucleotide sequences of synonym codons for arginine, valine, cysteine, and alanine. *Proc. Natl. Acad. Sci. U. S. A.* **54:** 954–960.
39. Kellogg D.A. et al. 1966. RNA codons and protein synthesis. IX. Synonym codon recognition by multiple species of valine-, alanine-, and methionine-sRNA. *Proc. Natl. Acad. Sci. U. S. A.* **55:** 912–919.
40. Nirenberg M.W. et al. 1966. The RNA code and protein synthesis. *Cold Spring Harb. Symp. Quant. Biol.* **31:** 11–24.
41. Marshall R.E. et al. 1967. RNA codewords and protein synthesis. XII. Fine structure of RNA codewords recognized by bacterial, amphibian, and mammalian transfer RNA. *Science* **155:** 820–826.
42. Crick F.H.C. 1966. Codon-anticodon pairing: The wobble hypothesis. *J. Mol. Biol.* **19:** 548–555.
43. Holley R.W. et al. 1965. Structure of a ribonucleic acid. *Science* **147:** 1462–1465.
44. Söll D. et al. 1965. Studies on polynucleotides, XLIX. Stimulation of the binding of aminoacyl-sRNA's to ribosomes by ribotrinucleotides and a survey of codon assignments for 20 amino acids. *Proc. Natl. Acad. Sci. U. S. A.* **54:** 1378–1385.
45. Clark B.F.C. and Marcker K.A. 1966. The role of N-formyl-methionyl-sRNA in protein biosynthesis. *J. Mol. Biol.* **17:** 394–406.
46. Brenner S. et al. 1965. Genetic code: The "nonsense" triplets for chain termination and their suppression. *Nature* **206:** 994–998.
47. Weigert M.G. and Garen A. 1965. Base composition of nonsense codons in *E. coli* evidence from amino-acid substitutions at a tryptophan site in alkaline phosphatase. *Nature* **206:** 992–994.
48. Brenner S. et al. 1967. UGA: A third nonsense triplet in the genetic code. *Nature* **213:** 449–450.
49. Nirenberg M. 1969. The genetic code. In *Les Prix W. Nobel en 1968. Nobel Foundation* (ed. P.A. Stockholm et al.), pp. 221–241, Norstedt and Söner.

The History of Deciphering the Genetic Code: Setting the Record Straight

AKIRA KAJI* AND HIDEKO KAJI†

*University of Pennsylvania, Philadelphia, Pennsylvania, USA;
†Thomas Jefferson University, Philadelphia, Pennsylvania, USA

We enjoyed immensely the interesting article by Marshall Nirenberg entitled "Historical review: Deciphering the genetic code—a personal account," which was published in the January 2004 issue of *TiBS*[1]; it brought back memories of the "good old days." We are extremely honored to be credited for one of our first papers as independent scientists. However, we were neither interested nor intending—and, indeed, did not show—that "poly(U) stimulated the binding of radioactive polyphenylalanine-tRNA to ribosomes" as the author stated.

Our paper, entitled "Specific interaction of soluble RNA with polyribonucleic acid induced polysomes,"[2] described mixing an extract containing ribosomes with poly(U) and a mixture of deacylated tRNAs. The reaction mixture was subjected to sucrose-gradient centrifugation, and the sedimentation of the ribosomes was examined. The presence of $tRNA^{phe}$(unesterified tRNA specific for phenylalanine) was estimated by subjecting each fraction to amino-acylation by aminoacyl-tRNA synthetase. We found that $tRNA^{phe}$ sedimented with polysomes complexed with poly(U). This was seen with poly(U), but not with poly(A). Conversely, $tRNA^{lys}$ bound to the complex of poly(A)–ribosomes, but not with poly(U)–ribosomes.

Reprinted from *Trends in Biochemical Sciences* June 2004, 29(6): 293.

We believe that, in this experiment, we provided the first confirmation of the adapter hypothesis proven by Chapeville et al. in a joint effort from the laboratories of Fritz Lipmann and Seymour Benzer.[3] More importantly, as pointed out correctly by Nirenberg, our paper helped to decipher the genetic code by providing a more appropriate method (specific binding of tRNA to programed ribosomes) over and above the earlier approach. Previous investigators examined amino acid incorporation into polypeptides under the direction of heteropolymers (polyribonucleic acids). However, these studies often gave ambiguous results.[4,5] We should add that ribosome-recycling factor (RRF; discovered shortly after the determination of the genetic code[6]) binds to the ribosome and shares the tRNA-binding sites that were identified during 1963–1966.[2,6–9] However, the mode of ribosomal binding of RRF is quite different from that of tRNA [10] despite its near-perfect structural mimic of tRNA.[11]

ACKNOWLEDGEMENTS

We thank F. Frankel (University of Pennsylvania, PA, USA) for his help in preparing this article.

REFERENCES

1. Nirenberg M. 2004. Historical review: Deciphering the genetic code—a personal account. *Trends Biochemical Sci.* **29:** 46–54.
2. Kaji A. and Kaji H. 1963. Specific interaction of soluble RNA with polyribonucleic acid induced polysomes. *Biochem. Biophys. Res. Commun.* **13:** 186–192.
3. Chapeville F. et al. 1962. On the role of soluble ribonucleic acid in coding for amino acids. *Proc. Natl. Acad. Sci. U. S. A.* **48:** 1086–1092.
4. Nirenberg M.W. et al. 1963. Approximation of genetic code via cell-free protein synthesis directed by template RNA. *Fed. Proc.* **22:** 55–61.
5. Wahba A.J. et al. 1962. Synthetic polynucleotides and the amino acid code. VI. *Proc. Natl. Acad. Sci. U. S. A.* **48:** 1683–1686.
6. Hirashima A. and Kaji A. 1970. Factor dependent breakdown of polysomes. *Biochem. Biophys. Res. Commun.* **41:** 877–883.
7. Kaji H. and Kaji A. 1964. Specific binding of sRNA with the template-ribosome complex. *Proc. Natl. Acad. Sci. U. S. A.* **52:** 1541–1547.
8. Kaji H. et al. 1966. Binding of specific sRNA to ribosomes: binding of sRNA to the template 30S subunits complex. *J. Biol. Chem.* **241:** 1251–1256.

9. Suzuka I. et al. 1966. Binding of specific sRNA to 30S ribosomal subunits: effect of 50S ribosomal subunits. *Proc. Natl. Acad. Sci. U. S. A.* **55:** 1483–1490.
10. Lancaster L. et al. 2002. Orientation of ribosome recycling factor in the ribosome from directed hydroxyl radical probing. *Cell* **111:** 129–140.
11. Selmer M. et al. 1999. Crystal structure of *Thermotoga maritima* ribosome recycling factor: a tRNA mimic. *Science* **286:** 2349–2352.

Biochemistry or Molecular Biology? The Discovery of "Soluble RNA"

MAHLON HOAGLAND

Academy Road, Thetford, Vermont, USA

In the late 1940s and early 1950s, biochemists and the vanguard of those who would come to identify themselves as molecular biologists operated in separate spheres. The biochemists, within whose ranks I came of age, were occupied with enzymology, with interconversions in energy metabolism and with biosynthetic mechanism—frequently using animal cell preparations as experimental systems. The molecular biologists-to-be, emerging from physics, genetics and microbiology, favored physicochemical studies of the structure of large molecules and exploring problems of genetic information transmission—using the relatively simple bacteria and viruses as models. In particular, with respect to the early stages of solving the problem of protein synthesis, one could say that the biochemists viewed it as a search for cellular machinery that could assemble a polymer from its constituent units; the success of this process would depend upon isolating and characterizing intermediate stages in a sequential series of chemical reactions. The molecular biologists came to the problem from the theoretical side, attempting to discern the genetic code and the broad principles underlying its translation into protein molecules. We biochemists tended to see the molecular biologists as imperious intruders from an alien culture and sensed that they viewed us as drab, if industrious, blue-collar laborers.

A merging of the interests and talents of the two groups was inevitable, as each came to value the contributions of the other. I think it is fair to say that the dual discovery of amino acyl-tRNA synthetases and of transfer RNA

Reprinted from *Trends in Biochemical Sciences* February 1996, 21(2): 77–80.

in the laboratories of Paul Zamecnik at the Massachusetts General Hospital (MGH) in the mid 1950s was the event that, by linking information theory and the study of machinery, unlocked the door between the two cliques and began the process that eventually made the distinction between them meaningless.

APPROACHING THE PROTEIN SYNTHESIS PROBLEM

I joined Paul Zamecnik's group at Harvard in 1952, in the Department of Medicine at MGH, after finishing medical school, where a two-year bout with tuberculosis had felicitously diverted me from medicine to biochemistry. I spent three postdoctoral years at the hospital in proximity to Paul's group, watching developments in protein synthesis with undisguised fascination, while working on a separate project.

With Paul's blessing and an understanding that I could join his group in two years' time, I spent a year with Kaj Linderstrøm-Lang at the Carlsberg Laboratory in Copenhagen, and another with Fritz Lipmann back at MGH. During the year in Denmark, a long and rewarding friendship with Herman Kalckar began (within a year he moved permanently to the USA). These three scientists had strongly influenced both Paul and me to see the energetics of peptide-bond formation as a problem of coupling with an exergonic energy source. It is hard to believe now that, even into the 1950s, in some biochemical quarters, thinking about the energetics of peptide-bond synthesis sought to implicate an intrinsic process—the reversal of proteolysis. Lipmann and Kalckar had, however, in prescient papers as far back as 1941, speculated that phosphorylated intermediates must be involved in the activation of amino acids for the synthesis of proteins.

Paul and his associates, notably Philip Siekevitz, Bob Loftfield, Betty Keller, John Littlefield and Mary Stephenson had, by 1952, established that ribosomes and a soluble cell fraction, along with ATP and an ATP-regenerating system, were essential for protein synthesis in cell-free rat liver preparations. One floor below, in Lipmann's lab, I suspected that those prepared minds would peek over the shoulders of Paul's group and begin to seek evidence for amino acid activation in those protein-synthesizing fractions. But no. To my growing incredulity, no one in Lipmann's group seemed to be interested in attacking this problem. Instead, their focus was on acetate activation and the mechanism of the biosynthesis of coenzyme A. (Lipmann won the Nobel Prize that year for his discovery of CoA.) I spent the year working on CoA biosynthesis, but was itching to get back with Paul and apply the techniques I was learning from Lipmann to what seemed to me to be the much more exciting problem of the energetics of protein synthesis.

BIOCHEMICAL EVIDENCE FOR AMINO ACID ACTIVATION

In 1953, upon returning to Paul's lab, I began applying the technique of phosphate–ATP exchange (which I learned from Werner Maas, who had been using it in Lipmann's lab to study CoA biosynthesis) to detect evidence of amino acid activation in the group's cell fractions. In the following months, it became apparent that the rat liver soluble fraction vigorously catalyzed an amino-acid-dependent incorporation of ^{32}P-labeled pyrophosphate (^{32}PP) into ATP—strong indication that activation was proceeding by the reversible adenylation of amino acid carboxyl groups, with the release of pyrophosphate. This effect could be obtained only after extensive dialysis of the fraction, as indigenous amino acids in the fraction produced a high background of ATP–PP exchange.

Furthermore, the fraction rapidly degraded PP and so it was necessary to add fluoride to inhibit the pyrophosphatase activity. We noted that, even when the intrinsic pyrophosphatase activity of the system was suppressed, there was no net accumulation of PP in the presence of a full complement of amino acids. This suggested that the amino-acylated intermediate did not dissociate to any extent from the enzyme that catalyzed its formation.

The addition of hydroxylamine (NH_2OH) to the system, as an artificial terminal acceptor of activated carboxyl groups, resulted in a stoichiometric hydrolysis of ATP and an accumulation of amino-acyl hydroxamic acids and pyrophosphate. The stimulation of ^{32}PP–ATP exchange and the accumulation of hydroxamic acids was additive for different amino acids, indicating independent activation sites. Our preliminary attempts at crude fractionation of the proteins present in the soluble fraction afforded suggestive evidence that different amino acids were activated by different proteins.[1,2]

Later, we were able to demonstrate that, in the enzymatic activation of tryptophan, ^{18}O from the carboxyl group of tryptophan ended up in the adenyl portion of ATP.[3] This experiment was done in collaboration with Paul Boyer and with Fritz Lipmann (who, upon learning of our findings, reproduced our results and then partially purified a tryptophan-activating enzyme), further corroborating the postulated formation of an amino-acyl–adenylate intermediate in the reaction.

It happened that I first announced the discovery of amino acid activation at a meeting of molecular biologists in Detroit in 1955. The palpable indifference with which the audience received the news showed how tightly closed the door between the biochemistry and molecular biology compartments was. The focus of the meeting, of course, was on how the polymerizing system *ordered* its units, not on how it *energized* them.

ARE PROTEIN SYNTHESIS AND RNA SYNTHESIS LINKED?

Paul got to wondering if the crude fractions involved in protein synthesis might also be involved in RNA synthesis, i.e. if there might be some coupling between the synthesis of the two polymers. It occurred to him that as the system catalysed the formation of AMP–amino acids, the AMP portion of the energized unit might get incorporated into RNA. If this were the case, the amino acid portion should *not* get attached to RNA. So, he and Mary Stephenson incubated ^{14}C-labeled ATP with ribosomes and soluble fraction, both together and separately, looking for incorporation of label into RNA and doing parallel incubations with ^{14}C-labeled amino acids as controls. To their surprise, ^{14}C-labeled amino acids were rapidly incorporated into the relatively small amount of low molecular weight RNA in the soluble fraction. Incorporation into the more abundant ribosomal RNA was negligible. Also, ^{14}C from ATP was incorporated into the RNA of this low molecular weight, or "soluble RNA" fraction, a finding later shown by Paul and another associate, Lisa Hecht, to be end-adenylation of transfer RNA (tRNA).

At this time (early 1956), Paul was appointed head of the Huntington Laboratories and his new duties, together with clinical responsibilities in the hospital, prompted him to ask me to pursue this lead.

The existence of any kind of chemical interaction between amino acids, on the one hand, and RNA, a molecule thought to have a template role, on the other—together with the fact that the reaction was occurring in the same fraction that activated amino acids—was, of course, enormously intriguing to us. The small quantity of RNA residing in the soluble fraction—some 10% of total cellular RNA—had been noted before by us and others, but no one had attached much significance to it. It was viewed as "junk"—fragments of the larger ribosomal RNA produced during the vigorous procedures used to break open cells. We soon found that the labeling of soluble RNA using ^{14}C-labeled amino acids was quite rapid—complete in two minutes at 37°C—and was specifically dependent on ATP.

Yeast RNA and rat liver ribosomal RNA were not similarly labeled when added to the reaction, and the labeling was completely inhibited by ribonuclease. The RNA–amino acid product was non-dialysable, acid stable and alkali labile, and it yielded an amino-acyl hydroxamic acid when incubated with hydroxylamine—presumptive evidence that the indigenous activating enzymes in the fraction were catalyzing the amino–acylation of RNA via an amino acyl–adenylate intermediate.

THE KEY EXPERIMENT: "SOLUBLE" RNA BEHAVES LIKE AN INTERMEDIATE

Then came the most exciting and memorable experiment of my career. I prelabeled the indigenous RNA in the soluble fraction with ^{14}C-amino acids and ATP. Then, I used our lab's procedure of re-precipitating the protein and RNA of the fraction twice at pH5 to remove free amino acids and ATP, redissolved the crude material at pH7.4 and incubated it with ribosomes. Parallel incubations were done with and without GTP. (This nucleotide was found to be essential for protein synthesis, in addition to ATP, by Betty Keller and Paul about a year previously.) To quote from an earlier account,[4] "It was night by the time the samples were dried, stacked, and ready to move automatically under the counter tube. I still can clearly see the dark windows of the lab, smell the organic solvents, hear the buzzing of a defective fluorescent lamp in the next room. In front of me were the transfixing flashing lights on the Geiger counter as the samples began to be counted. First, the unincubated complete system: 489 counts of labeled amino acid in RNA, 30 counts in protein. Good—the few counts in protein are expected background in such a relatively crude system. Next, the incubated system, GTP omitted: 111 counts in RNA, 40 counts in protein. Okay—RNA loses amino acid, but it is not getting into protein. Next, the complete incubated system: 180 counts in RNA, 374 counts in protein. Those little numbers caused a shiver to go down my spine: amino acids had left the RNA and entered protein! From that night on, I had little doubt that this small soluble RNA was the physical link between activated amino acids and their ordered arrangement in protein molecules."

At this time, two years before the discovery of mRNA, it was generally assumed by workers in the field that ribosomes contained DNA-derived RNA, which served as a template for ordering amino acids. We interpreted our findings to mean that "soluble" RNA was receiving activated amino acids from cognate activating enzymes and carrying them to a template on ribosomes.

ACCUMULATING EVIDENCE ATTRACTS MOLECULAR BIOLOGISTS!

Paul, Mary and I spent the remaining months of 1956 refining the transfer reaction: we showed that purified amino-acylated soluble RNAs, extracted from the soluble fraction by the phenol method, rapidly and quantitatively transferred amino acids to protein in the presence of GTP and that different amino acids became linked to the RNA additively—indicating separate binding sites for each amino acid. Mary also demonstrated that the same events occurred in comparable fractions derived from another animal system—mouse ascites tumor cells.[5,6]

Not unexpectedly, others were hot on the trail. Evidence for the existence of an intermediate between activation and polymerization was emerging at the time from three other laboratories. Hultin and Beskow[7] were finding that the incorporation of labeled amino acid into protein in a cell-free system was not affected, i.e. not diluted by an excess of free amino acid. Holley[8] reported that the activation of alanine (but, oddly, not of other amino acids) was inhibited by ribonuclease, and in Japan, Ogata and Nohara[9] were discovering evidence for an RNA intermediate.

Now we had attracted the attention of the molecular biologists! The first encounter took the shape of a visit to my lab in late 1956 by Jim Watson, who had just taken an academic appointment at Harvard. As I excitedly related our story, I nervously sensed that Jim shared my enthusiasm, but was impatiently waiting for me to finish—did he know something I didn't? Finally, had I not heard of Francis Crick's adaptor hypothesis? No—apprehensively—I had not. He went on: a year previously, Francis had written a private communication to his fellow molecular biologists "On Degenerative Templates and the Adaptor Hypothesis: A Note to the RNA Tie Club." In it, after a pre-amble quoting a little-known Persian writer, "Is there anyone so utterly lost as he that seeks a way where there is no way?," he sets forth the idea that amino acids could only be recognized by an ordering template if they were first linked to small "adaptor" molecules that had a complementary chemical relationship to the template. "Each amino acid would combine chemically, at a special enzyme, with a small molecule which, having a specific hydrogen-bonding surface, would combine specifically with the (RNA) template. . . . In its simplest form there would be 20 different kinds of adaptor molecules, one for each amino acid, and 20 different enzymes to join the amino acid(s) to their adaptors."

I was intrigued by the prescience of Francis' idea, but also a bit deflated and miffed at having the theoretical framework for our discovery foisted on us by an outsider—indeed, by a molecular biologist—after we had revealed transfer RNA and correctly interpreted its significance! A vision rose before me: we explorers sweating and slashing our way through a dense jungle, finally rewarded by the discovery of a beautiful long-lost temple—and looking up to see Francis, circling above our goal on gossamer wings of theory, gleefully pointing it out to us.

Oddly, Francis never published the adaptor paper. When he learned of our results he was initially skeptical that "our" soluble RNAs could be "his" adaptors—mainly because of their large size (some 60 to 70 nucleotides). After all, he had already suggested that a triplet of nucleotides was sufficient to code for an amino acid. However, we were aware that the RNA had to interact with a cognate enzyme to become amino-acylated, and lacking involvement with the coding problem, were not as uncomfortable with its size (for discussion on the adaptor hypothesis in relation to our discoveries, see Ref. 11).

So, while we could not help but admire the power of Francis' imagination, we could take satisfaction in the knowledge that, in this instance, a fine theory neither substituted for, nor guided the successful analytical dissection of the machinery of protein synthesis (see also the perspectives of Paul Zamecnik[10]).

Well, the door had opened! Francis visited in the spring of 1957 and invited me to spend the next academic year at the Medical Research Council Unit for the Study of Molecular Structure of Biological Systems in Cambridge. Indicative of the sea-change, he wanted to try his hand at isolating and experimenting with transfer RNA. We entertained the hope—in retrospect, a forlorn one—that the association of each amino acid with a unique RNA might be an approach to cracking the genetic code. I, of course, was happy to represent biochemistry at the epicenter of molecular biology, whose members at the time were occupied with protein structure and the co-linearity of gene and protein. Francis got more than he had bargained for. One day, I entered the lab we had set up together and found him under the bench trying to capture a rat, who was eluding his efforts to use its liver as a source of

FIGURE 1. A laboratory scene from 1957. Mahlon Hoagland enters the lab to find Francis Crick attempting to catch a rat whose liver was destined to be the source of transfer RNA. Artwork by Bert Dodson.

transfer RNA—a scene that aptly portrayed the demise of the distinction between biochemistry and molecular biology (see Fig. 1).

REFERENCES

1. Hoagland M.B. 1955 *Biochim. Biophys. Acta* **16:** 288–289.
2. Hoagland M.B., Keller E.B., and Zamecnik P.C. 1956. *J. Biol. Chem.* **218:** 345–358.
3. Hoagland M.B. et al. 1957. *Biochim. Biophys. Acta* **26:** 215–217.
4. Hoagland M.B. 1990. In *Toward the Habit of Truth: a Life in Science,* W.W. Norton & Co.
5. Hoagland M.B., Zamecnik P.C., and Stephenson M.L. 1957. *Biochim. Biophys. Acta* **24:** 215–216.
6. Hoagland M.B. et al. 1958. *J. Biol. Chem.* **231:** 241–256.
7. Hultin T. and Beskow G. 1956. *Exp. Cell Res.* **11:** 664.
8. Holley R.W. 1957. *J. Am. Chem. Soc.* **79:** 658.
9. Ogata K. and Nohara H. 1957. *Biochim. Biophys. Acta* **25:** 659.
10. Zamecnik P.C. 1969. *Cold Spring Harbor Symp. Quant. Biol.* **34:** 6.
11. Judson H. 1979. *The Eighth Day of Creation,* Simon & Schuster.

Back to Camelot: Defining the Specific Role of tRNA in Protein Synthesis

BERNARD WEISBLUM

University of Wisconsin Medical School, Madison, Wisconsin, USA

After graduating from Medical School in 1961, I went to work in Seymour Benzer's laboratory at Purdue University, where I was privileged to participate in a series of exciting experiments on the then emergent genetic code. One study that received some notoriety was a critical test of the "adaptor hypothesis" proposed by Francis Crick in 1958. Crick had postulated that a small oligonucleotide, possibly soluble RNA (sRNA, as it was then known; tRNA as it is known today), functions as an adaptor for the incorporation of amino acids into protein.[1] Thus, it followed that once an amino acid is attached to sRNA, the specificity with which it is incorporated into protein resides solely in the sRNA adaptor to which it is attached.

The Raney-nickel experiment[2] (Fig. 1), as it came to be called, is often cited as the critical experiment that proved the adaptor hypothesis. It allowed us to demonstrate that, in an *in vitro* protein-synthesis system, alanine from alanyl-$tRNA^{Cys}$ is incorporated into protein at positions normally occupied by cysteine rather than at those occupied by alanine. The Raney-nickel experiment, however, was only one of a series of experiments that confirmed the adaptor hypothesis. As interesting as the results of the experiments themselves was the way in which these experiments came to be done and what followed.

Reprinted from *Trends in Biochemical Sciences* June 1999, 24(6): 247–250.

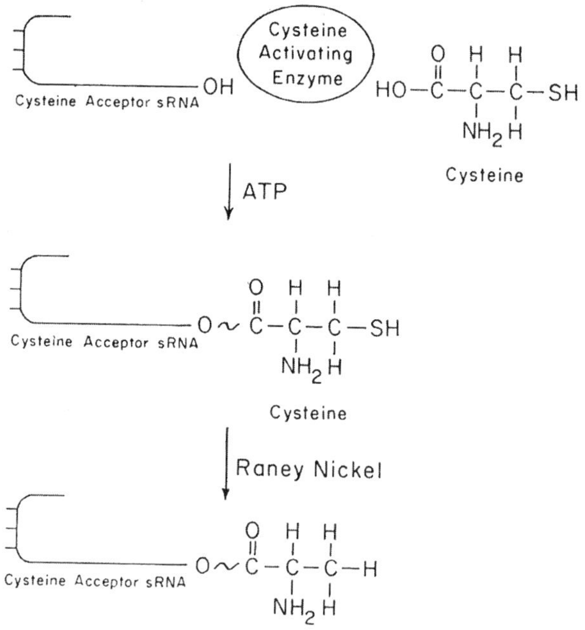

FIGURE 1. "Plan of experiment. Cysteine is attached to its normal acceptor sRNA through the mediation of the cysteine activating enzyme. By the action of Raney Nickel, the cysteine, while still attached, is converted to alanine. The coding properties of the hybrid molecules are then investigated."[2] Figure reproduced, with permission, from Ref. 2. sRNA, soluble RNA (tRNA).

GENETIC STUDIES OF ALLELE-SPECIFIC SUPPRESSION THAT LED TO THE RANEY-NICKEL EXPERIMENT

This trail of research did not start out as an attempt to test the adaptor hypothesis, but developed from allele-specific (i.e. mutant-specific) genetic-suppressor studies of the phage-T4 *rII* system by Benzer and Champe, which began around 1959.[3] While these studies were in progress, a series of papers from Francois Gros' laboratory[4–6] revealed that *Escherichia coli* grown in the presence of 5-fluorouracil (5-FU) made abnormal proteins. For example, alkaline phosphatase and β-galactosidase were shown to have altered amino acid compositions and altered thermostability, but conserved antigenicity. We now believe that 5-FU exerts its suppressor activity because, although it is incorporated into mRNA as uracil, it base pairs with guanine in amino-acyl-tRNA anticodons (i.e. it exhibits the incorporation specificity of cytosine). The observation that the amounts of proline and tyrosine incorporated into total protein, as well as into tRNA-containing fractions, were markedly increased, suggested that the effect was informationally specific. In parallel with the 5-FU studies, yet a third system, *E. coli* tryptophan synthetase, provided insight into allele-specific suppression. Yanofsky and St Lawrence, in a review entitled "Gene Action,"[7] suggested that some forms of allele-specific suppression that they had seen in their studies might be caused by alterations in the specificity with which amino acids are incorporated into protein.

Members of the Benzer lab decided to attempt to explain allele-specific suppression of *rII* mutants of phage T4. The fluorouracil effect suggested a biochemical variable that they could include in their studies. The most striking aspect of the fluorouracil effect was its high degree of specificity: it restored enzyme activity very effectively for some *rII* mutants but not at all for others. This suggested that there was a relationship between the fluorouracil effect and the apparently altered specificity of amino acid incorporation reported by the Gros and Yanofsky labs.[4–7]

By 1960, five years of intensive genetic mapping by the Benzer lab had saturated the *rII* region with mutations to a degree unprecedented in any other genetic system. We therefore hoped that the patterns of suppression by 5-FU at specific sites might be correlated with the wealth of detailed information about those sites. The problem of allele-specific suppression became even more interesting when it was noted that certain strains of *E. coli* K carry genetic suppressors (which eventually turned out to be mutant tRNA genes, as predicted) whose action mimics the phenotypic-suppressor activity of 5-FU.

An indicator strain is one on which phage can grow and multiply. By measuring the extent to which an indicator is consumed after infection with a phage-containing sample, one can quantify the number of phage present. Some *rII* mutants that, as expected, did not grow in *E. coli* KB (the indicator K strain from the Benzer lab) were, however, able to grow in *E. coli* KT (the indicator K strain from Irwin Tessman's lab) and vice versa. This observation was unexpected because all *E. coli* K strains originated as *E. coli* K-12 from Edward Tatum's strain collection at Yale. We now know that *E. coli* KB and KT carried different sets of allele-specific suppressors of *rII* mutations, whose actions were mimicked by inclusion of 5-FU in the plating medium. Thus, 5-FU provided a chemical dimension in which to study allele-specific suppression.

BIOCHEMICAL STUDIES OF ALLELE-SPECIFIC SUPPRESSION

There appeared to be two roads to altering the specificity of incorporation of amino acids into protein. The first road, based on 5-FU experiments, was phenotypic and led to the notion that 5-FU suppresses debilitating mutations through its uracil-specific incorporation into mRNA. The presence of 5-FU induces low-level miscoding at the mutated site by mimicking cytosine in base pairing with aminoacyl-tRNA. The second road, based on *E. coli* strain differences, was genotypic and led to the notion that inherited alterations in tRNA (or of amino-acid-activating enzymes) also induce miscoding and therefore lead to a partial restoration of wild-type activity. With these ideas "in the air," we reached for a direct biochemical handle on the problem by studying the enzymatic charging of tRNA with amino acids.

Because of my medical training (and therefore lack of squeamishness about doing the type of biochemistry that required grinding up of biological specimens), I became the designated biochemist. For starters, I prepared the tRNA (first from yeast, and later from rat liver and *E. coli*), whereas Seymour prepared the activating-enzyme extracts and did the charging experiments. In a three-way comparison that involved the charging of tRNA from *E. coli*, yeast and rabbit liver, we found that certain activation enzymes were active only with homologous tRNA but that others were active both in homologous and in heterologous combinations.[8] Given that the charging reaction involved a key form of informational specificity in protein synthesis, we asked whether nonrecognition seen in certain heterologous charging reactions might indicate non-universality of the genetic code. (The structural bases for these observations are beginning to be elucidated a third of a century later.[9,10]) Similarly, the nonsense suppression of phage-T4 *rII* mutations[11,12] suggested that given that the fidelity of reading of the genetic code can be modified by 5-FU, we should expect that mechanisms that suppress nonsense mutations also evolved.

SYNTHESIS OF RABBIT HEMOGLOBIN WITH AMINO-ACYL-tRNA FROM *ESCHERICHIA COLI*

Günther von Ehrenstein, a post-doc in Fritz Lipmann's lab at the Rockefeller Institute,[13] had synthesized rabbit hemoglobin *in vitro* by charging *E. coli* tRNA with a mixture of amino acids containing [^{14}C]leucine and synthesizing rabbit hemoglobin with rabbit reticulocyte ribosomes. The success of the experiment posed a dilemma: is the genetic code universal? On the one hand, the genetic code could apparently be altered phenotypically (by 5-FU, or genetically, as in the case of the "natural" variation between *E. coli* strains KB and KT). On the other hand, charged *E. coli* tRNA could be used to synthesize a protein from rabbit hemoglobin mRNA that was indistinguishable in two-dimensional fingerprints of tryptic digests from hemoglobin synthesized in rabbit cells.

Fortunately, Günther was about to leave the Lipmann lab and was preparing to move on to his new position as Assistant Professor in the Biophysics Department at the Johns Hopkins University School of Medicine. Seymour invited him to spend a part of his transition time during 1962 in our lab at Purdue. This provided me with a unique opportunity to become Günther's apprentice and to learn how to perform his sorcery. It was a thrilling experience.

The Benzer lab lunched together daily. At lunch, which was a 1 1/2-hour period that began at noon, we got together to talk about anything. The focus was mostly science, but not always. Anything went, as long as it was interesting, and especially if it involved exotic foods. During these lunch periods, under Seymour's prodding, we discussed whether we could perform a critical test of the adaptor hypothesis by charging a tRNA with the wrong amino acid. If so, we could then try to do Gunther's experiment again and ask whether the amino acid retained its specificity of incorporation into protein or whether the amino acid was incorporated into protein with the specificity of the tRNA that carried it. A formal possibility was that incorrectly charged tRNAs were "dead" for purposes of protein synthesis.

We decided to attach an amino acid to its cognate tRNA and to modify the amino acid side chain by some chemical or enzymatic trick. None of us possessed the expertise necessary to do such creative organic chemistry. Luckily, Bill Ray had recently joined the Department of Biological Sciences at Purdue; previously, he had worked with Dan Koshland at the Brookhaven National Laboratory. Bill's areas of expertise included organic chemistry, protein chemistry and enzymology. He suggested that the reactivity of the sulfhydryl group in cysteine was unique, and that Raney nickel could probably remove it under conditions that did not appreciably modify either the tRNA adaptor to which the cysteine was attached or the ester bond that linked the two species.

CYSTEINE INTO CYSTEIC ACID, CYSTEINE INTO ALANINE

My task was to perform the magic transformation that Bill suggested. I was able to show that, indeed, free cysteine in aqueous solution was easily converted to alanine; however, performing this magic on cysteine, while the amino acid was bound to tRNA, proved to be much more difficult. Francois Chapeville, then a post-doc in the Lipmann lab, became impressed by the possibilities of cysteine chemistry and volunteered his services. He attached [^{35}S]cysteine to cysteine's cognate tRNA and converted the resultant cysteinyl-tRNA^{Cys} to cysteic-acid-tRNA^{Cys} by treatment with periodic acid. He then showed that, *in vitro*, the cysteic acid was incorporated into protein by ribosomes with the same specificity as cysteine. In our lunchtime deliberations, Bill had considered various chemical modifications of amino acid side chains that would yield unnatural amino acids, but Seymour did not consider any of these suggested modifications, or the conversion of cysteine into cysteic acid, to be a proof of the adaptor hypothesis. He insisted that we should convert one natural L-amino acid into another. Nevertheless, Chapeville had obtained an interesting result and presented his findings at the annual meeting of the Federation of American Societies of Biology—the occasion for presenting "blockbuster" findings—in the spring of 1962.[14]

A factor that made the cysteic acid experiment feasible was that [^{35}S]cysteine that had maximum possible specific activity could be obtained easily. The cysteine–alanine interconversion, however, would remove the ^{35}S, which left us at the mercy of the availability of ^{14}C-labelled cysteine. Although many ^{14}C-labelled amino acids were available at specific activities approaching the theoretical maximum, uniformly labelled [^{14}C]cysteine was not one of them (because of the low yields with which cysteine could be obtained from [^{14}C]-uniformly labelled *Chlorella* protein hydrolyzate).

By this time, I had shown that Raney nickel would convert free or tRNA-charged cysteine to alanine; however, the only [^{14}C]cysteine available commercially was specifically labelled at the carboxyl group and had a specific activity of only 10 Ci mol^{-1}. Owing to the low specific activity, the counts from the [^{14}C]alanine produced were too low for us to perform an experiment that would give a measurable amount of incorporation into protein. Chapeville made high-specific-activity [^{14}C]cysteine by treating uniformly labelled [^{14}C]serine with serine sulfhydrase, which he had purified from yeast. The resultant uniformly labelled [^{14}C]cysteine had the necessary specific activity, and the Cys–Ala conversion gave enough radioactive label to make the experiment possible.

The experiment proved the adaptor hypothesis. Translating a synthetic messenger (random polyribouridylateguanylate, which contained many cysteine codons but no alanine codons), the *E. coli* protein-synthesis machinery incorporated alanine into protein but with the specificity of cysteine.[2] Given that we had come this far, we asked whether synthesis of rabbit globin in which alanine substituted for cysteine also was feasible: it was, although we had pushed the incorporation to the point at which only a few precious alanine counts were incorporated.[15] The presence of thiolated bases (e.g. 4-thiouridine and 2-methylthio-isopentenyladenosine) in tRNA came to light only after the Raney-nickel experiments were completed. The assumption that tRNA would remain structurally unaffected by Raney-nickel treatment is, therefore, not generally true, and we were fortunate that $tRNA^{Cys}$ was not inactivated by a treatment that resulted in the desulfuration of these two sulfur-containing bases.

BOB HOLLEY VISITS PURDUE

Bob Holley adapted the countercurrent-distribution method to the fractionation of tRNA and, using homogeneous tRNAs purified by the method, he determined the first tRNA sequences.[16] Countercurrent distribution is a fractionation technique that is based on the ability of individual components present in a complex mixture of compounds to dissolve at different relative ratios in two immiscible phases (which are mechanically separable). For each compound, the ratio with which it separates into the two phases is called its partition coefficient. Partition and separation of the phases is a technique that is taught in elementary organic chemistry. The power of countercurrent distribution derives from an ingenious automated device, invented by Lyman Craig. The device performs thousands of sequential partition steps that compound small differences in partition coefficients; this allows complete separation of individual components present in a complex mixture. Holley noted that some acceptor activities could be separated into multiple peaks that varied in number and degree of resolution. This heterogeneity made such tRNA less attractive as a candidate for sequencing; however, tRNA heterogeneity was exactly what we needed for the next experiment.

At Purdue, Holley presented a seminar on his tRNA studies, and he agreed to fractionate about 200 mg (most of the batch) of one of my purest tRNA preparations. When the fractionated tRNA came back, we had a division of labor: Seymour performed the analytical studies and tested fractions for acceptor activity, searching in particular for optimally separated multiple peaks for a given amino acid; I charged multiple tRNA peaks with amino acids and determined the specificities with which they were incorporated into protein. Given that only homopolymer RNA (namely, poly rU, poly rA, poly rC and poly rI) was available commercially, I had to make all 11 possible ribo-copolymer combinations, using polynucleotide phosphorylase.

With hindsight, the choice of the multiple leucine-accepting fractions of *E. coli* tRNA as the focus of our studies was a fortunate one. Leucine (along with arginine and serine) has six codons, which maximizes the number of cognate tRNAs. Indeed, in the countercurrent fractionation, it showed two well-resolved peaks. These two peaks were later separated into at least five distinct peaks of leucine-acceptor activity. The resolution of the two original leucine-acceptor peaks that we reported[17] sufficed to show that one incorporated leucine into protein when poly rUG was the messenger, whereas the other incorporated leucine when poly rUC was the messenger.[17]

The leucine-degeneracy experiment complemented the Raney-nickel experiment. In the latter, two different amino acids were attached to the same tRNA adaptor. By contrast, in the leucine-degeneracy experiment, the same amino acid was attached to two different tRNA adaptors and had two different specificities. The only intervention necessary for separation of the multiple acceptors was countercurrent distribution, which used gentle reagents [ethylene glycol monomethylether ("methyl cellosolve") and phosphate buffer] that were far less harsh than Raney nickel.

In follow-up amino-acid-incorporation studies, the results obtained for random polyribonucleotide messengers poly rUC and poly rUG were confirmed: two countercurrent-fractionated leucine-specific *E. coli* tRNAs transferred their respective leucines *in vitro* into mutually exclusive positions in the α chain of rabbit hemoglobin synthesized by reticulocyte ribosomes.[18]

A CREATIVE IDEA CONTINUES TO LIVE

Lyman Craig invented countercurrent distribution to fractionate antibiotic peptides and other small molecules; Bob Holley modified the method to fractionate macromolecular tRNA. This line of development was further extended when Seymour, imaginatively, fractionated even larger macromolecules: fruit flies—"as if they were molecules of behavior."[19] This new approach, which was designed to enrich a population of mutagenized fruit flies for visually impaired mutants, signaled Seymour's ever-creative approach to research and, in this instance, his move into neurobiology. Seymour Benzer's contributions to neurobiology from the 1960s to date were discussed in a popular article, "The Lord of the Flies" by Jonathan Weiner, that recently appeared in *The New Yorker*,[20] and in a book by the same author entitled *Time, Love, Memory*.[21]

The "cysteine–alanine" experiment is alive and well; in its new incarnation, however, it is known as stop-codon-suppressor methodology. Sidney Hecht's lab and Peter Schultz's lab developed the approach[22,23] to allow mischarging of any tRNA with almost any amino acid—including designer tRNAs and unnatural amino acids. This technology has been used for synthesis of a wide range of proteins that carry substituted amino acids or that can test models of enzyme action.[24]

The question of how the structure of tRNA determines the specificity with which it is charged has been pursued productively in the labs of Bill McClain, now at the University of Wisconsin-Madison, and Paul Schimmel, now at the Scripps Institute. Bill was a graduate student in Sewell Champe's lab at Purdue during the mid-1960s and became infected with the tRNA-recognition problem at that time. Recent publications from these two laboratories[9,10] have begun to unravel the structural basis for the recognition of a tRNA by its cognate activating enzyme. And the work goes on. We can look back with satisfaction on the simple experiments that we did, the questions that they provoked and the knowledge that we have acquired.

ACKNOWLEDGEMENTS

I thank Seymour Benzer for accepting me as a post-doctoral research fellow in his laboratory and for the inspiring mentorship that he provided. I also thank my Purdue colleagues (Sewell Champe, Bill Ray and Irwin Tessman) and my Wisconsin colleagues (Brian Kay, Bill McClain and Millard Susman) for their friendship and for many helpful comments.

REFERENCES

1. Crick F.H.C. 1958. *Symp. Soc. Exp. Biol.* **12:** 138–163.
2. Chapeville F. 1962. *Proc. Natl. Acad. Sci. U.S. A.* **48:** 1086–1092.
3. Benzer S. and Champe S.P. 1961. *Proc. Natl. Acad. Sci. U. S. A.* **47:** 1025–1038.
4. Naono S. and Gros F. 1960. *C.R. Acad. Sci. (Paris)* **250:** 3527–3529.
5. Naono S. and Gros F. 1960. *C.R. Acad. Sci. (Paris)* **250:** 3889–3891.

6. Bussard A., Naono S., Gros F., and Monod J. 1960. *C.R. Acad. Sci. (Paris)* **250:** 4049–4051.
7. Yanofsky C. and St Lawrence P. 1960. *Annu. Rev. Microbiol.* **14:** 311–340.
8. Benzer S. and Weisblum B. 1961. *Proc. Natl. Acad. Sci. U. S. A.* **47:** 1149–1154.
9. Gabriel K., Schneider J., and McClain W.H. 1996. *Science* **271:** 195–197.
10. Beuning P.J., Yang F., Schimmel P., and Musier-Forsyth K. 1997. *Proc. Natl. Acad. Sci. U.S. A.* **94:** 10150–10154.
11. Champe S.P. and Benzer S. 1962. *Proc. Natl. Acad. Sci. U. S. A.* **48:** 532–546.
12. Benzer S. and Champe S.P. 1962. *Proc. Natl. Acad. Sci. U. S. A.* **48:** 1114–1121.
13. von Ehrenstein G. and Lipmann F. 1961. *Proc. Natl. Acad. Sci. U. S. A.* **47:** 941–950.
14. Chapeville F. 1962. *Fed. Proc.* **21:** 414.
15. von Ehrenstein G., Weisblum B., and Benzer S. 1963. *Proc. Natl. Acad. Sci. U. S. A.* **49:** 669–675.
16. Holley R.W. et al. 1963. *Cold Spring Harbor Symp. Quant. Biol.* **28:** 117–121.
17. Weisblum B., Benzer S., and Holley R.W. 1962. *Proc. Natl. Acad. Sci. U. S. A.* **14:** 1449–1454.
18. Weisblum B., Gonano F., von Ehrenstein G., and Benzer S. 1965. *Proc. Natl. Acad. Sci. U. S. A.* **53:** 328–334.
19. Benzer S. 1967. *Proc. Natl. Acad. Sci. U. S. A.* **58:** 1112–1119.
20. Weiner J. 1999. *The New Yorker* **75:** 45–51.
21. Weiner J. 1999. *Time, Love, Memory,* Alfred A. Knopf, New York.
22. Heckler T.G. et al. 1984. *Biochemistry* **23:** 1468–1473.
23. Robertson S.A. et al. 1989. *Nucleic Acids Res.* **17:** 9649–9660.
24. Mendel D., Cornish V.W., and Schultz P.G. 1995. *Annu. Rev. Biophys. Biomol. Struct.* **24:** 435–462.

The Crystallization and Structural Determination of tRNA

BRIAN F.C. CLARK

Aarhus University, Aarhus, Denmark

Transfer RNA (tRNA) is a central class of macromolecules in molecular biology. It translates the genetic message imprinted in DNA into protein by decoding the mRNA, and thus ensures accuracy in the interpretation of genetic information. The mechanism by which this is performed involves correctly charging a tRNA species with a specific amino acid, a reaction catalysed by the cognate aminoacyl-tRNA synthetase, and correctly decoding the genetic message into mRNA, the responsibility of the charged tRNA. Each charged tRNA is carried to the aminoacyl-tRNA binding site (A-site) of the ribosome where the mRNA is decoded. (Normally, each tRNA can be charged with only one of the usual 20 amino acids.)

Because of its importance in decoding, its small size (~25 000 Da or 73–93 nucleotides) and its availability, tRNA became a favoured example of a nucleic acid molecule in the 1960s for both sequence determination and 3D-structural elucidation. The first primary structural determination of a tRNA, that of yeast alanine tRNA by Robert Holley's group at Cornell University (Ithaca, NY, USA) in 1965, earned Holley a share in the 1968 Nobel Prize for Physiology or Medicine. By contrast, despite its great importance for establishing the foundation of structural nucleic acid chemistry at atomic resolution and for giving insight into the mechanism of protein synthesis and transfer of genetic information, the determination of the 3D structure of the tRNA (in 1974) has not been recognized with such distinction. It was probably overshadowed by the intense competition among many research groups

Reprinted from *Trends in Biochemical Sciences* August 2001, 26(8): 511–514.

that ended in strong rivalry between the two groups that first solved successfully the crystal structure: these being led by Aaron Klug at the MRC Laboratory of Molecular Biology (LMB), Cambridge, UK, and Alex Rich at MIT, Dept of Biology, Cambridge, MA, USA. Indeed, in December 1968 the *New Scientist* wrote an article called "The race for transfer RNA" reporting that any one of several research groups could bring off one of the biggest coups in "classical" molecular biology by obtaining crystals of tRNA large enough and ordered enough to allow a structural determination by X-ray crystallography.[1] How did I become involved?

DECODING AND SEQUENCING

After a PhD working on the chemistry of phosphinositides with Dan Brown in the Dept of Organic Chemistry at Cambridge University, UK (1961), I learned biochemistry during two postdoctoral periods: one with Jack Buchanan at MIT's Division of Biochemistry (1961–1962) and the other with Marshall Nirenberg at the NIH's National Heart Institute, Bethesda, MD, USA (1962–1964). It was in the laboratory of Nirenberg that he and Heinrich Matthaei synthesized polyPhe, directed by poly U, using a cell-free system, thus deciphering the first codon for an amino acid and starting the race for the elucidation of the genetic code. The experience I gained in Nirenberg's laboratory led, in 1964, to a staff position at the MRC LMB's Division of Molecular Genetics, then co-headed by Francis Crick and Sydney Brenner. I returned to Cambridge with alacrity as the Americans thought that the coding problem was essentially solved. However, researchers in Cambridge realized that there should be signals for both initiation and termination of protein synthesis. Therefore, I started working on nonsense codon (a codon specifying chain termination) elucidation and separation of suppressor tRNAs (tRNAs which could decode nonsense codons) with John Smith and Brenner. However, my attention was soon captured by the "strange" tRNA, shown to be formylmethionyl-tRNA (fMet-tRNA), recently discovered by Kjeld Marcker and Fred Sanger. I was able to put my experience of decoding and cell-free protein synthesis to good use in a close collaboration with Marcker over the next six years.

My task was to purify triplet oligonucleotides and nonradioactive initiator tRNA with the aim of elucidating the codons for initiation. Our work showed that fMet-tRNA was a prokaryotic initiator because, on analysis, fMet was found at the N-terminal end of polypeptides synthesized under the direction of synthetic and natural mRNAs.

Because Marcker was in Sanger's division at the LMB, he was conversant with the new rapid tRNA sequencing methods developed by Sanger, George Brownlee and Bart Barrell. A severe sequencing problem with tRNAs was the high occurrence of modified bases that could only sometimes be detected by the radioactive method. Thus Marcker's group and my group emphasized different aspects of tRNA biochemistry. His group was concerned with the rapid sequencing of methionine tRNAs whereas my group was developing large-scale purification methods to allow identification of any modified bases in specific tRNAs, which required milligrams of purified tRNA species.

Fortunately, the sequencing of the $tRNA_f^{Met}$ species proceeded quickly as this initiator tRNA did not contain many modified bases. After a few months, the primary structure was determined. However, we were still in the process of developing large-scale purification methods for tRNAs, especially for the initiator tRNA. We were becoming less interested in identifying new modified bases and more interested in understanding the molecular mechanism of protein biosynthesis. With the splendid structurally inclined atmosphere of the LMB in the mid-1960s we knew that it would be necessary to determine 3D structures of protein synthesis components. This area has spurred me on for the past 30 years.

PURIFICATION AND CRYSTALLIZATION

Actually, it was in the Autumn of 1966 that we began to work on the project of separating different tRNA species on a large enough scale to obtain a single species of methionine tRNA with the aim of crystallizing this tRNA as an intact molecule, or even as a fragment, for crystallographic analysis and tertiary structure determination. The crystallization project had, in particular, the enthusiastic support of Crick.

The large-scale purification of pure species of tRNA was no mean feat in the mid-1960s, particularly considering that >50 tRNA species with similar properties needed to be separated. We noted that the method of countercurrent distribution (CCD) used by Holley's group in the purification of $tRNA^{Ala}$ separated two methionine tRNAs. With my chemist's background I soon set up this method in the laboratory. Marcker and I then showed that the two types of $tRNA^{Met}$ were the initiator $tRNA_f^{Met}$ and the elongator

$tRNA_m^{Met}$.[2] The large-scale purification of the methionine tRNAs was published in 1969 in collaboration with Bhupendra P. (Doc) Doctor, Brian Wayman, Suzanne Cory and Philip Rudland.[3] We produced >100 mg of pure initiator tRNA from 10 g of unfractionated tRNA.

My initial crystallization attempts did not produce identifiable crystals of tRNA but rather crystals of various salts. My simple techniques involved the precipitation of tRNA with standard oligonucleotide- or nucleotide-precipitating agents such as ethanol, acetone or dioxan. My interest was faltering when, in the early Spring of 1968, a procedure by a German postdoctoral worker, Hasko Paradies, purporting to have crystallized yeast $tRNA^{Ser}$, was given to me by Ken Holmes at LMB. These claims were never verified by X-ray data, and neither ourselves nor Paradies successfully used this method to crystallize the initiator tRNA. It was discovered later that his published tRNA crystals were, in fact, protein.[4]

Nevertheless, his crystallization claim stimulated my group to put more systematic efforts into crystallization of the initiator tRNA. At the same time, our group and potential collaborators got together to discuss more systematic approaches. Indeed, I have notes from these days that our discussions with crystallographers such as John Finch, Tom Steitz and David Blow in the laboratory were continued with Holmes, his colleague Shirley Morris and my PhD student Philip Rudland, in the Prince Regent pub.

Meanwhile, my systematic crystallization attempts using specific salt forms of tRNA succeeded. At first, I obtained small spherical crystal aggregates called spherulites using initiator tRNA. Morris also obtained spherulites using a sample of impure $tRNA^{Lys}$. A solution of the purified uncharged $tRNA_f^{Met}$ as a mixed magnesium potassium salt was dialysed against water to remove the excess of inorganic cations. This was then lyophilized and redissolved in water to make a 2–5% solution that was then equilibrated with an atmosphere of dioxan (35% v/v with water) in a desiccator. This procedure produced spherulites in ~16 hours, and these spherulites were indicative of disordered crystalline forms, thereby strongly suggesting that single crystals were possible.

A microscopic study of the spherulites revealed that they became more ordered in time, although the overall shape of the spherulite was maintained. Microcrystals could also be directly produced if the tRNA solution was equilibrated for several days with a lower concentration of dioxan. The microcrystals disappeared in 30 minutes when a trace of ribonuclease A was added to the mother liquor, indicating the ribonucleic acid nature of the microcrystals. These microcrystals, although far too small for single-crystal analysis, did allow powder X-ray photographs to be taken and analysed by Holmes and Klug. The analysis yielded probable unit cell dimensions and space group identifying crystal symmetry characteristics, thus giving evidence that the microcrystals were true 3D crystals and not liquid crystals. Thus, the first documented crystals of tRNA were authenticated and published in *Nature* in September 1968.[5] Further progress now needed large single crystals of a pure tRNA species.

Following the demonstration that a tRNA molecule could be crystallized, the tRNA field became very competitive. The Oak Ridge National Laboratory (Oak Ridge, TN, USA), under the aegis of the US Atomic Energy Commission, developed new chromatography methods based on reversed phase chromatography, purified separate species of tRNA on a large scale and offered them to anyone interested in crystallization.

COMPETITION

In the Autumn of 1968, the race for tRNA structure determination involved Hans Zachau's group in Munich working on yeast $tRNA^{Ser}$, Fritz Cramer's group in Göttingen collaborating with Holmes's group (by that time moved to Heidelberg) on yeast $tRNA^{Phe}$, a collaboration between the late Bob Bock's group in Madison and the late Paul Sigler's in Chicago, a collaboration between Rich at MIT and Don Caspar at the Boston Children's Cancer Research Foundation, and Jacques Fresco's group in Princeton. In the UK we also had competition from the Dept of Biophysics at King's College, London.

After publication of the Cambridge crystallization conditions I was soon able to obtain single crystals of both $tRNA^{Val}$ and the initiator tRNA. With the advent of single crystals of tRNA 50 x 20 x 20 µm in dimensions, Crick encouraged me to join forces with Klug's crystallography group, thus starting a friendly and productive collaboration. The crystals were extremely fragile and sensitive to environmental conditions so I grew them in quartz capillary tubes, which could be sealed and used directly for X-ray analysis. That we collected any data at all was a result of the persistence of John Finch, who worked heroically in the cold room trying to obtain good resolution pictures from the crystals.

By the end of 1968, there were already publications in *Science* and *Nature* of single tRNA crystals from the groups of Cramer, Bock, Rich and Fresco. Another article in the *New Scientist* in January 1969 described the competitive nature of the field and made the point that the 3D structure of tRNA would not be solved in the near future because of the various problems several experienced groups had in producing and handling a variety of tRNA crystals. Indeed, Arnold Hampel and Bock grew yeast $tRNA^{Phe}$ crystals as large as 0.5 × 2.0 mm while, in Cambridge, I was obtaining only small crystals 0.15–0.2 mm in diameter. Furthermore, Rich and Sung-Hou Kim at MIT grew crystals of the initiator *E. coli* tRNA obtained from Oak Ridge from a chloroform–water mixture. These crystals were as large as 1.0 × 1.7 × 0.6 mm. However, none of these publications showed that the crystals were "crystallographers" crystals capable of giving high resolution X-ray data. We were soon in the same situation as the other laboratories with respect to crystal size by obtaining a variety of large single crystals, as my group was purifying its own tRNAs and producing an increasing number of tRNA species. Unfortunately, things then became more and more depressing. First, our lead in tRNA production was wiped out by the US National Laboratory at Oak Ridge, and then our crystals proved to be somewhat disordered, giving resolution no better than 7 Å. I called these crystals mere bags of water.

Thus, there was a lull in our crystallographic data production for three years. However, the unstinting support of the British Medical Research Council provided space, people, equipment and materials to help us try to solve the tRNA structure and win the race. My group grew to include five research assistants (Bill Whybrow, Bob Coulson, Ray Brown, Daniela Rhodes and Margaret Prentice) and several PhD students, who purified tRNAs and helped with the crystallization, and postdoctoral worker Jane Ladner from Caltech, who soon converted from an NMR spectroscopist to a crystallographer.

At the later stages of tRNA production and crystallographic analyses, there was more and more structural collaboration with Klug's group, in particular with John Finch and a new postdoctoral worker Jon Robertus, for structural analysis. It was to my chagrin that we never produced ordered crystals of initiator tRNA to better than ~10 Å resolution.

STRUCTURAL DETERMINATION

Because of the lack of production of good crystals from *E. coli* tRNAs for analysis, we started obtaining samples from outside the laboratory (e.g. from either Oak Ridge, Strasbourg or commercial sources). By 1972, we were able to obtain beautiful crystals of several *E. coli* tRNA species but still the resolution was no better than 7 Å.[6] We were therefore depressed to find that Kim and Rich had been able to obtain X-ray reflections to 3 Å resolution from yeast $tRNA^{Phe}$ bought from Boehringer Mannheim.[7]

We had also bought yeast $tRNA^{Phe}$ from Boehringer, which Ladner cocrystallized with *E. coli* $tRNA^{Val}$ to check whether tRNAs with complementary anticodons would crystallize better (Fig. 1). She obtained some extremely thin crystals by vapour diffusion in sitting drops. Despite their appearance, these crystals (X-rayed in 1972 by Brown) gave X-ray diffraction patterns that extended beyond 3 Å. It turned out that only the y $tRNA^{Phe}$ was present in the crystal. As these crystals possessed a smaller unit cell than that published by the MIT group, it gave us confidence that our crystals would ultimately give better diffraction data. Hence, we switched to working with y $tRNA^{Phe}$ and the race was really on.[8]

In 1972, the solution of the y $tRNA^{Phe}$ structure was still a tough problem. The y $tRNA^{Phe}$ crystals that diffracted well were thin and twinned (i.e. composed of two intergrown crystals), making good data collection very difficult. Large crystals were grown by dialysis as used successfully in Ieuan Harris's group by a postdoctoral worker for *Bacillus stearothermophilus* protein crystallization. Indeed, Rhodes used this method to grow most of the large y $tRNA^{Phe}$ crystals for data collection. Out of thousands of crystals few were suitable. Thus Finch and Brown took close to 500 precession photographs to find heavy atom derivatives.

FIGURE 1. Large (~1 mm long), flat crystals of yeast $tRNA^{Phe}$.

Certainly, at the beginning of the race the MIT group had a short lead, particularly when they published in *Science*[9] a 4 Å resolution structure of yeast tRNA^{Phe} with a general shape they called an "L," which has stuck over the years. We went straight for a 3 Å resolution structure published in 1974.[10] Ladner, Robertus and Finch were mainly involved with the data collection, and Rhodes and Brown with the production of the necessary crystals. My group thus concentrated on growing crystals and searching for compounds suitable for heavy atom derivatives. We also embarked on a study using chemical probing to show that the three-dimensional structure of y tRNA^{Phe} was the same in solution as in the crystal.[11] In particular, Rhodes and Brown were leading players for our part of the work, which exemplified a complementary integration of the work of Klug's crystallography group and my biochemistry group (Fig. 2).

The 3D structure of y tRNA^{Phe} in the monoclinic crystal form was solved to 3 Å resolution in Cambridge using isomorphous replacement with five heavy atom derivatives.[10] One of these, Pt, was located by chemical methods to a site in the anticodon loop and served to assign unambiguously residues in the electron density map.[12]

Why did the bitterness between MIT and Cambridge occur? As I mentioned previously, we were disheartened when the MIT group published in *Science* a 4 Å resolution structure that looked very similar to our structure with respect to the backbone tracing. Our horizon then brightened when they published a 3 Å resolution structure in *Nature*[13] in the Spring of 1974.

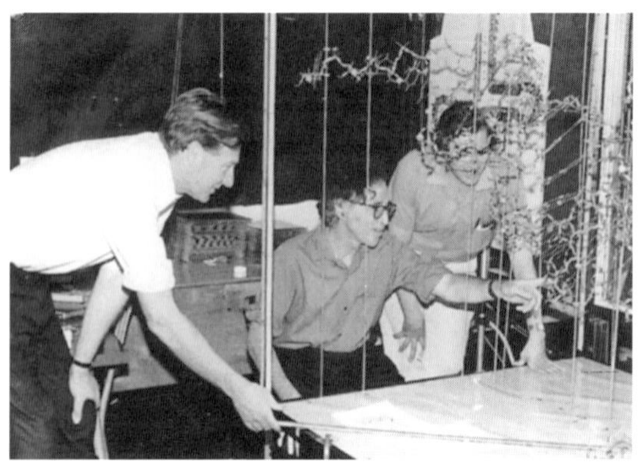

FIGURE 2. Aaron Klug, John Finch and Brian Clark discussing finer points of the skeletal model, built in Cambridge, UK, to represent the 3D structure of yeast tRNA^{Phe}.

At this time we were just finishing our structure determination and were astonished to find that, according to our structure, the MIT group had made a serious mistake in tracing the polynucleotide chain. We sailed on and wrote the paper for *Nature*, submitting it in June 1974. In addition, Robertus and I gave talks on the new structure at a Steenbock Symposium in Madison, also in June 1974. At this time we reported detailed information on our structure. We called our structure a "T" because of the two long helices forming a "T" junction. These helices were composed of the acceptor stem that was stacked on the TψC-stem, and the anticodon stem stacked on the D-stem. We also described, for the first time, several base triplets that helped hold the structure together as well as some crucial non Watson–Crick base pairs that had not been presented by the MIT group. Our "T" for tRNA structure did not catch on. Imagine our chagrin when the MIT group who had attended at the Steenbock Symposium wrote a revised paper on the 3 Å structure and submitted it to *Science*. This was published so quickly (within one month) that it came out two weeks before our *Nature* paper in August 1974.

POSTSCRIPT

Further work by both the Cambridge and MIT groups pushed the resolution to 2.5 Å resolution and refined the structure. Interestingly, many investigations of tRNA structure in which the tRNA is either alone (naked) or in complexes with aminoacyl-tRNA synthetases and translational elongation factors, have encountered difficulties in reaching a higher resolution for tRNA structures. However, with modern technology, Rhodes, still at the LMB, and Peter Moore's group at Yale, have recently re-examined y $tRNA^{Phe}$ crystals (Rhodes' were ~20 years old) and obtained resolution structures better than 2 Å, so it could be worthwhile to return to investigating the tRNAs we studied before.

Elucidation of the y $tRNA^{Phe}$ structure was a milestone in the development of our knowledge about the molecular mechanisms governing protein biosynthesis. It enabled various functional parts of the structure to be assigned and gave a framework for later work in the tRNA field, for example, determination of interaction sites between tRNA and aminoacyl-tRNA synthetases and elongation factors, and assignment of tRNA binding sites on the ribosomal subparticles. Moreover, the y $tRNA^{Phe}$ structure was the first nucleic acid structure to be solved to 3 Å and gave a picture of the ordered complexity of a folded RNA molecule—a complexity as great as that of a protein—and also gave details of metal ion binding sites. The types of nucleotide conformation, bonding and interactions, including metal-binding sites seen in the folded $tRNA^{Phe}$, have been shining examples for other RNA structures. An interesting by-product of the $tRNA^{Phe}$ structure was the first detailed chemical picture of a G–U base pair in a double helical stem, in which the pairing was that predicted by Crick's "Wobble" hypothesis. It was also satisfying that our group was the first to confirm that the y $tRNA^{Phe}$ structure was the same in solution as in the crystal.[10]

So, the 3D structural determination of a tRNA did not gain anyone a Nobel Prize, possibly because too many people were involved and the methods used turned out to be relatively standard, but it gave us enormous satisfaction at the time. And, I am happy to add that Klug, who led the crystallographic analysis at the LMB, did win the Nobel Prize in 1982 for many other important, innovative contributions to structural biology.

ACKNOWLEDGEMENTS

I thank Ray Brown, Aaron Klug, Kjeld Marcker, Daniela Rhodes and Jan Barciszewski for critical reading of the manuscript and helpful suggestions.

REFERENCES

1. Chedd G. 1968. The race for transfer-RNA. *New Sci.* **40:** 606–607.
2. Clark B.F.C. and Marcker K.A. 1966. The role of N-formyl-methionyl-sRNA in protein biosynthesis. *J. Mol. Biol.* **17:** 394–406.
3. Doctor B.P. et al. 1969. Studies on the *Escherichia coli* methionine tRNAs. *Eur. J. Biochem.* **8:** 93–100.
4. Hendrickson W.A. et al. 1983. True identity of a diffraction pattern attributed to valyl tRNA. *Nature* **303:** 195–196.
5. Clark B.F.C. et al. 1968. Crystallization of transfer RNA. *Nature* **219:** 1222–1224.
6. Brown R.S. et al. 1972. Crystallization of pure species of bacterial tRNA for X-ray diffraction studies. *Eur. J. Biochem.* **31:** 130–134.
7. Kim S.H. et al. 1971. High resolution X-ray diffraction patterns of crystalline transfer RNA that show helical regions. *Proc. Natl. Acad. Sci. U. S. A.* **68:** 841–845.
8. Ladner J.E. et al. 1972. High resolution X-ray diffraction studies on a pure species of tRNA. *J. Mol. Biol.* **72:** 99–101.
9. Kim S.H. et al. 1973. Three-dimensional structure of yeast phenylalanine transfer RNA: folding of the polynucleotide chain. *Science* **179:** 285–288.

10. Robertus J.D. et al. 1974. Structure of yeast phenylalanine tRNA at 3 Å resolution. *Nature* **250:** 546–551.
11. Robertus J.D. et al. 1974. Correlation between three-dimensional structure and chemical reactivity of transfer RNA. *Nucleic Acids Res.* **1:** 927–932.
12. Rhodes D. et al. 1974. Location of a platinum binding site in the structure of yeast phenylalanine tRNA. *J. Mol. Biol.* **89:** 469–475.
13. Suddath F.L. et al. 1974. Three-dimensional structure of yeast phenylalanine transfer RNA at 3.0Å resolution. *Nature* **248:** 20–24.

What Was the Message?

ELLIOT VOLKIN

Oak Ridge National Laboratory, Oak Ridge, Tennessee, USA

"At this point Francis and Sidney leaped to their feet . . . Began to gesticulate. . . . What had set off Francis and Sidney was, once again, a connection between the lactose system and phage. . . . Exactly the properties required for what we called X, the unstable intermediary we had postulated for galactosidase. Why, in Paris, when we were looking for a support material for X, had we not thought of this phage RNA? Why had we not thought of it?" from François Jacob's *The Statue Within.*[1]

"Where, then, is the message? At this point Sidney Brenner let out a loud yelp—he had seen the answer. (So had I, for that matter, but nobody else had.) . . . It is difficult to convey two things. One is the flash of enlightenment when the idea was first glimpsed. . . . The other is the way it cleared away so many of our difficulties. . . . Why had we not seen it before?" from Francis Crick's *What Mad Pursuit.*[2]

These were the recollections of an informal meeting at Cambridge on Good Friday, 1960, from which Jacob, Crick and Brenner gave birth to the revolutionary concept of messenger RNA. The key, then, was the RNA that Larry Astrachan and I first described in 1956[3] and whose properties we studied over the subsequent three years. What was the history of this nucleic acid that we called DNA-like RNA?

In the late 1940s and early 1950s a large body of circumstantial evidence implicated RNA synthesis with active protein synthesis and cellular growth in a diverse variety of organisms, even though at that time no method existed for the isolation of individual RNA species. Of all the present-day organisms, lytic bacteriophage were particularly attractive to study, being the most simple examples of a complete biological entity.[4] Their physical properties and genetic pathways were the subject of intense study in many eminent lab-

Reprinted from *Trends in Biochemical Sciences* May 1995, 20(5): 206–209.

oratories. Lytic phage infection had the unique capacity to divert the host's biochemical mechanisms immediately to become synthetic systems for phage production instead. However, despite the demonstration of highly active synthesis of protein and DNA,[5] it was unambiguously shown, especially by members of S.S. Cohen's lab,[6] that RNA synthesis came to an abrupt halt upon phage infection. Using ^{32}P to study the course of its incorporation into phage DNA, A.D. Hershey[5] was the first to note that a minor metabolically active RNA species might exist following T2-phage infection. This elegant experimental system begged for the resolution of the role of RNA in phage nucleic acid metabolism. I was confident that we had the methodology to determine definitively whether or not some RNA was synthesized in these infected cultures.

THE EXPERIMENTAL BACKGROUND

In the early 1950s, two approaches were used to separate the degradation products of nucleic acids: ion-exchange chromatography and paper chromatography. The emergence of gel chromatography and electrophoresis came some years later. (At this point, it may be worth recalling that Erwin Chargaff, a pioneer in the use of the paper chromatographic method, clearly demonstrated the A = T and G = C relationship in DNA from a variety of biological sources.[7] This was an instrumental factor in Watson and Crick's[8] resolution of the structure of DNA.) Our laboratory, notably through the efforts of Waldo Cohn, was at the forefront of the development and utilization of the ion-exchange methodology. During the Second World War Cohn had been part of a group at the Oak Ridge National Laboratory (ORNL) that exploited ion-exchange for the heroic task of quantitatively separating highly radioactive masses of fission products. One such product, carrier-free ^{32}P, was readily available to us from ORNL and was a necessary component of our bacteriophage experiments. (Cohn, incidentally, holds a patent for the preparation of carrier-free ^{32}P.) Our group, consisting of Cohn, D.G. Doherty, J.X. Khym and myself, took advantage of the ion-exchange methodology to separate essentially all of the important alkaline and acid degradation products of RNA and DNA, as well as separating the unbound nucleoside mono-, di- and triphosphates.[9] Cohn and I first demonstrated that the 5′-nucleotides could be derived from RNA enzymatically.[10] This result, together with the identification of the 2′- and 3′-products of alkaline hydrolysis, was important in Brown and Todd's proposal[11] that the linkage between the nucleosides in RNA (i.e. 3′ to 5′) was the same as in DNA, and that the 2′- and 3′-nucleoside phosphates arose by way of a triphosphorylated cyclic intermediate.

THE EXPERIMENTAL APPROACHES

Our experiments with bacteriophage were based on two fundamental observations ascertained by ion-exchange chromatography. First, the sole products of alkaline hydrolysis of RNA were the 2′- and 3′-nucleoside phosphates of all four bases.[12] Second, these nucleotides could be separated cleanly from the corresponding 5′-ribonucleoside phosphate precursor molecules. Thus, any activity recovered with the 2′,3′-nucleotides would eliminate any ambiguity in defining its source as RNA.

The initial experimental approach with bacteriophage was to infect the host *Escherichia coli* with a high multiplicity of T2R+ phage particles to ensure complete infection. Then, after a few minutes to allow for phage adsorption, $^{32}PO_4$ was added to the culture medium. The experiment was carried out in both broth and synthetic media. At various times after this, we isolated total nucleic acid. It was important to use a method that produced a quantitative separation of the RNA in polymeric form. For this purpose, we applied the perchloric acid precipitation method proposed by Tyner et al.[13] This procedure removes acid-soluble constituents, lipids and protein. The resulting total nucleic acid product is retained in a completely nondialysable form.

After hydrolysis of the nucleic acid with NaOH under conditions that convert RNA completely to the 2′- and 3′-mononucleotides, these products were separated quantitatively by anion-exchange chromatography and all fractions recovered from the column were counted for ^{32}P. The data made it quite clear that there was a small but significant uptake of ^{32}P into RNA after phage infection and that this was a genuine property of the infected host. Radioactivity resided solely under the ultraviolet absorbency peaks corresponding to the 2′- and 3′-nucleoside phosphates of the four bases. Although at early times after infection the uptake of ^{32}P into RNA even exceeded that in phage DNA—T2-infected bacteria behave like a DNA-producing factory—the amount of radioactivity in the RNA nucleotides leveled off quickly. This raised the question of whether the synthesis of this RNA stopped shortly after infection or, instead, whether it never accumulated because of constant synthesis and turnover. More on that point later. What gave me an immediate jolt was the observation that the total counts and the specific activities associated with the individual nucleotides did not resemble in any way the nucleotide composition of the host *E. coli* RNA (C : G : U : A ratios of 1 : 1.4 : 1 : 1). Instead, the distribution of radioactivity mimicked the composition of the DNA of the infecting phage. Thus, the ^{32}P associated with the 2′,3′-ribonucleoside phosphates in T2-infected cells had a C : G : U : A of about 1 : 1 : 1.9 : 1.9. The related deoxyribonucleotides in T2 DNA have a hydroxymethylcytosine (HMC) : G : T : A ratio of 1 : 1 : 1.95 : 1.95.[14]

At this point, I realized that this was hot property. I urged Larry Astrachan, who was completing an experiment on proflavine DNA,[15] to join me in experiments designed to resolve fundamental questions about the properties of this DNA-like RNA. The following summarizes the findings and conclusions of these experiments, some of which were done together with Joan Countryman, who joined us as a postdoctoral investigator.

It occurred to us that, since the isotopic nucleotide composition of this RNA resembled so closely the base ratios of the phage DNA, the RNA could be a component of the phage particle itself. Estimates of the RNA phosphorus content of these phages, usually done colorimetrically, were reported to be less than 2.4% of the total phosphorus.[16] Since these estimates were subject to some error, and since the molecular weight of the RNA may be only about 1% that of the DNA, phage containing an approximately equal number of molecules of each would exhibit a corresponding difference in phosphorus content. To resolve this question we mixed phage that were highly labeled with ^{32}P with an excess of carrier RNA and subjected the mixture to alkaline hydrolysis and ion-exchange chromatography.[17] An upper limit of 0.025% of the ^{32}P was recovered from the column and this radioactivity was not consistent with the elution of the mononucleotides. We concluded that T2 bacteriophage contain no RNA.

Although synthetic medium was required for some experiments, peptone broth[5] was still the ideal growth medium for most. However, because of the high concentration of inorganic phosphate in broth, this medium presented challenges when we used $^{32}PO_4$, both to allow reasonable uptake of the isotope and to avoid exposing the laboratory to large amounts of the β-emitter. We modified the medium initially by using HCl to precipitate some of the material that interfered with ultraviolet absorbencies of the nucleotides. Then, with the controlled addition of $MgCl_2$ and NH_4OH, the inorganic phosphate content of the medium was reduced to less than 3 $\mu g ml^{-1}$ by removing the insoluble $MgNH_4PO_4$ complex. *E. coli* itself and the phage-infected bacteria not only thrived on this medium but had the fortunate property of sopping up the bulk of the medium's radioactive inorganic phosphate before using the large store of the culture's organically bound phosphorus.

Initially, it was of vital importance to resolve the question of whether there was metabolic turnover of the DNA-like RNA. For this purpose, the obvious pulse–chase experiments, carried out at various times after T2 infection, unequivocally showed that, indeed, highly active turnover of this RNA takes place. As mentioned above, the experiments of Hershey et al.[5] were the first to indicate the existence of a physiologically active RNA after T2 infection. We carried out a series of pulse–chase experiments at various times after T2 infection.[18,19] The differences in the synthesis and turnover of the RNA labeled early and late after infection are shown in Fig. 1. RNA labeled 1–3 min after infection can have more than ten times the ^{32}P content found in the phage DNA, and the RNA produced very early exhibits the highest turnover rate. Interestingly, there is a gradual change in the ^{32}P ratio of the mononucleotide as the isotope leaves RNA. Thus, the radioactivities remaining in the constituent four nucleotides approach uniformity. In addition, a definite residue of this RNA always remains. It is interesting to note that

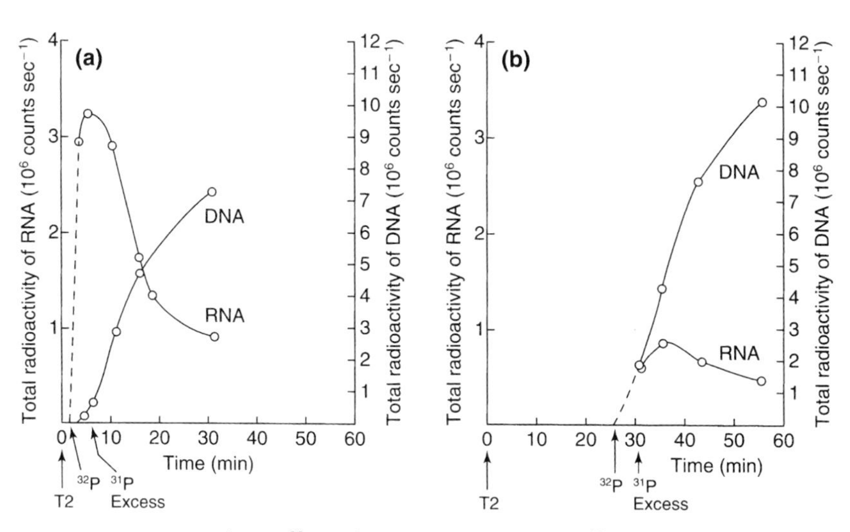

FIGURE 1. Turnover of RNA ^{32}P and accumulation of DNA ^{32}P at (a) early and (b) late pulse-chase times after T2 infection. Adapted from Ref. 19.

when $^{32}PO_4$ is added to the medium late after T2 infection (26 min), the RNA labeled late still has an isotope composition mimicking that of T2 DNA, but its turnover is slower.

From this series of pulse–chase experiments we concluded that two or more species of RNA are labeled in the host. One type is synthesized at a high rate shortly after infection, is characterized by a high metabolic turnover, and has a composition similar to that of the infecting phage's DNA. The other class or, more likely, classes of RNA contain at least one type whose nucleotide composition approaches equality. It is labeled more slowly and exhibits little turnover.

In light of the identification, some years later, of both "early" and "late" phage-specific enzymes and other proteins that are necessary components for phage production,[20] we can speculate that some of the more slowly labeled RNA species are transcribed to carry the information necessary for the synthesis of the "late" proteins.

If pulse–chases were performed very early after phage infection, an almost typical RNA to DNA precursor–product pattern was observed. We asked the question of whether the labeled RNA was a mandatory precursor of the DNA. From the experiments of RNA turnover carried out at various times after infection,[19] we observed that there was a definite lack of correspondence in the rate of RNA synthesis with the rate of DNA synthesis. We concluded that the RNA could not be an obligatory precursor of DNA. I mention this because it became popular in some quarters to assume that we missed the boat, or perhaps misled ourselves by insisting that the RNA's function was only as a precursor to phage DNA. Certainly, the biochemical fate (but not the function) of the degraded RNA is to serve as precursor material for the enormous synthesis of T2 DNA.

We compared the T2 system with that of the unrelated virulent phage, T7.[21] This has molar base ratios that are widely different from those of T2 DNA and from its host's (*E. coli*) DNA. Although T7, unlike T2, converts the host's DNA almost exclusively, with little utilization of precursors, for its own DNA synthesis, a definite quantity of $^{32}PO_4$ could be taken up from the medium and incorporated into T7-infected cells. The labeled RNA in this system also underwent turnover, and the mononucleotide labeling of C, G, A and U closely paralleled the base composition of T7 DNA (C, 23.5%; G, 23.8%; A, 26.5%; T, 26.5%).

A legitimate question may be raised about the use of alkali as a means of hydrolysing and analysing $^{32}PO_4$ incorporation into RNA. Thus, the 2′,3′-

mononucleotide products are actually nearest neighbors to the 5′-nucleotide intermediates that are the true precursors. This is not an issue if, as we believed, the minor species of DNA-like RNA were the only RNA synthesized after phage infection. But if ^{32}P incorporation involves a heterogeneous uptake into some pre-existing RNA, then the probability is high that the isotope associated with the 5′-precursor nucleotide will be found with a different nucleotide. To resolve this point, we turned to the use of [^{14}C]formate,[22] known to be a precursor for purine base synthesis,[23] and thus independent of the mode of phosphate transfer. The results with [^{14}C]formate were in complete accord with the data obtained by using ^{32}P as the tracer. [^{14}C]formate pulse–chase experiments with T2, of necessity carried out in synthetic medium, were virtual copies of those with $^{32}PO_4$, in which the A : G ratio of the RNA was about 2 : 1, the same as in T2 DNA, not the approximate ^{14}C ratio of 1 : 1 found in uninfected bacteria. The experiments with T7 phage, although not as clear-cut, definitely pointed to an A : G ratio of ^{14}C that was closer to the base ratio found in T7 DNA than that in *E. coli* DNA. In the T2 system there was an unanticipated conversion of some [^{14}C]formate to pyrimidines. This was not seen in uninfected *E. coli* or in T7-infected bacteria. Apparently, there is a distinctive metabolism of this compound in T2-infected cells. But most important, from our point of view, was the observation that, here again, the ^{14}C ratio of U : C in RNA was about 2 : 1, comparable to the T : HMC ratio in T2 DNA. Some of these data are shown in Table 1.

TABLE 1. *Labeling of RNA mononucleotides after [^{14}C]formate assimilation*

	Specific activity (counts per min per μmole of nucleotide)					
Mononucleotide	T2 infection			Uninfected bacteria		
Min after [^{14}C]formate	7	12[a]	49	5	10[a]	50
Adenylic acid	86	133	45	128	366	375
Guanylic acid	41	67	33	138	382	368
A : G ratio	2.10	1.99	1.36	0.92	0.96	1.02
Uridylic acid	42	68	37	–	–	–
Cytidylic acid	16.5	32	26	–	–	–
U : C ratio	2.54	2.12	1.42			

[a]At 10 min a 100-fold excess of unlabeled sodium formate was added to the cultures. No significant loss of RNA occurred in the T2 system, whereas RNA continued to accumulate in the uninfected *E. coli*. From Ref. 22.

Early on we did some experiments to study the approximate intracellular distribution of labeled RNA in infected[24] and uninfected *E. coli.*[25] We used the rather crude method[26] of disrupting the cells by grinding them with alumina. After low-speed centrifugation to remove unbroken cells and alumina, particulate fractions at 20 000 g (P1) and 140 000 g (P2) and the subsequent supernatant (S) were collected. Surprisingly, especially in the T2 and T7 phage systems but even in uninfected *E. coli,* the P1 fraction (membranes?), which contained a minimum molar amount of RNA, incorporated and turned over ^{32}P into its RNA at a high rate. The other two fractions, P2 (mostly ribosomal) and S (mostly transfer RNA), remained metabolically stable. These experiments, their deficiencies notwithstanding, confirmed our notion that the physiologically active species of RNA was a minor fraction of the cellular total.

From experiments using the known protein synthesis inhibitor chloramphenicol,[27] we sought correlations among the syntheses of RNA, protein and DNA.[28] Chloramphenicol was added before T2, very shortly afterwards and at later times when T2 had been adsorbed; and the course of ^{32}P uptake and turnover was followed by pulse–chasing. We observed an interesting relationship between RNA turnover and DNA synthesis: when chloramphenicol was added shortly before or shortly after infection, although DNA synthesis was inhibited and the DNA-like RNA accumulated but with no turnover, late addition of chloramphenicol inhibited neither DNA synthesis nor RNA turnover. Thus, it appears that blocking protein synthesis at an early time after T2 infection does not slow the synthesis of RNA but stops its turnover (biological utilization?). From these experiments it appears that RNA metabolism and protein synthesis are closely interrelated.

In summary, this series of experiments disclosed that a previously unrecognized kind of RNA is synthesized *de novo* immediately upon lytic phage infection. The RNA, as evidenced by its incorporation of precursors labeled with ^{32}P and ^{14}C, has a base composition virtually identical to that of the infecting phage's DNA. It undergoes rapid metabolic turnover, is associated with a quantitatively minor subcellular component, and its activity appears to be interconnected with protein synthesis.

THE DELAYED REALIZATION THAT DNA-LIKE RNA IS THE MESSENGER

Why then was there a four-year hiatus between this series of papers enumerating the properties of the DNA-like RNA and the resolution of its vital role in the transfer of genetic information? In his autobiographical account,[2] Crick states "the irony was, of course, . . . we had not recognized it till that fateful Good Friday morning." Was it largely unrecognized by everyone? Not really. Our work actually created quite a stir in biochemical circles. Jukes,[29] in writing of the benefits and excitement of attending FASEB meetings, states "Some years later, I think in 1956, I had squeezed my way into a doorway of a packed room to hear a paper by Volkin and Astrachan on DNA-like RNA!" I gave invited presentations on these findings at the prestigious McCollum-Pratt Symposium in 1956,[18] and at the International Microbiology meeting in Stockholm and the International Biochemical Society meeting in Vienna in 1958. In fact, as part of the 1958 conference tour, I was asked to discuss our research to a select, relatively small group of phage researchers at an abbey near Paris. I believe the session was organized by Monod's group at the Pasteur Institute.

Part of the explanation in the relative nonrecognition of DNA-like RNA may reside in the unfortunate circumstance that, at that time, there seemed to be a self-imposed chasm between many leading biochemists/chemists on the one hand and their geneticist/virologist counterparts on the other, whereby the accomplishments of either group were largely discounted by the other. This quote from Jukes' highly critical review[30] of H.F. Judson's book[31] may illustrate the point. "The story of the messenger concept is developed in detail. I well remember the excitement caused by the 'DNA-like RNA' finding of Volkin and Astrachan in 1956. The statement that this work 'could not quite be dismissed as the inconsequential and self-admittedly sloppy work of people not well known at a lab not highly regarded' is an unjustified slur on the lab where Cohn and Volkin pioneered on phosphate linkages in nucleic acids." (For a very different, more complimentary, view of Judson's book, one should refer to the review by Edsall.[32])

But it is easy to put aside these personal surmises and accept the probable conclusion that it is only after a critical mass of important data and key discoveries has accumulated that the ultimate postulate can be formulated. I believe such was the case with the beautifully conceived concept of messenger RNA.[1,2,33]

REFERENCES

1. Jacob F. 1988. *The Statue Within: An Autobiography*, Basic Books.
2. Crick F. 1988. *What Mad Pursuit: a Personal View of Scientific Discovery*, Basic Books.

3. Volkin E. and Astrachan L. 1956. *Virology* **2:** 149–161.
4. Adams M.H. 1950. Methods *of Medical Research* (Vol. 2) p. 1, Year Book Publishers.
5. Hershey A.D. 1953. *J. Gen. Physiol.* **37:** 1–3.
6. Cohen S.S. 1948. *J. Biol. Chem.* **174:** 281–293.
7. Chargaff E. 1955. *The Nucleic Acids* **1:** 307–371.
8. Watson J.D. and Crick F.H.C. 1953. Cold *Spring Harbor Symp. Quant. Biol.* **18:** 123–131.
9. Volkin E. and Cohn W.E. 1953. *J. Biol. Chem.* **205:** 767–782.
10. Cohn W.E. and Volkin E. 1952. *Arch. Biochem. Biophys.* **35:** 465–467.
11. Brown D.M. and Todd A.R. 1955. *Am. Rev. Biochem.* **24:** 311–338.
12. Cohn W.E. 1950. *J. Am. Chem. Soc.* **72:** 1471–1478.
13. Tyner E.P., Heidelberger C., and LePage G.A. 1953. *Cancer Res.* **13:** 186–203.
14. Wyatt G.R. 1953. *Cold Spring Harbor Symp. Quant. Biol.* **18:** 133–134.
15. Astrachan L. and Volkin E. 1957. *J. Am. Chem. Soc.* **79:** 130–134.
16. Putnam F.W. 1953. *Adv. Protein Chem.* **8:** 177.
17. Volkin E. and Astrachan L. 1956. *Virology* **2:** 594–598.
18. Volkin E. and Astrachan L. 1957. *The Chemical Basis of Heredity,* pp. 686–694, Johns Hopkins Press.
19. Astrachan L. and Volkin E. 1958. *Biochim. Biophys. Acta* **29:** 536–544.
20. Geiduschek E.P. 1991. *Annu. Rev. Genet.* **25:** 437–460.
21. Volkin E., Astrachan L., and Countryman J.L. 1958. *Virology* **6:** 545.
22. Volkin E. 1958. *Fourth Int. Congress Biochem.* **7:** 212–224.
23. Reichard P. 1955. *The Nucleic Acids,* pp. 277–306, Academic Press.
24. Volkin E. and Astrachan L. 1956. *Virology* **2:** 433–437.
25. Countryman J.L. and Volkin E. 1959. *J. Bacteriol.* **78:** 41–48.
26. McIlwain H. 1948. *J. Gen. Physiol.* **2:** 288–291.
27. Crawford L.V. 1958. *Biochim. Biophys. Acta* **28:** 208–209.
28. Astrachan L. and Volkin E. 1959. *Biochim. Biophys. Acta* **32:** 450–456.
29. Jukes T.H. 1977. *Nature* **267:** 8.
30. Jukes T.H. 1979. *Nature* **281:** 505–506.
31. Judson H.F. 1979. *The Eighth Day of Creation: The Makers of the Revolution in Biology,* Simon and Schuster.
32. Edsall J.T. 1980. *J. Hist. Biol.* **13:** 141–158.
33. Jacob F. and Monod J. 1961. *Cold Spring Harbor Symp. Quant. Biol.* **26:** 193–201.

An RNA Heresy in the Fifties

HENRY HARRIS

Sir William Dunn School of Pathology, University of Oxford, Oxford, UK

I drifted into the world of RNA in all innocence. From 1952 to 1954 I had been working on chemotaxis of leucocytes and had become a dab hand at manipulating populations of granulocytes, macrophages and lymphocytes *in vitro.* In 1955, I turned to fibroblasts and was busy studying the incorporation of radioactive amino acids into the proteins of these and other cells when Jacques Monod and his colleagues[1] announced to the world that protein turnover, as defined by Rudolf Schoenheimer, was an artefact. The biochemistry of my undergraduate years had been dominated by Schoenheimer's *The Dynamic State of Body Constituents,*[2] which appeared in 1942. By the use of isotopes, Schoenheimer had shown that the proteins of the body were in a constant flux of synthesis, breakdown and resynthesis. But Monod claimed, on the basis of observations made on bacteria growing exponentially, that the proteins of bacterial cells were stable and, being a great generalizer, he concluded that this must also be true for somatic cells in animals. What Schoenheimer had observed was, according to Monod, simply the concomitant multiplication and death of cells. This struck me as very unlikely for a number of reasons, but especially because the amount of cell multiplication and cell death that could be detected in organs such as adult liver or adult kidney was nothing like enough to account for the extent of the protein turnover that was actually observed. Now the macrophages that I was working with could be maintained *in vitro* for many days without undergoing either cell multiplication or cell death, and they continued to exhibit all their normal physiological functions, such as locomotion, chemotaxis and phagocytosis. It seemed to me that if one wanted to know what was going on in the animal body, one should not look at cells multiplying exponentially. In the body,

Reprinted from *Trends in Biochemical Sciences* July 1994, 19(7): 303–305.

exponential cell multiplication is seen only in the terminal stages of a few highly malignant tumours. John Watts, my first PhD student, and I therefore measured the incorporation and release of radioactive valine in the proteins of macrophages, and we found extensive protein turnover. So Monod was wrong, and Schoenheimer was right.

I wouldn't have bothered about RNA if it hadn't been for the appearance of a paper by Siminovitch and Graham[4] in which it was claimed that all the RNA in multiplying animal cells was, as Monod had proposed for protein, completely stable. I thought that, while I was about it, I might as well have a look at the stability of the RNA in macrophages. The result was dramatic. Although the total amount of RNA and protein in these nonmultiplying cells remained constant over the course of the experiment, a rapid and large-scale turnover of RNA was observed. There was obviously some fraction of RNA in the cell that was not only rapidly synthesized, but also rapidly broken down, and we could recover the end products of this breakdown as acid-soluble components in the medium.[5] Naturally, I wanted to know where in the cell this RNA fraction was to be found, so I turned to autoradiography and quickly saw that the rapidly labelled RNA component that I was interested in was located in the cell nucleus.[6] There were several autoradiographic studies in the literature in which it had been observed that, when cells exposed to a short pulse of radioactive RNA precursor were transferred to nonradioactive medium (a pulse–chase experiment as it was then called), radioactivity left the nucleus and appeared in the cytoplasm. This result was universally interpreted as evidence that the RNA initially labelled by the radioactive pulse was the precursor of the radioactive RNA that subsequently appeared in the cytoplasm. However, when I did this experiment on macrophages, I found that the radioactivity, as expected, disappeared from the nucleus, but this disappearance was not accompanied by any measurable increase in the amount of radioactive RNA in the cytoplasm[6] (Fig. 1). And thus my heresy began.

I decided next to have a look at fibroblasts isolated from newborn rat heart in order to see what happened when a pulse–chase experiment of this kind was done on multiplying tissue. I found, as others had done, that when cells briefly labelled with a radioactive RNA precursor were transferred to non-radioactive medium, the disappearance of radioactivity from the nucleus was indeed accompanied by the appearance of radioactivity in the cell cytoplasm. But when I measured the two processes carefully, I found that there was no correspondence between the amount of radioactivity lost from

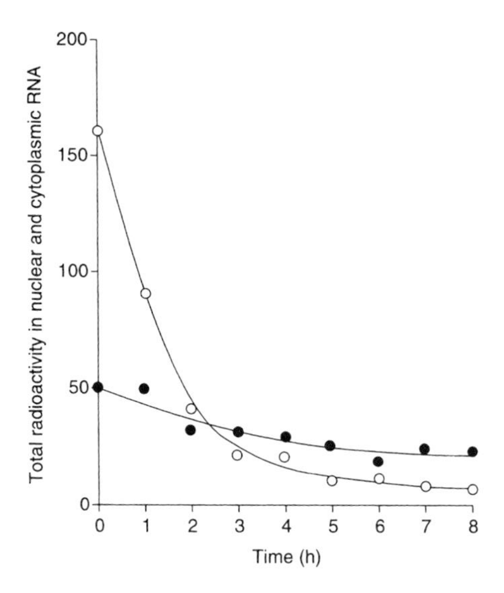

FIGURE 1. A pulse–chase experiment in nonmultiplying macrophages. Radioactivity is rapidly lost from the nuclear RNA (○) during the chase, but there is no corresponding rise in the amount of radioactivity in the cytoplasmic RNA (●).

the nucleus and the amount that appeared in the cytoplasm. It was evident that the two processes, even in a multiplying cell, did not simply reflect the passage of RNA from nucleus to cytoplasm. We then turned to bulk cultures of HeLa cells, which we had trained to grow in suspension so that we would have enough material for detailed quantitative studies. Harold Fisher and I devised a rapid technique for isolating cell nuclei that enabled us to separate the cytoplasmic RNA from the rapidly labelled nuclear RNA without measurable contamination of the former by the latter.[7] This and other improvements in our techniques permitted us to make a very thorough study of the passage of radioactivity through the RNA of the cell in pulse–chase experiments.[8] We found, among other things, that even in suspension cultures in which the cells were growing exponentially, up to 13 times as much radioactivity was lost from the nuclear RNA per unit time as appeared in the cytoplasmic RNA.[8,9] Moreover, by analysing the distribution of radioactivity in the individual RNA nucleotides, we could show that the distribution of radioactivity in the nucleotides of the rapidly labelled nuclear RNA was quite different from that of the labelled RNA in the cytoplasm.[8,9] This provided additional strong support for our conclusion that the redistribution of radioactivity between nucleus and cytoplasm could not be accounted for simply by translocation of the rapidly labelled nuclear RNA to the cytoplasm.

Our results were strongly resisted. Indeed, at one point a mathematical assault on our data appeared in *Nature* claiming that all our findings could easily be accommodated by a simple precursor–product relationship between the labelled nuclear and the labelled cytoplasmic RNA.[10] With Neil Gilbert's help, I penned an appropriately mathematical riposte,[11] but the exchange of mathematics served merely to obfuscate our results rather than clarify them. So I found myself, for the first time, embattled in a rather hostile world. The advent of actinomycin D made matters worse. Pulse–chase experiments were done in which *de novo* synthesis of RNA was blocked during the chase by the administration of actinomycin D; but radioactivity was found, none the less, to pass from nuclear to cytoplasmic RNA. I repeated these actinomycin D experiments with great care and was again able to show that the distribution of radioactivity within the individual nucleotides of the cytoplasmic RNA was quite different from that found in the pulse-labelled nuclear RNA.[12] Again, I found that there was massive breakdown of the RNA that had been labelled during the pulse. Then I was told that actinomycin D induced this breakdown. This scholastic trench warfare went on for some years, but eventually I managed to get most of my adversaries to admit that there was, under physiological conditions, a large-scale turnover of RNA in the cell. What remained far from clear was whether or not the breakdown took place within the nucleus or after the pulse-labelled RNA had been transferred to the cytoplasm.

Although the idea made no immediate biological sense, our experiments continued to yield increasingly strong evidence that the breakdown took place within the nucleus. Although the separation of nuclear from cytoplasmic RNA fractions in our experiments was as clean as you could wish, I could not find any RNA component in the cytoplasm that showed the rapid turnover that we had measured. When isolated nuclei containing pulse-labelled RNA were incubated at 37°C in solutions whose ionic composition approximated to that of intracellular fluid, breakdown of the labelled RNA to acid-soluble components occurred with kinetics that mimicked what we had observed in the intact cell.[14] No such rapid breakdown of labelled RNA occurred in isolated cytoplasmic fractions. Moreover, the breakdown of the labelled nuclear RNA released 5′-nucleotides that could rapidly be reincorporated, a recycling that we knew to occur in the intact cell. The much slower breakdown of cytoplasmic RNA released only 3′-nucleotides that could not be reutilized. So, reluctantly, I was driven to the conclusion that the rapid RNA turnover that I had discovered was taking place within the nucleus. I was encouraged to receive a note from Jesse Scott in Paul Zamecnik's department at the Massachusetts General Hospital, telling me that he had repeated our experiments with a different cell type, could confirm our data and agreed with my interpretation of it; Howard Hiatt also published some findings that were closely consonant with ours[15,16]; and Jan-Erik Edström, studying the RNA of nuclear sap isolated by direct micro-dissection from the nuclei of fixed starfish oocytes, noted the presence there of heterogeneous polynucleotides of bizarre base composition that were obviously intermediates of RNA breakdown.[17]

I think that, by the early 1960s, I had convinced most people that there was a rapid turnover of RNA in animal cells, and some people that this turnover was an intranuclear process. But in 1961 my life was greatly complicated by the appearance of two celebrated papers, one by Sydney Brenner, François Jacob and Matt Meselson,[18] and the other by François Gros and his colleagues.[19] These papers launched the idea that there was a short-lived RNA that functioned as a vector of information from the genes to the cytoplasm of the cell, an RNA that was aptly named "messenger" RNA. I had been monitoring pulse–chase experiments for years, and I knew only too well that when cells in which the RNA had been pulse-labelled were transferred to unlabelled medium, the labelled cells did not at once begin to draw on a pool of unlabelled precursors for the synthesis of their RNA but, on the contrary, continued for a long time to use a substantial pool of labelled precursors that had accumulated before the transfer; and similarly for cells transferred from ^{15}N to ^{14}N. So, I had some pretty serious reservations about the probative value of the pulse–chase experiments on which these two papers were based, but I had more than technical grounds for my scepticism. I knew from many experiments on the behaviour of enucleate cells, and especially from the classical experiments of Hämmerling on *Acetabularia,* that the templates for protein synthesis (the messenger RNA) in higher cells could not be short-lived on the timescale that the messenger RNA hypothesis envisaged, and although I had better evidence than anyone else for the existence of a short-lived RNA in animal cells, to my infinite regret I could not find it in the cell cytoplasm, where protein synthesis took place. My scepticism became well known and was treated pretty roughly. In fact, in some quarters I was relegated to the lunatic fringe which, in my experience, is a distinctly uncomfortable place for a young man to be, although, no doubt, an excellent regime for the development of moral fibre. Since our studies on RNA had been meticulously done, and our evidence had not been breached at any point, I simply dug in my heels. Indeed, I dug them in too deep, for I maintained a perfectionist scepticism about the messenger RNA hypothesis long after a great deal of circumstantial evidence had accumulated in its favour. It was not until Jean-Pierre Ebel and his colleagues[20,21] finally succeeded in sequencing a large part of the ribosomal RNA that I was finally won over. This obduracy on my part cost me a bottle of very good whisky which I lost in a bet with John Paul.

None the less, the tide gradually turned, and increasing numbers of people began to take an interest in intranuclear RNA turnover. Wherever I went, I was asked what its possible function could be. We were, by then, getting used to futile metabolic cycles, but a futile cycle for so much of the RNA made in the cell nucleus was a bit hard to take. It was often suggested to me that the short-lived RNA had a regulatory role in gene expression, but I didn't like that much because I couldn't see then how RNA molecules could recognize, by base-pairing, genes with which they had no significant homology. My thoughts began instead to turn around the observation, well known at the time, that there was much more DNA in higher organisms than was necessary to code for any plausible estimate of the number of genes that these organisms possessed. I found myself being drawn to the notion that the RNA that broke down in the cell nucleus was made on noncoding DNA; and, if so, it was not difficult to suppose that it had some evolutionary role. My thoughts along these lines were prodded by an invitation from Henry Vogel to talk about intranuclear RNA turnover at a symposium that he and Vernon Bryson were organizing at Rutgers on "Evolving Genes and Proteins." Henry made it a condition of his invitation that I had to drag evolution into my talk somehow. That suited me well enough, for that was the direction in which my thoughts were, in any case, taking me. In my lecture, I reviewed all the evidence I had for rapid turnover of RNA within the cell nucleus and then aired my views on what I thought it all meant.[22] Here is the resumé with which I concluded:

> Only a small proportion of the RNA made in the nucleus of animal and higher-plant cells serves as a template for the synthesis of protein. This RNA is characterized by its ability to assume a form which protects it from intranuclear degradation. Most of the nuclear RNA, however, is made on parts of the DNA which do not contain information for the synthesis of specific proteins. This RNA does not assume the configuration necessary for protection from degradation and is eliminated within the cell nucleus. It plays no role in the

synthesis of cell protein, but serves as a background on which mutation and selection may operate to produce new templates for protein synthesis.

Thirteen years later, when introns and RNA splicing were discovered, Walter Gilbert[23] wrote an article in *Nature* entitled *Why genes in pieces?* in which he advanced evolutionary arguments very similar to those I had proposed at Rutgers. An editorialist in the same journal did, indeed, make mention of the fact that I had first described intranuclear RNA turnover many years previously, but only a few old friends to whom I had sent reprints of the Rutgers paper recalled that I had also come pretty close to the mark in my assessment of the biological significance of the apparently pointless process that I had stumbled upon.

Perhaps you would like to know what I did next. Well, I dropped RNA altogether and did not come back to it for several years. My reason for doing this was that I was convinced that no deep understanding of intranuclear RNA turnover would be achieved until methods were devised for sequencing nucleic acids, and I knew that I would not be the one to provide them. Instead, I thought it would be of interest to see what might happen if you fused together cells from different animal species. But that is another story, and I have told it many times.

REFERENCES

1. Hogness D.S., Cohn M., and Monod J. 1955. *Biochim. Biophys. Acta.* **16:** 99.
2. Schoenheimer R. 1942. *The Dynamic State of Body Constituents,* Harvard University Press.
3. Harris H. and Watts J.W. 1958. *Nature* **181:** 1582.
4. Siminovitch L. and Graham A.F. 1956. *J. Histochem. Cytochem.* **4:** 508.
5. Watts J.W. and Harris H. 1959. *Biochem. J.* **72:** 147.
6. Harris H. 1959. *Biochem. J.* **73:** 362.
7. Fisher H.W. and Harris H. 1962. *Proc. R. Soc. London Ser. B* **156:** 521.
8. Harris H. and Watts J.W. 1962. *Proc. R. Soc. London Ser. B* **156:** 109.
9. Harris H. et al. 1963. *Proc. R. Soc. London Ser. B* **157:** 177.
10. Singh U.N. and Koppelman R. 1963. *Nature* **198:** 181.
11. Harris H. 1963. *Nature* **198:** 183.
12. Harris H. 1963. *Nature* **198:** 184.
13. Harris H. 1964. *Nature* **202:** 1301.
14. Harris H. 1963. *Proc. R. Soc. London Ser. B* **158:** 79.
15. Hiatt H.H. 1962. *J. Mol. Biol.* **5:** 217.
16. Hiatt H.H. and Lareau J. 1962. In *Regulation of Enzyme Activity and Synthesis in Normal and Neoplastic Liver* (ed. G. Weber), p. 77, Pergamon.

17. Edström J.-E. 1964. In *Role of Chromosomes in Development* (ed. M. Locke), p. 137, Academic Press.
18. Brenner S., Jacob F., and Meselson M. 1961. *Nature* **190:** 576.
19. Gros F. et al. 1961. *Nature* **190:** 581.
20. Fellner P., Ehresmann C., and Ebel J.P. 1970. *Nature* **225:** 26.
21. Fellner P. and Ebel J.P. 1970. *Nature* **225:** 1131.
22. Harris H. 1965. In *Evolving Genes and Proteins* (ed. V. Bryson and H.J. Vogel), p. 469, Academic Press.
23. Gilbert W. 1978. *Nature* **271:** 501.

The Discovery of "Giant" RNA and RNA Processing: 40 Years of Enigma

KLAUS SCHERRER

Institut Jacques Monod, CNRS and University of Paris, Paris, France

RNA processing is a primordial paradigm of gene expression. Iconoclastic when discovered, after 40 years there is still no general rationale for this apparent "wasting" of up to 90% of RNA transcripts. This article tells the story of the discovery of RNA in the laboratory of J.E. Darnell. The discovery of "giant" RNA and its conversion into rRNA revealed the phenomenon of RNA processing and pre-rRNA. Genuine mRNA was also identified, but the majority of DNA-like nuclear RNA was also found to be giant and unstable. In spite of early evidence, pre-mRNA processing was only accepted in 1977 when the discovery of gene fragmentation in DNA made it obvious.

The years 1961–1963 were most exciting in the field that had just become molecular biology. News regarding the latest experiments and results came in almost every day and they were discussed over coffee in enthusiastic, and sometimes almost sacral, terms; late one night I remember Dan Nierlich saying: "François Gros identified the repressor; it is . . . (deep silence) . . . RNA!" Across the Charles River at the Massachusetts General Hospital (www.mph.harvard.edu) Paul Zamecnik led the research team of Mary Stephenson, Bob Loftfield and Mahlon Hoagland. They had worked out the tRNA (then "soluble" RNA) story, which made it possible to formulate the adaptor hypothesis, suggesting how DNA might be translated into polypeptides.[1] James Watson was leading a school of young pioneers in modern biology at Harvard

Reprinted, with corrections, from *Trends in Biochemical Sciences* October 2003, 28(10): 566–571.

(www.harvard.edu). His DNA double helix[2]—which he had worked out with Francis Crick in 1953, taking into account experiental evidence provided by Maurice Wilkins and Rosalind Franklin—had become the founding stone of the "central dogma" of molecular biology: DNA makes RNA and RNA makes protein. The concept of mRNA, a hypothesis based on the Jacob–Monod Model[3] and bacterial genetics, was about to be proven biochemically by Sydney Brenner, François Jacob and Matthew Meselson[4] at the California Institute of Technology (www.caltech.edu), and separately by François Gros[5] from the Pasteur Institute (www.pasteur.fr/externe) who was working part-time with Howard Hiatt and Waly Gilbert in Watson's laboratory. Pulse-labelling experiments on *Escherichia coli* showed the existence of a short-lived RNA species. In contrast to rRNA, which represented the majority of RNA in bacterial and animal cells, bacterial mRNA was supposed to be highly unstable.

The ribosome was recognized as being the factory for protein synthesis.[6] In the beginning, Hildegard Lamfrom thought it might carry the message by itself,[7] then Ian Matthaei and Marshall Nirenberg[8] demonstrated that poly(U) could program the ribosomes to synthesize poly-phenylalanine. This major breakthrough was the start of the race to solve the genetic code. Alfred Tissieres brought ideas and techniques from the team led by Jean Weigle at the University of Geneva (www.unige.ch) and, in Watson's laboratory, found that fast-sedimenting ribosomes were involved in protein biosynthesis.[9] In New York, Paul Marks found the same to be true of reticulocytes.[10] Thus, the flow of data that was either published or about to be published in 1962 was very exciting. Interestingly, the new results were more or less as expected; intellectual concepts based mainly on genetics were about to be neatly translated into biochemical reality from the DNA double helix to the genetic code.

Unexpected results obtained in James Darnell's laboratory at Massachusetts Institute of Technology (MIT; http://web.mit.edu) rapidly led to the discovery of "giant" RNA and its processing, on the model case of rRNA biosynthesis[11,12] and, eventually, to the pre-mRNA concept.[13] Controversially, the evidence we produced was at odds with the most efficient and intellectually satisfying bacterial schemes already established, and I was taken for a crackpot!

DIGGING IN

I first met Jim Darnell when he was about to set up a laboratory as a newly appointed Assistant Professor at MIT, under the wing of Salvador Luria (warmly, but respectfully, called "Salva"). Applying a method established for *E. coli*, Jim and Cyrus Levintal ("Cy") were grinding up HeLa cells with alumina powder, which I had often used in my pre-biochemistry years as an organic chemist prior to my thesis with Carl Martius in Zurich. After centrifugation, Jim showed me a tube and drew my attention to a luminescent glow suspended in the aqueous phase. One of them announced, "that's the RNA!" However, being familiar with alumina, I knew it was, in fact, colloidal particles that always stay in the liquid phase. And I got the slightly reassuring feeling that, possibly, my training at the Swiss Institute of Technology (www.ethz.ch) would be useful in the adventures yet to come.

Jim had spent several years with Harry Eagle, a virologist and pioneer in cultivating cells and, in particular, human HeLa cells. At that time, only a few investigators were interested in the molecular biology of animal cells; the "serious" research was with *E. coli* and bacteriophages. James Watson visited MIT frequently, and would discuss our strange results. One day, he told me "To work with animal cells, you've got to be a hero or a fool!"

In spite of Watson's ambiguous comment, I continued to extract RNA from HeLa cells to be analyzed on sucrose gradients. But the "hydraulic" system of collection, to which Pablo Amati had introduced me, yielded accordion-type gradient patterns; the drops collected were never the same size. Finally, I resorted to a "finger pump," to regularly obtain the same number of fractions. Looking over my shoulder, Jim said "terrific!" And within two days, all the finger pumps in the MIT biolabs were busy!

I had started to play with phenol extraction techniques, first used by the German chemist Otto Westphal, to replace alumina powder grinding, and also began using sodium dodecyl sulfate (SDS) to dissolve animal cells totally.[14] Walter Vielmetter introduced me to the "historic" DNA and RNA literature going back, via Schramm, to the Miescher tradition. It was then that I came across the reports by Eberhard Wecker[15] who extracted Tobacco Mosaic Virus (TMV) RNA from plant cells at 60°C, yielding fair amounts of infectious RNA.[16] Applying phenol extraction at 60°C after dissolution with SDS yielded significant amounts of RNA from HeLa cells.

Everyone who did this type of experiment—we were not the only ones in the race to discover animal-cell mRNA—was confronted with the high viscosity of the solution being handled; this was due to the presence of DNA which had to be destroyed by DNase. The DNase was heavily contaminated by RNase and, hence, fatal for RNA as well. I read somewhere that a dramatic shift in the solubility of DNA occurs as a function of pH and temperature. At pH 5, but only below 4°C, the DNA goes into the phenol phase, whereas RNA stays in the supernatant. By labelling HeLa cell DNA, I found that the DNA content was reduced to <0.1%. But although we retrieved DNA-free undegraded RNA and beautiful 28S and 18S rRNA peaks, as well as a 4S sRNA on the sucrose gradients, there was no sign of mRNA.

MIRACLES IN THE WAKE OF CHRISTMAS EVE

In the late afternoon a few days before Christmas Eve 1961, I was quite desperate. After three months, the RNA pulse-labelled with radioactive uridine remained consistently at the bottom of the centrifuge tube (Figure 1a). The mRNA should have shown up on a sucrose gradient in François Gros' "classical" 10S position.[5] Somehow, the RNA in human HeLa cells seemed to behave different from that in bacteria.

Before joining MIT, Jim Darnell had spent a year with François Jacob at the Pasteur Institute; he brought back the technique of using sucrose gradients. My aim was to apply RNA isolated from HeLa cells to the gradients, thereby mimicking François Gros' *E. coli* mRNA experiment. The Pasteur method was: prepare your RNA, load it onto a gradient, start the ultracentrifuge at 6 p.m.; go home, sleep well, and come back the following morning at 9 a.m. to collect the gradient; determine the positions of small and large rRNA in the spectrophotometer; and then precipitate the RNA with acid to locate the pulse-label supposedly marking the mRNA.

After seeing my labelled RNA "precipitated" to the bottom of the centrifuge tubes, Jim was convinced that something must be wrong with my methods: "I always told you that heat destroys life and, possibly, RNA." He would not accept the argument of a chemist: phosphodiester bonds are quite stable at 65°C, at least for a while. I tried many tricks to improve matters, but without success. However, I had just read a German paper by Wolfgang

FIGURE 1. Sucrose gradient analysis of RNA from HeLa cells. (a,b) RNA pulse labelled by Uridine-^{14}C, extracted by the hot phenol technique as described and centrifuged for various times at 25 000 rpm in a Spinco ultracentrifuge.[11,12] (a) Labelled for 60 min. and spun for 14 h. (b) As (a) but centrifuged for only 6 h. (c) The cells were pulse labelled by ^{32}P for 30 min, lysed by Dounce homogenization in reticulocyte standard buffer (RSB), treated with 0.5% desoxycholate (DOC) and centrifuged for 60 min at 25 000 rpm on a sucrose gradient. Fractions 15–25 (of 42) were pooled and extracted by hot phenol as described and centrifuged for 9 h at 25 000 rpm. Radioactivity is shown in gray, absorbency at 260 nm is shown in black.

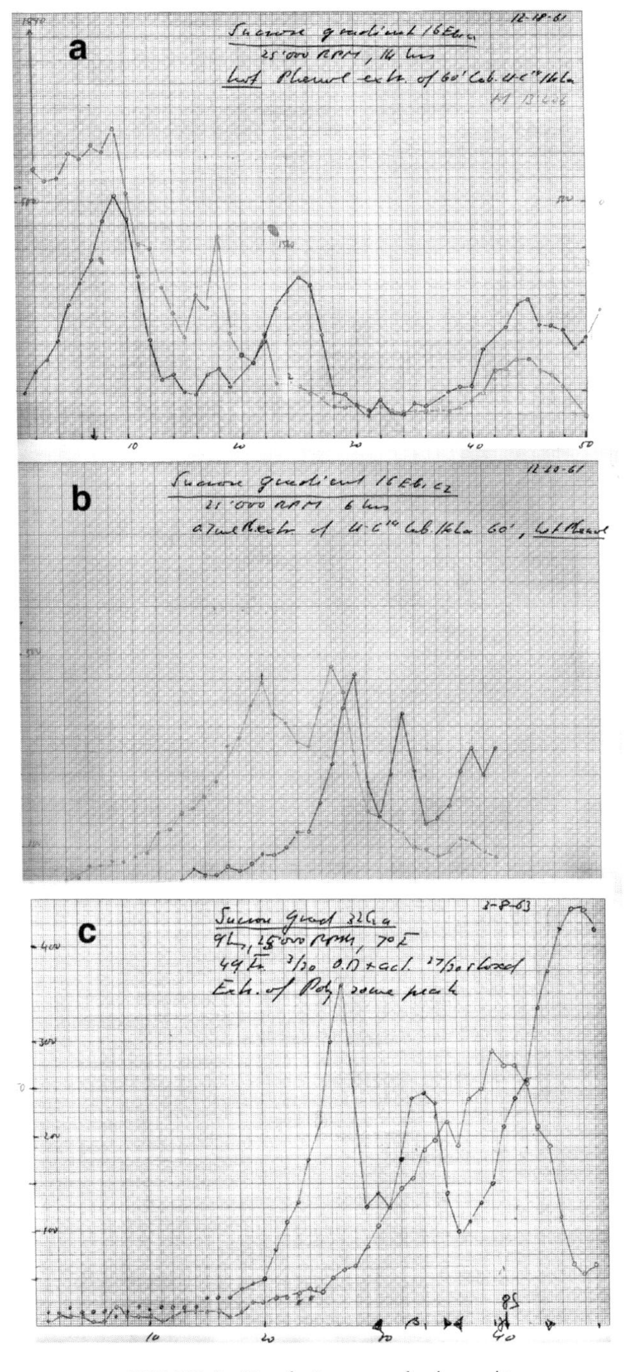

FIGURE 1. (*See facing page for legend.*)

Beerman (cited in Ref. 17) on activation of polytene chromosome bands, which could be observed by microscopy in the salivary glands of *Chironomus* larvae; cytogenetics had shown that these bands correspond to meiotic genes. I began to consider that transcription in higher eukaryotes might be quite different from that in *E. coli* given the huge "puffs" that sprung up from such genetically marked bands when the genes were activated. It occurred to me that, possibly, my radioactive "precipitate" at the bottom of the tube could correspond to something real.

Thus, I decided to break the sacred routine and, the labelled RNA being ready, went home for dinner with my lonely wife (already a laboratory widow!), and came back to start the centrifuge at midnight. I stopped the centrifuge at six the next morning and collected the gradient. This time, the large 28S rRNA peak had not even moved to the middle of the tube, very little absorbancy was picked up at 260 nm in the spectrophotometer at the bottom of the tube. Excitedly, I precipitated the fractions and counted them one by one.

And there it was; still no 10S mRNA but, most amazingly, two clear-cut peaks of radioactivity stood out, much heavier than the 28 S rRNA, juxtapositioned like the "Eiger and Jungfrau" in the Alps (Figure 1b). Beyond them, there was sort of hilly region, extending to the bottom of the tube, reminding me of the "Uetliberg" placed between the Alps and the lake of Zurich.

Everybody, and in particular Jim, was stunned. At that time no RNA bigger than that of the large ribosomal subunit was thought to exist in any type of uninfected cell. Later on, to brake this "credo," I spent more than ten years proving that the "Uetliberg" also represented giant RNA of particular significance. Fortunately, in a pulse-chase experiment the 45S and 35S peaks turned out to be metabolic precursors of rRNA. That Christmas Eve "giant" unstable RNA and, hence, RNA processing was born!

Unfortunately, we then fell into the same trap as before, and routinely sedimented the "Uetliberg" hilly zone to the bottom of the tube, favouring high resolution of the 45S and 35S peaks. (I didn't publish the pattern of the original gradient run in December 1961 until 1979 in a review[18] [Figure 1b].) In hindsight, that gradient zone contained practically pure pre-mRNA, which sedimented faster than the 45S RNA. This mistake delayed the discovery of pre-mRNA by several years.

In the early spring of 1962, a symposium was held in New York. The highlight of this meeting was a monumental confrontation between Severo Ochoa and Erwin Chargaff. Chargaff, on the basis of his solid background in chemistry, condemned the talk of unstable mRNA as mere dreams of

French intellectuals. I can still see tall Severo Ochoa getting up and hearing him saying: "If by now I have not lost faith, I certainly may have lost face!" Attending that meeting with my fellow post-doc in the Darnell laboratory, Ernie Zimmerman, we heard a talk by Richard Franklin who told how to use the drug actinomycin to stop transcription. At the end of the talk, Ernie and myself turned to each other and said: "that's what we've got to do!" Back in Cambridge, we started to label the cells for 15–30 min, and stopped incorporation of label with actinomycin, thus enabling us to study the RNA decay.

Jim Darnell rated our work as highly interesting. To boost the output, he gave me technical help in the person of lovely Harriet Latham (today the well-known scientist Ms H. Robertson), then a working student at MIT. She took care of the RNA analysis, and the three of us headed on with these pulse-chase experiments. Soon it became evident that ~50% of the 45S peak, labelled for 30 min, was transformed after ~20 min into 35S RNA, and by ~4 h into 28S and 18S rRNA (Figure 2). Thus, "Eiger" and "Jungfrau" were the metabolic precursors of rRNA; pre-rRNA processing was born! To my great pride, our work was reported in the *Scientific American.*

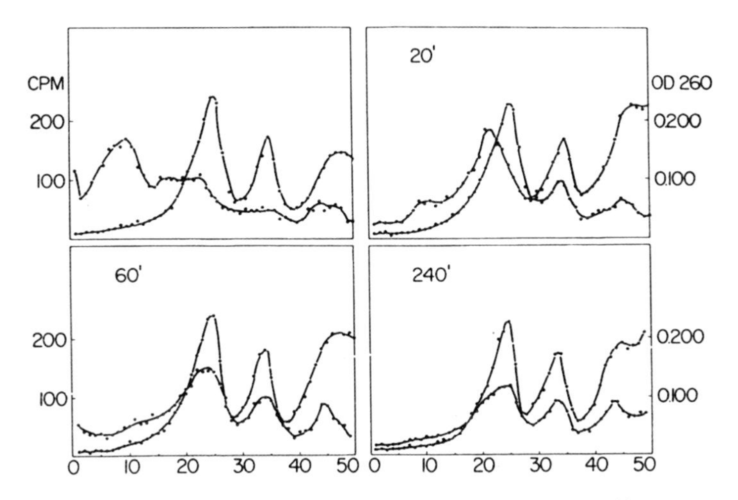

FIGURE 2. Fate of 45S RNA after actinomycin treatment. Uridine-2-^{14}C (0.01 mM and 1 μC/μM) was added to 600 ml of HeLa cells in suspension culture. After 30 min 150 ml was removed 20, 60, and 240 min after the addition of actinomycin. RNA from each sample was extracted and analyzed by sucrose gradient sedimentation. (o = CPM, o = OD_{260}). (Reproduced, with permission, from Ref. 12.)

LOOKING BACK AND FORTH

At about the same time, Robert P. Perry ("Bob"), working at the Fox Chase Cancer Center (www.fccc.edu), obtained similar results.[19,20] Thus, the pre-rRNA story was easily accepted in the world of science. Later, David Schlesinger found pre-rRNA in *E. coli*, which introduced RNA processing to the bacterial world.

For me however, the excitement of having discovered RNA processing with Jim Darnell, in the case of rRNA, was still unsatisfactory in the absence of any evidence for mRNA in animal cells. By base-composition analysis I found pulse-labelled A + U-rich RNA, as was expected for mRNA; but this RNA was "giant" as well and, hence, too big to be bona fide mRNA.

Some time before, Jonathan R. Warner, a student of Alex Rich working with Paul M. Knopf from Howard Dintzis team, came to work part time in the Darnell laboratory. Isolating cytoplasmic extracts from rabbit reticulocytes producing up to 90% haemoglobin, they came across heavy cytoplasmic ribosomal complexes that turned out to be the polyribosomes, the sites of protein biosynthesis.[21,22] The complexes that Jon found appeared as a thin thread linking approximately five ribosomes under the electron microscope.[23] Sedimenting faster than the 80S ribosomes on the gradient, they collapsed to the 80S position when treated by RNase. The actual nature of that RNase-labile thread was not known, but it was suggestive of mRNA. Radioactive labelling was, however, not possible in reticulocytes devoid of nuclei.

Thus, collaborating with Sheldon Penman (a nuclear physicist then "escaping" into biology) and Yechiel Becker, I labelled these structures in uninfected HeLa cells with ^{32}P. This enabled us to observe finally a 10S RNA peak (Figure 1c) and to determine the base composition of the pulse-labelled RNA. To our delight, the base composition of the pulse-labelled polyribosomal RNA was not ribosomal, but high in A + U and, thus, "DNA-like" as mRNA should be.[24] At last we had got hold of 10S mRNA of the animal cell. I also realized that most of the AU-rich RNA was not in the cytoplasm, but in the nuclear high molecular weight RNA.[12] Eventually, this project led to the formulation of the pre-mRNA concept, which we first proposed in 1966.[13]

During 1961 and 1962, Georgii Georgiev was working with Olga Samarina and Nika Mantieva on the same problems in Moscow. Unfortunately, they were confronted with endless obstacles because they were confined behind the Iron Curtain. As a former collaborator of Ilia Zbarsky, the earliest pioneer of Nuclear Matrix,[25,26] Georgiev eventually joined the Institute of the venerable Academician Vladimir Engelhart. Not only were we ignorant of each other's work owing to the communication barrier, but not belonging to the "nomenklatura," Georgiev initially had no access to ultracentrifuges. Thus, he had no chance of observing giant messenger-like RNA (mlRNA) or pre-rRNA. But they devised an extraction procedure going stepwise from 40–80°C, getting consecutively cytoplasmic, nuclear (pre-) ribosomal- and DNA-like RNA.[27] Later, with access to ultracentrifuges, Georgiev and Samarina found 35S and 45S rRNA peaks, as well as high molecular weight heterogeneous "DNA-like RNA" (dRNA). These and other data on cytoplasmic AU-rich RNA (presumably mRNA) were published in *Nature* in 1963.[28] Most importantly, Georgiev shared my view that the mlRNA/dRNA was probably related to mRNA in some way. When we met in Moscow in 1967, he was one of the few persons abroad with whom results and ideas concerning pre-mRNA could be discussed. Fortunately, this was also the case in Paris where I had joined the laboratory of François Gros in the spring of 1963. François' intellectual input and criticism, as well as the interest and trust he had in what our small "eukaryote team" was doing, was crucial in the determination of the pre-mRNA concept between 1963 and 1967.

On the basis of cytogenetics and the RNA puffs observed in *Chironomus* and *Drosophila*, and our data showing the existence of giant AU-rich RNA, I considered the existence of pre-mRNA to be as likely, by analogy to pre-rRNA. However, it took several agonizing years on the battlefield, with arguments over association and degradation, to convince colleagues of the reality of giant RNA. This we finally achieved thanks to the collaboration of Nicole Granboulan, an expert electron microscopist who later discovered the "prosomes." She showed by EM the specific pre-rRNA metabolites as well as the giant mlRNAs, by finding uninterrupted linear molecules of the expected size.[29] But the conceptual predominance of the *E. coli* model of mRNA and protein biosynthesis was still the main obstacle.

I started the analysis of transcription in duck erythroblasts, a system introduced to the Gros laboratory by Irving London, collaborating with Lise Marcaud and Lislott Voegelin. In contrast to rabbit reticulocytes, these red cells have nuclei. Not dividing any more in the peripheral blood of anaemic birds, immature red cells form no ribosomes and, hence, little pre-rRNA, but actively synthesize globin as well as its mRNA. For the first time, this enabled us to label a specific messenger of an animal cell, the chicken globin mRNA.[30] It was also less difficult to show the existence of the AU-rich high molecular weight mlRNA[12] and its processing into smaller size molecules.[13,30] Such RNA was also observed across the Atlantic by Rui Soeiro, who was working in Jim Darnells laboratory, and was given the alternative name of "heterogeneous nuclear RNA" (HnRNA), which was supposed to replace "mlRNA" and "dRNA."[31] But, although the Scherrer, Georgiev and Darnell laboratories had demonstrated the existence of the same type of pulse-labelled RNA, it was thought that the "artefact" explanation was easier and little credit was given to our interpretations linking mlRNA/dRNA to mRNA.

Processing was obviously the main paradigm of rRNA synthesis. But it took ten more years to be generally accepted that it might also apply to mRNA. At the Cold Spring Harbor symposium in 1970, Georges Spohr and myself presented competition hybridization experiments, which indicated that the giant nuclear RNA contained the same sequences as pulse-labelled bona fide mRNA isolated from HeLa cell polyribosomes.[32] However, the pre-mRNA concept was only accepted in 1977 after the more "noble" discovery of splicing.[33,34] The fragmentation of genes in the DNA made it all evident that RNA had to be processed to glue the pieces of genes together before their expression. And the iconoclastic pre-mRNA became trivial.

CONCLUDING REMARKS

Looking back over this 40-year-long story, it is my conviction that the giant transcripts and processing of pre-mRNA, often the results of gene fragmentation in the genome, might reflect the requirements of the DNA-based nuclear topological organization. Furthermore, the RNA processing is the basis of the dynamic nuclear architecture, organizing selective transport and control of individual transcripts on the nuclear matrix. It is known that specific genes are placed in specific nuclear sectors and that specific mRNA is transported to specific cytoplasmic sites in which proteins are made and assembled locally. This process leads to sub-cellular and cellular morphogenesis, as I proposed some years ago in the "unified matrix hypothesis."[35] In my opinion, all the insertion sequences found are beneficial and provide a selective advantage. A more rational explanation of the giant RNA discovered 40 years ago might, thus, make eventually redundant intellectual crutches, such as references to "waste," "junk," "archaeological," "futuristic," or, in short, "nonsense" RNA and DNA. Not only is the significance of intergenic and intervening sequences in pre-rRNA and pre-mRNA at stake, but also, more importantly, the problem of the 95% of unexplained DNA—the apparent nonsense DNA or, better, "mystery" DNA thought by some to be still hidden in the human genome.

Today, 40 years after the original discovery of ribosomal and non-ribosomal giant RNA subject to processing, the *raison d'être* of such high molecular weight RNA is still being discussed. Exon shuffling seems to be a consequence rather than an originator of gene fragmentation in DNA and RNA processing. In the early days, nobody would have dared to propose a mechanism like splicing; experimental evidence enforced it onto the scientific community. Similarly, the *raisons d'être* of "giant" RNA might spring from lucid interpretation of bona fide data. And the evidence that nuclear RNA, and in particular giant globin RNA,[36,37] is a structural component of the nuclear matrix[35,38,39,40] (and A. Rynditch, unpublished observations) might lead us in the right direction. It revives the concept of "architectural RNA" proposed by Penman years ago.[41] This interpretation is not popular as yet, but sometimes courage pays off—even in science!

I would like to conclude by citing a paragraph from Ernest Borek's book *The Code of Life* published in 1965,[42] which gives a pertinent definition of the molecular biology at that time and describes the intellectual and social atmosphere well. "In superficial hindsight it always appears that science moves from point to advanced point in a direct line to ever-increasing understanding of the world around us . . . it must be remembered that in science as in any other human endeavour the participants are all too human. There are those who are ready to jump on a rolling bandwagon and bask in the security of working on a popular hypothesis approved by the pillars of the profession. But there are those who, endowed either with real imagination or just with a perverse streak, will refrain from joining the crowd and endeavour to hack out a path of their own . . ." One might conclude that, in science, "perverse" cracking of pots is sometimes necessary to design new experiments, and to imagine new ideas that could help one to forge ahead, beyond the actual jungle of dogmas and preconceived opinions.

ACKNOWLEDGEMENTS

The author wishes to thank most cordially all colleagues and friends who assisted the writing of this review with personal recollections, corrections and advice; impossible to mention all of them. Particular thanks are due to Mariann Bienz, Martin Billeter, Sidney Brenner, Georgii Georgiev, François Gros, François Jacob, Lennart Philipson and of course, Jim Darnell. Please note the historical account by James Darnell: Darnell, J.E. (2002) the surprises of mammalian molecular cell biology. Nat. Med. 8, 1068–1071.

REFERENCES

1. Hoagland M.B. et al. 1958. A soluble ribonucleic acid intermediate in protein synthesis. *J. Biol. Chem.* **231:** 241–257.
2. Watson J.D. and Crick F. 1953. A structure for deoxyribose nucleic acid. *Nature* **171:** 737.
3. Jacob F. and Monod J. 1961. Genetic regulatory mechanisms in the synthesis of proteins. *J. Mol. Biol.* **3:** 318–356.
4. Brenner S. et al. 1961. An unstable intermediate carrying information from genes to ribosomes for protein synthesis. *Nature* **190:** 576–581.
5. Gros F. et al. 1961. Unstable ribonucleic acid revealed by pulse labelling of *Escherichia coli. Nature* **190:** 381–585.
6. Siekevitz P. and Zamecnik P. 1981. Ribosomes and protein synthesis. *J. Cell Biol.* **91:** 53s–65s.
7. Lamfrom H. 1961. Factors determining the specificity of hemoglobin synthesized in a cell free system. *J. Mol. Biol.* **3:** 241–252.
8. Nirenberg M. and Matthaei J.H. 1961. The dependence of cell-free protein synthesis in *E. coli* upon naturally occurring or synthetic polynucleotides. *Proc. Natl. Acad. Sci. U. S. A.* **47:** 1588–1602.
9. Tissieres A. et al. 1960. Protein synthesis. *Proc. Natl. Acad. Sci. U. S. A.* **46:** 1450.
10. Marks P. et al. 1962. Protein synthesis in erythroid cells. I. Reticulocyte ribosomes active in stimulating amino acid incorporation. *Proc. Natl. Acad. Sci. U. S. A.* **48:** 2163–2171.
11. Scherrer K. and Darnell J.E. 1962. Sedimentation characteristics of rapidly labelled RNA from HeLa cells. *Biochem. Biophys. Res. Commun.* **7:** 486–490.
12. Scherrer K. et al. 1963. Demonstration of an unstable RNA and of a precursor to ribosomal RNA in HeLa cells. *Proc. Natl. Acad. Sci. U. S. A.* **49:** 240–248.
13. Scherrer K. et al. 1966. Patterns of RNA metabolism in a differentiated cell: a rapidly labeled, unstable 60S RNA with messenger properties in duck erythroblasts. *Proc. Natl. Acad. Sci. U. S. A.* **56**: 1571–1578.
14. Philipson L. 1961. Chromatographic separation, and characteristics of nucleic acids from HeLa cells. *J. Gen. Physiol.* **44:** 899.
15. Wecker E. 1959. The extraction of infectious virus nucleic acid with hot phenol. *Virology* **7:** 241–243.
16. Gordon M.P. et al. 1963. Heat inactivation of TMV. *Virology* **19:** 416.
17. Beermann W. 1962. *Protoplasmologia, Handbuch der Protoplasmaforschung,* Springer-Verlag.

18. Scherrer K. et al. 1979. On pre-messenger RNA and transcriptions. A review. *Mol. Biol. Rep.* **5:** 5–28.
19. Perry R.P. 1962. The cellular sites of synthesis of ribosomal and 4S RNA. *Proc. Natl. Acad. Sci. U. S. A.* **48:** 2179–2186.
20. Perry R. et al. 1964. Hybridization of rapidly labeled nuclear ribonucleic acids. *Science* **145:** 504–507.
21. Warner J.R. et al. 1963. A multiple ribosomal structure in protein synthesis. *Proc. Natl. Acad. Sci. U. S. A.* **49:** 122–129.
22. Warner J. and Knopf P. 2002. The discovery of polyribosomes. *Trends Biochem. Sci.* **27:** 376–380.
23. Slayter H.S. et al. 1963. The visualization of polyribosomal structure. *J. Mol. Biol.* **7:** 652–657.
24. Penman S. et al. 1963. Polyribosomes in normal and poliovirus-infected HeLa cells and their relationship to messenger-RNA. *Proc. Natl. Acad. Sci. U. S. A.* 49: 654–661.
25. Zbarsky I. and Debov S. 1948. On the proteins of cell nuclei. *Doklady Acad. Nauk SSSR* **62:** 785–798.
26. Zbarsky I. and Georgiev G. 1959. Cytological characteristics of protein and ribonucleoprotein fractions of cell nuclei. *Biochim. Biophys. Acta* **32:** 301–302.
27. Georgiev G. and Mantieva V. 1962. The isolation of DNA-like RNA and ribosomal RNA from the nucleo-chromosomal apparatus of mammalian cells. *Biochim. Biophys. Acta* **61:** 153–154.
28. Georgiev G. et al. 1963. Biosynthesis of messenger and ribosomal ribonucleic acids in the nucleochromosomal apparatus of animal cells. *Nature* **200:** 1291–1294.
29. Granboulan N. and Scherrer K. 1969. Visualisation in the electron microscope and size of RNA from animal cells. *Eur. J. Biochem.* **9:** 1–20.
30. Scherrer K. and Marcaud L. 1965. Remarques sur les ARN messagers polycistroniques dans les cellules animales. *Bull. Soc. Chim. Biol.* **47:** 1697–1713.
31. Soeiro R. et al. 1966. Rapidly labelled hela cell nuclear RNA II. Base composition and cellular localisation of heterogeneous RNA fraction. *J. Mol. Biol.* **19:** 362–372.
32. Scherrer K. et al. 1970. Nuclear and Cytoplasmic Messenger-like RNA and their Relation to the Active Messenger RNA in Polyribosomes of HeLa Cells. In *Cold Spring Harbor Symposium* (Vol. 35) (Laboratory, C.S.H., ed.), pp. 539–554.
33. Berget S. et al. 1977. Spliced segments at the 5′ terminus of adenovirus 2 late mRNA. *Proc. Natl. Acad. Sci. U. S. A.* **74:** 3171–3175.
34. Chow L. et al. 1977. An amazing sequence arrangement at the 5′ ends of adenovirus 2 messenger RNA. *Cell* **12:** 1–8.
35. Scherrer K. 1989. A unified matrix hypothesis of DNA-directed morphogenesis, protodynamism and growth control. *Bioscience Reports* **9(2):** 157–188.
36. Imaizumi M.-T. et al. 1973. Demonstration of globin messenger sequences in giant nuclear precursors of messenger RNA of avian erythroblasts. *Proc. Natl. Acad. Sci. U. S. A.* **70:** 1122–1126.
37. Broders F. et al. 1990. The chicken α-globin gene domain is transcribed into a 17-kilobase polycistronic RNA. *Proc. Natl. Acad. Sci. U. S. A.* **87:** 503–507.
38. Fey E.G. et al. 1986. The non-chromatin substructures of the nucleus: the ribonu-

cleoprotein (RNP)-containing and RNP-depleted matrices analysed by sequential fractionation and resinless section electron microscopy. *J. Cell Biol.* **102:** 1654–1665.

39. Maundrell K. and Scherrer K. 1979. Characterization of pre-messenger-RNA containing nuclear ribonucleoprotein particles from avian erythroblasts. *Eur. J. Biochem.* **99:** 225–238.
40. De Conto F. et al. 2000. In mouse myoblasts the nuclear prosomes are associated with the nuclear matrix and accumulate preferentially in the peri-nucleolar areas. *J. Cell Sci.* **113:** 2399–2407.
41. Penman S. et al. 1982. Cytoplasmic and nuclear architecture in cells and tissue: form, functions, and mode of assembly. In *Cold Spring Harb Symp Quant Biol.* (Vol. 46; Pt 2), pp. 1013–1028, Cold Spring Harbour Lab. Press.
42. Borek E. 1965. *The Code of Life,* Columbia University Press.

The Discovery of "Split" Genes: A Scientific Revolution

J.A. WITKOWSKI

Cold Spring Harbor Laboratory, Cold Spring Harbor, New York, USA

In 1962 T.S. Kuhn[1] published a seminal book entitled *The Structure of Scientific Revolutions* in which he suggested that changes in scientific thought come about as revolutions in which an existing theory is replaced by a more or less radical theory, rather than by a gradual cumulative process. Kuhn's work itself sparked something of a revolution in studies of the history of science and the general theme of his work is largely accepted. However, it is striking that most examples of revolutions have been drawn from the physical sciences, a peculiarity recognized by Kuhn himself and by Cohen[2] in his magisterial book *Revolution in Science.* I think that there have been many revolutions in the biological sciences and in the early part of 1977 a quite stunning example occurred, the discovery of "split" genes.

Papers describing phenomena attributable to the presence of non-coding, intervening sequences had been published as early as 1974, but it was not until the first half of 1977 that splicing was proposed as the mechanism to account for these phenomena. It was a remarkable feature of this revolution that once the phenomenon had been recognized, results obtained in a wide variety of systems were understood. In alphabetical order these included the adenovirus late region, the chick ovalbumin gene, *Drosophila* ribosomal genes, mouse, and rabbit globin genes, Simian Virus 40 (SV40) and yeast tRNA. I shall describe these various discoveries, but some disclaimers are required. I have used only published materials as my sources so that this is not a definitive account of the discovery of "split" genes. I am *not* concerned with trying to establish a precise chronology for these events or with trying

Reprinted from *Trends in Biochemical Sciences* March 1988, 13(3): 110–113.

to assign priority for the discovery. Instead I want to use the discovery of intervening sequences to illustrate the various aspects of a scientific revolution as defined by Kuhn and Cohen.

A SCIENTIFIC REVOLUTION

What constitutes a scientific revolution? Kuhn[1] emphasized that scientists work at solving puzzles within the framework of the existing theory or "paradigm" and that this "normal" science is what occupies most scientists most of the time. As research progresses, an increasing number of findings cannot be accounted for by the current theory and eventually a scientist or group of scientists proposes a radical modification or even a complete replacement of the current theory. The new theory encompasses the older findings as well as those that had been anomalies under the old theory, and in addition the new theory suggests lines of research that were formally undreamt of. The new theory is accepted with varying degrees of enthusiasm but, if successful, it eventually becomes the new paradigm until it in its turn is replaced by further developments.

How does one recognize a scientific revolution? Cohen[2] suggests four tests that can be used to determine if a change in scientific thought counts as a revolution. Firstly, there is what observers of the scene thought at the time; secondly, there is the later documentary evidence such as is found in reviews and textbooks; thirdly, there is the judgement of historians of science; and finally there is the opinion of scientists working in the field today.

I shall treat the topic system by system, but there is the difficulty that I cannot cover all the work that was done and I shall have to be selective. I shall deal with papers published up to and including 1977, and choose papers illustrating Kuhn's and Cohen's ideas of a scientific revolution.

MESSENGER RNA

The prediction of the existence of mRNA is generally regarded as one of the intellectual breakthroughs in the history of molecular biology. The classic experiment was performed by Brenner, Jacob and Meselson[3] although previously others, in particular Volkin and Astrachan,[4] had obtained results that were highly suggestive of the existence of mRNA. In their discussion, Brenner et al. remarked that: "It is a prediction that the messenger RNA should be a simple copy of the gene, and its nucleotide composition should therefore correspond to that of the DNA." This was the unquestioned paradigm that had guided the way people thought about the relationship between the nucleotide sequences of a gene and its mRNA, and the amino acid sequence of the protein. (The discovery of mRNA qualifies in its own right as a scientific revolution and exciting descriptions by Brenner, Crick and Jacob will be found in Ref. 5). How was the paradigm overthrown?

DROSOPHILA RIBOSOMAL DNA

In early 1977 *Cell* devoted 57 pages to reports from three laboratories on the organization of the ribosomal genes of *Drosophila.* The studies had been carried out using restriction mapping but principally by exploring the R loop technique of White and Hogness.[6] In this technique, single-stranded RNA is hybridized to double-stranded DNA under conditions that favor RNA–DNA hybrids. The RNA displaces one of the DNA strands of the duplex which forms a loop that indicates the site of transcription of the RNA. The conclusions of three groups[6–9] were the same. In *Drosophila,* rDNA exists in two forms of repeating units, 17 kb and 11.4 kb in length. What was curious was that the 28S rDNA was in two parts, separated by an insert of variable length ranging from 0.5kb to 6.5kb. However, some 55% of the 28S genes were not interrupted in this way and it could be argued that the interrupted genes were not transcribed, or if they were transcribed they were not functional; there was no evidence that the presence of inserted sequences was a functional feature of these genes.

ADENOVIRUS

It was hoped that study of the relatively small adenovirus (Ad2) genome and its transcription would shed light on the processes of gene regulation in eukaryotes. It was the subject of intense investigation in the 1960s and 1970s (and remains so today) and work on adenovirus is particularly well documented.[10] By the mid-1970s, anomalies were observed in three separate investigations using adenovirus.

One approach was to map the transcriptional units by isolating mRNAs and determining what they coded for by *in vitro* translation and polyacrylamide gel electrophoresis. The first results[11] using mRNAs fractionated on the basis of size showed that some viral mRNAs were much longer than predicted from the amino acid sequences of their protein products. This approach was refined by using *Eco*RI restriction fragments of the adenovirus late genes to select specific RNAs; as the order of the *Eco*RI fragments was known, the gene products could be assigned to particular regions of the adenovirus genome.[12] There were two puzzling features. Firstly, there were minor protein bands that were explained away by the presence of contaminating non-homologous mRNA (a justifiable assumption at that time and one that became questionable only with hindsight). Secondly, it was found that significant quantities of P-VIII were synthesized by RNAs selected by *Eco*RI fragments A, and F and D that were widely separated in the adenovirus genome. It was suggested that "The information for the two polypeptides is contained in separate mRNA molecules, but these molecules contain some common sequences."

Another curious result had been obtained by Gelinas and Roberts[13] who analysed the sequences at the 5′ ends of the adenovirus mRNAs and then mapped these in the genome. Remarkably, they found that there was a single 5′ capped oligonucleotide. This was "unexpected" because many adenovirus mRNAs were present late after infection, and Gelinas and Roberts had ". . . anticipated one 5′ capped oligonucleotide DNA for each mRNA species." There was one further puzzling feature; Gelinas and Roberts were "surprised" to find that the 5′ oligonucleotide was very sensitive to nuclease digestion in mRNA–DNA hybrids when it should have been protected from attack. They suggested that the 5′ terminus of mRNA or of its DNA complement was able to make a hair-pin loop. The 5′ oligonucleotide might have some regulatory function for which it was distributed at a number of sites in a large transcript that was subsequently "processed" to give individual mRNAs.

Finally in this pre-revolution period, groups at Cold Spring Harbor Laboratory (Chow et al.[14]) and at the National Institute of Health (Westphal et al.[15]) were using electron microscopy and R loop mapping to determine the sites of transcription of late mRNAs. Chow et al. noticed that about 25% of loops made with late mRNAs had short free 5′ ends that did not hybridize to the DNA. They referred to the results of Gelinas and Roberts, remarking that the 5′ tails might be due to ". . . the presence of sequences on the mRNA which are not encoded in the region flanking the hybridized transcript." Westphal's group observed similar tails but suggested that they represented 3′ poly(A) tails or were artefacts resulting from displacement of RNA by re-annealing DNA.

These various anomalies were resolved in the early part of 1977 when Berget et al.[16] and the Cold Spring Harbor Laboratory groups[17–20] submitted papers to *Proceedings of the National Academy of Sciences* and *Cell* respectively.

Berget et al.[16] had been using the R looping technique to map the hexon mRNA to the Ad2 genome. When hexon mRNA was hybridized to the *Hin*dIIIA fragment, an R loop was formed, but there were also short 5′ and 3′ tails. A similar result was obtained when the mRNA was hybridized to single-stranded *Hin*dIII-digested DNA; double-stranded hybrids were formed, again with short single-stranded 5′ and 3′ tails. The most spectacular result came from the hybridization of hexon mRNA to single-stranded *Eco*RI A fragment. No fewer than three loops of single-stranded DNA formed at the 5′ end of the hybrid molecule, indicating that the 5′ end of the mRNA was transcribed from map positions 16.8, 19.8 and 26.9. These sequences, Berget et al. suggested, were "spliced" to the main body of the mRNA during processing of a long precursor mRNA.

Chow et al.[17] continued their studies by identifying the origins of the free 5′ tails that they had observed using single-stranded DNA restriction fragments as probes. Their results were identical to those of Berget et al.; sequences from map positions 16.6, 19.6 and 26.6 were complementary to the 5′ tails of mRNA that were transcribed to the right of position 36. With a degree of understatement, Chow et al. remarked that their results were ". . . not directly consistent with any mechanism previously suggested for the biosynthesis of mRNA in eukaryotic cells."

What of the anomalies in the earlier studies using *in vitro* translation of mRNA selected with restriction fragments? Lewis et al.[18] used other enzymes to produce smaller restriction fragments and found that the *Hpa*I C fragment selected mRNA for IVa2, 15k and IX polypeptides, but it and the neighbouring fragment selected for almost all late mRNA at levels that were substantially higher than those obtained with λ DNA used as a negative control. These findings ". . . were initially difficult to explain" but interpretation had been ". . . greatly facilitated recently by several different investigations. . .." In short, it was now clear that several late mRNAs had sequences that were transcribed from a common site away from the main transcription region of each mRNA.

Finally, Klessig[19] performed a refined analysis of the 5′ oligonucleotide studied by Gelinas and Roberts,[13] concentrating on producing highly purified fiber and 100K mRNA. The purified mRNAs had the same 5′ oligonucleotide, confirming the results of Gelinas and Roberts. Klessig tried to purify the mRNAs by hybridizing them to the *Eco*RI DNA fragments and then treating with RNase. The DNA fragments used coded for the 5′ ends of the mRNAs but failed to protect the 5′ oligonucleotide from digestion. Furthermore, the 5′ oligonucleotide was protected in DNA–RNA hybrids using the *Bam*HI (0–29.1) and *Hin*dIII (7.5–17.00) DNA fragments. Klessig wrote: "The scheme that emerges for the biosynthesis of Ad2 mRNAs is unlike any hitherto described or postulated."

These studies followed on from previous lines of investigation, but Dunn and Hassell[20] presented new data from an analysis of the transcription of SV40–Ad2 hybrids. Their results showed that the transcripts included sequences that were distant from the site of insertion of the SV40 DNA in the Ad2 genome. They suggested that what was needed was a direct sequence comparison between a gene and the mRNA transcribed from it.

YEAST tRNA GENES

The first sequence comparisons seem to have been carried out not using an mRNA but rather using yeast $tRNA^{Tyr}$ genes. Goodman, Olson and Hall[21] identified an *Eco*RI fragment that contained the *SUP4* locus and used this as a probe to screen an ochre suppressor *SUP4-o* yeast λ library. They then determined the sequences of the mutant gene and three wild-type genes and found that the genes contained a stretch of 14 nucleotides to the 3′ side of the anticodon triplet that was not present in $tRNA^{Tyr}$. The insert was present in functional genes and it was unlikely to be a cloning artefact. As they remarked then, ". . . it is difficult to muster any real evidence on how such genes or gene systems might work."

THE RABBIT AND MOUSE β-GLOBIN GENES

The experiment suggested by Dunn and Hassell had already been done in part by Jeffreys and Flavell[22] using the rabbit β-globin mRNA, and their report appeared in the same historic issue of *Cell* as did the Ad2 work (see Figure 1). In fact Jeffreys and Flavell had compared restriction maps rather than sequence data but the results were convincing nevertheless. Two examples of their anomalous results must suffice here: (1) the enzyme *Hae*III liberated a 333 bp fragment from the cDNA but an 800 bp fragment from the genomic DNA; (2) a *Bam*HI site known to be 67 bp from an *Eco*RI site was

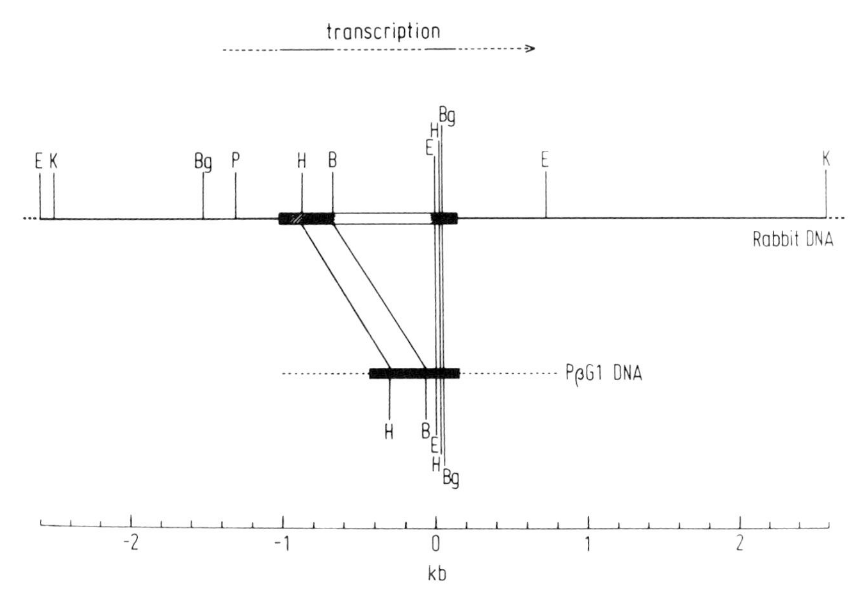

FIGURE 1. This figure is reprinted from Jeffreys and Flavell (Ref. 22), and compares the restriction enzyme map of the rabbit beta-globin gene (above) with that of a rabbit beta-globin cDNA. The distance between the *Bam*HI (B) and *Eco*RI (E) sites in the genomic DNA is 700 nucleotides compared with 67 nucleotides in the cDNA. The empty box in the genomic DNA is what became known as an intron.

found to be in a "strange" position some 700 bp from the *Eco*RI site in the genomic DNA. Jeffreys and Flavell thought that it was remarkable that there was a 600 bp fragment within the gene that did not code for globin. With considerable understatement, echoing a paper published 23 years earlier, they went on that if this was a general phenomenon, ". . . then there are certain implications for recombinant DNA research."

Tilghman et al.[23] worked on the mouse β-globin gene and obtained similar results. They found that *Hin*dIII and *Sst*I which *failed* to cut the cDNA, *readily cleaved* the cloned gene (their italics). R looping was also used and "suprisingly" two loops instead of the expected one were found, indicating the presence of an inserted fragment. Tilghman et al. sequenced 40 bp across the suspected junction of the inserted DNA and found that the DNA sequence agreed with the known amino acid sequence for residues 93–104, but that it then diverged and the DNA sequence did not code for at least the next five amino acids. Here was direct evidence for a non-translated region within a eukaryotic gene.

THE CHICKEN OVALBUMIN GENE

In the early days of gene cloning it was tremendously advantageous to work with genes that were highly expressed, for example the globin genes and the chicken ovalbumin gene. Breathnach, Mandel and Chambon[24] "unexpectedly" found that the ovalbumin gene contained inserts. For example, the cDNA contained no *Hin*dIII or *Eco*RI sites, yet the cDNA probe detected several bands in genomic DNA cut with these enzymes. Comparisons of the lengths of fragments obtained with other enzymes produced a "striking anomaly," once again demonstrating the presence of inserts within a gene. It was possible that the unusual arrangement of the ovalbumin gene in these cells was related in some way to its regulation; the cells of the tubular gland of the oviduct respond to hormonal stimulation by synthesizing very large amounts of ovalbumin. However, Breathnach et al. showed that the same insert was present in DNA from erythrocytes and from 5-day embryos.

SV40 TRANSCRIPTION

By 1977 the SV40 genome was being mapped and functions were being assigned to its various regions. However, the sequences between 0.67 and 0.76 that were transcribed late in infection had not been ascribed a function. Aloni et al.[25] set out to remedy this and began by isolating mRNA hybridizing to DNA in the 0.67–0.76 region and also RNA hybridizing to DNA fragments from the region 0.0–0.17 that coded for 16S and 19S RNA. They found that these two sets of RNAs hybridized to both sets of restriction fragments and that analysis on sucrose gradients showed that they all contained 16S and 19S RNA. Mapping the 5′ ends of these RNAs showed that they were derived from the 0.67 region. Aloni et al. concluded that leader sequences (0.67–0.76) of the 16S RNAs were added to transcripts from the 0.95–0.17 region and they speculated that the deleted region might have some regulatory function.

"SPLIT GENES" AS A KUHNIAN SCIENTIFIC REVOLUTION

The first stage as described by Kuhn is the accumulation of anomalies that cannot be explained by existing theory. It is clear that the scientists working in this field were all too aware that they were finding anomalies. Words such as "surprised," "unexpected," "strange," "ambiguity" and "anomaly" recur repeatedly in papers published in this period. The results were not easy to explain for the old ideas on the relationship between the gene and its mRNA had failed. It had been recognized that hnRNA must be processed but it was not thought that the processing might involve the removal of sequences from within coding regions. (Although studies of hnRNA were beginning to suggest this might occur.) As Weinstock et al.[26] later remarked, their results using the ovalbumin gene were "not predictable from an examination of the restriction map of mRNA." Crick wrote: "What is remarkable is that the possibility of splicing had not at any time been seriously considered before it was forced upon us by the experimental facts. . . . Lacking evidence, we had become overconfident in the generality of our ideas."[27] This idea of being "forced" is explicit in some writings. O'Malley and Woo's laboratory wrote of their work[28] on the ovalbumin gene that ". . . the existence of intervening sequences within structural genes was unpredicted by our prior understanding of eukaryotic genes." They were "*obligated*" to conclude that the coding sequences were non-contiguous in the gene and Dunn and Hassell[20] wrote of their work that this conclusion seemed "*unavoidable*" (my italics). Yet once it was admitted that mRNAs ". . . possess a structure fundamentally different from any mRNAs hitherto described,"[9] the observed anomalies became excellent data to support the new theory.

Splicing was not only compatible with earlier results but clearly it opened up a whole new area for research. Paper after paper closed with references to the consequences of splicing for obtaining transcription of cloned genes, and to speculations on the functions of the inserted sequences and the 5′ leader sequences, and of course on the mechanism of splicing itself. It is not necessary to elaborate how splicing has become the new paradigm and in its turn has led to another revolution, the recognition that the RNA possesses catalytic activity once thought to be the exclusive property of proteins.[29] The implications for studies of gene evolution are only just being explored.

COHEN'S CRITERIA

The RNA splicing story seems to fit the Kuhnian model quite well; how well does it satisfy Cohen's four criteria? Firstly, it is obvious that the participants recognized that they were in the midst of a remarkable scientific event. The language used in the papers expresses astonishment at what was being discovered, exemplified by the title of the paper by Chow et al.: "An Amazing

Sequence Arrangement at the 5′ ends of the Adenovirus 2 Messenger RNA."[17] Watson[30] wrote in the 1977 *Annual Report of Cold Spring Harbor Laboratory* that "Splicing is likely to be the most important observation in molecular genetics since the 1960 discovery of messenger RNA" and with uncanny prescience, the 1977 Cold Spring Harbor Quantitative Symposium was devoted to chromatin. The summaries of the meeting written by Chambon[31] and Sambrook[32] again show that these events were recognized at the time as being revolutionary, and Watson[33] remarked that the participants left "feeling that they had been part of an historic occasion." Cohen lays great emphasis on contemporary opinion, particularly if the word "revolution" is used. At least three references in the early years referred to splicing as revolutionary; "In the last 2 years there has been a mini-revolution in molecular genetics,"(Crick[27]); ". . . revolutionize our ideas of genes and the proteins for which they code" (Watson[30]); "Our picture of the organization of genes in higher organisms has recently undergone a revolution" (Gilbert[34]).

Secondly, the later documentary evidence from reviews and textbooks is overwhelming. In his 1979 review Darnell[35] wrote that it was a "most astonishing discovery." All undergraduate text books on molecular biology describe splicing and Lewin devotes a chapter of 21 pages in *Genes*[36] to a discussion of splicing. It is ironic that the third edition of Watson's classic text *The Molecular Biology of the Gene*[37] appeared in 1976, just prior to the discovery of splicing, but in the fourth edition[38] splicing is covered extensively, including a chapter dealing with the evolutionary implications.

It is difficult to satisfy Cohen's third criterion that the opinions of competent historians of science should be consulted because the events of only ten years ago have not yet attracted analysis by professional historians. I hope that there will be an increasing trend for historians to deal with recent events in areas like molecular biology where changes occur with disturbing but exciting speed.

I have discussed this topic with many scientists who were not directly involved in the discovery of splicing, and there is a unanimous opinion that the discovery was indeed a revolution. To a degree, events of that period have taken on a mythical quality, with stories of how word of the American findings first reached London and were received with frank disbelief, and discussions of whether a Nobel Prize could be awarded for the work. (The discussion does not center on whether the work justifies a prize, but who would receive it!)

CONCLUSION

The discovery of non-coding, intervening sequences in genes qualifies as a scientific revolution according to the tests of both Kuhn and Cohen. It was perceived as such at the time and it has opened up new areas of research that were undreamt of in the pre-revolutionary period. The paradigm that it overthrew was that there was a direct correspondence at the nucleotide level between a gene and its RNA, and that the coding sequence of a gene was continuous without interruption. As Broker and Chow wrote: "This startling fact [splicing] has abruptly changed our conception of what a eukaryotic gene is and how its expression is controlled."[39]

REFERENCES

1. Kuhn T.S. 1970. *The Structure of Scientific Revolutions,* The University of Chicago Press.
2. Cohen I.B. 1975. *Revolution in Science,* The Belknap Press of Harvard University Press.
3. Brenner S., Jacob F., and Meselson M. 1961. *Nature* **190:** 576–581.
4. Volkin E. and Astrachan L. 1956. *Virology* **2:** 149.
5. Judson H.F. 1979. *The Eighth Day of Creation.* Jonathan Cape.
6. White R.L. and Hogness D.S. 1977. *Cell* **10:** 177–192.
7. Glover D.M. and Hogness D.S. 1977. *Cell* **10:** 167–176.
8. Wellauer P.K. and Dawid I.B. 1977. *Cell* **10:** 193–212.
9. Pellegrini M., Manning J., and Davidson N. 1977. *Cell* **10:** 213–224.
10. Broker T.R., Chow L.T., Dunn A.R., Gelinas R.E., Hassell J.A., Klessig D.F., Lewis J.B., Roberts R.J., and Zain B.S. 1978. *Cold Spring Harbor Symp. Quant. Biol.* **XLII:** 531–553.
11. Anderson C.W., Lewis J.B., Atkins J.F., and Gesteland R.F. 1974. *Proc. Natl. Acad. Sci. USA* **71:** 2756–2760.
12. Lewis J.B., Atkins J.F., Anderson C.W., Baum P.R., and Gesteland R.F. 1975. *Proc. Natl. Acad. Sci. USA* **72:** 1344–1348.
13. Gelinas R.E. and Roberts R.J. 1977. *Cell* **11:** 533–544.
14. Chow L.T., Roberts J.M., Lewis J.B., and Broker T.R. 1977. *Cell* **11:** 819–836.
15. Meyer J., Neuwald P.D., Lai S.-P., Maizel J.V., and Westphal H. 1977. *J. Virol.* **21:** 1010–1018.
16. Berget S.M., Moore C., and Sharp P.A. 1977. *Proc. Natl. Acad. Sci. USA* **74:** 3171–3175.
17. Chow L.T., Gelinas R.E., Broker T.R., and Roberts R.J. 1977. *Cell* **12:** 1–8.
18. Lewis J.B., Anderson C.W., and Atkins J.F. 1977. *Cell* **12:** 37–44.
19. Klessig D.F. 1977. *Cell* **12:** 9–21.

20. Dunn A.R. and Hassell J.A. 1977. *Cell* **12:** 23–35.
21. Goodman H.M., Olson M.V., and Hall B.D. 1977. *Proc. Natl. Acad. Sci. USA* **74:** 5453–5457.
22. Jeffreys A.J. and Flavell R.A. 1977. *Cell* **12:** 1097–1108.
23. Tilghman S.M., Timeier D.C., Seidman J.G., Peterlin B.M., Sullivan M., Maizel J.V,. and Leder P. 1978. *Proc. Natl. Acad. Sci. USA* **75:** 725–729.
24. Breathnach R., Mandel J.L. and Chambon P. 1977. *Nature* **270:** 314–319.
25. Aloni Y., Dhar R., Laub O., Horowitz M., and Khoury G. 1977. *Proc. Natl. Acad. Sci. USA* **74:** 3686–3690.
26. Weinstock R., Sweet R., Weiss M., Cedar H., and Axel R. 1978. *Proc. Natl. Acad. Sci. USA* **75:** 1299–1303.
27. Crick F. 1979. *Science* **204:** 264–271.
28. Lai E.C., Woo S.L.C., Dugaiczyk A., Catterall J.F., and O'Malleyu B. 1978. *Proc. Natl. Acad. Sci. USA* **75:** 2205–2209.
29. Cech T. 1986. *Cell* **44:** 207–210.
30. Watson J.D. 1977. *Cold Spring Harbor Laboratory Annual Report,* p. 7.
31. Chambon P. 1978. *Cold Spring Harbor Symp. Quant. Biol.* **XLII:** 1209–1234.
32. Sambrook J. 1977. *Nature* **268:** 101–104.
33. Watson J.D. 1978. *Cold Spring Harbor Symp. Quant. Biol* **XLII:** XV.
34. Gilbert W. 1978. *Nature* **271:** 501.
35. Darnell J.E. 1979. *Proc. Nucleic Acid Res. Mol. Biol.* **22:** 327–353.
36. Lewin B. 1983. *Genes.* John Wiley & Sons.
37. Watson J.D. 1976. *Molecular Biology of the Gene.* W.H. Freeman.
38. Watson J.D., Hopkins N.H., Roberts J.W., Steitz J.A., and Weiner A.M. 1987. *Molecular Biology of the Gene, 3rd edn., Vol. 2.* Benjamin/Cummings.
39. Broker T.R. and Chow L.T. 1977. *Cold Spring Harbor Laboratory Annual Report,* p. 43.

The *In Vitro* Biosynthesis of an Authentic Protein

NORTON D. ZINDER

Rockefeller University, New York, New York, USA

"However, it must be pointed out that formal proof of the structure-determining function of 'm-RNA' will be obtained only when the synthesis of a specific protein, known to be controlled by an identified structural gene, is shown to take place in a reconstructed system containing messenger-RNA from genetically competent cells while all other fractions were prepared from cells that lack this structural gene."

In their summary of the 1961 Cold Spring Harbor Laboratory Symposium on "Cellular Regulation," Monod and Jacob[1] stated the above as necessary to prove the central dogma that had been presented at that symposium: DNA makes RNA, RNA makes protein—the mRNA hypothesis. Some of us were already hoping to bypass part of the stringency of the experimental system implied in that statement; the DNA component. With the discovery of the RNA-containing bacteriophages of *Escherichia coli*,[2] it seemed that the *in vitro E. coli* protein synthesizing system might well use their RNA as a message because it clearly must do so in the cell. (In fact, it was about six months later before we explicitly showed there was no DNA intermediate.[3])

INITIAL STIMULUS

I had already made contact with Dan Nathans, who was then in Fritz Lipmann's laboratory, studying the details of protein biosynthesis in crude *E. coli* extracts. Little did I know that at that time Marshall Nirenberg was already getting some stimulus of protein synthesis by various RNAs including tobacco mosaic virus (TMV) and ribosomal RNA in *E. coli* extracts. If I had

Reprinted from *Trends in Biochemical Sciences* August 1997, 22(8): 318–320.

known, perhaps I would not have been so casual that summer of 1961.

In any event, I spent the summer of 1961 teaching the bacterial genetics course at Cold Spring Harbor, as I did both several years before and several years after. But then, there occurred one of those events that defy statistics. That summer, two of the students attending the course were from Nirenberg's laboratory, H. Matthaei and Oliver Jones. Naturally, I asked them, as I asked all of the students, what they were doing in their labs. From Matthaei, I learned that they were studying protein synthesis in crude extracts of *E. coli* with and without addition of nucleic acids. Nothing more. From Jones, however, I learned the important role of β-mercaptoethanol in keeping protein synthesizing extracts alive. I discussed with both of them our planned phage RNA experiments. Let us note that this was July of 1961. In the October issue of the *PNAS,* two papers that had been communicated on 3 August 1961 appeared and revolutionized the coding problem. The first by Matthaei and Nirenberg[4] was on the preparation and stabilization of the extracts. The second by Nirenberg and Matthaei[5] was entitled "The dependence of cell-free protein synthesis in *E. coli* upon naturally occurring or synthetic ribonucleotides." Also, at the Biochemistry Congress in Moscow that August, Nirenberg "electrified" the audience with the announcement that polyphenylalanine was formed under the direction of polyuridylic acid.

OUR FIRST EXPERIMENTS

Returning to Rockefeller in September, we started experiments using phage RNA. About a gram of *E. coli* in late log phase was harvested by centrifugation, ground in a mortar and pestle with alumina, extracted in buffer, treated with DNase and spun at 20 000 r.p.m. in an ultracentrifuge. To this supernatant (S30) was added sources of energy such as ATP, amino acids with one or more radioactively labeled and phage RNA.[5] After incubation at 37°C, all protein was precipitated with hot trichloroacetic acid (TCA) and counted for new biosynthesis.

Somehow things went wrong. Although we could occasionally reduce the background incorporation by pre-incubating the extracts and have the system stimulated two- to threefold by added RNA, it was not reproducible. Moreover, the only things we knew at the time about the phage's coat protein were its amino acid composition, approximate size of about 120 (129) amino acids and that it did not contain any histidine. Thus, in order to analyse the product, we needed a fairly large amount of labeled material with which we could develop an analytical procedure. During this period, we also became aware of the ubiquity and stability of RNAses, because they were on everyone's fingers and necessitated the wearing of rubber gloves (so-called "finger nuclease"). It was not until about December that we could reproducibly get tenfold or greater phage-RNA-stimulated incorporation.

PRODUCT ANALYSIS

To analyse the product, we decided to fingerprint tryptic peptides with high voltage, two-dimensional electrophoresis at pH 4.7 and pH 1.9. To label the peptides, we took advantage of the fact that trypsin cleaves proteins after arginine and lysine residues. We therefore labeled the protein product, defined by hot TCA precipitability with ^{14}C-labeled arginine and lysine, so that all soluble peptides except for the carboxy-terminal peptide would be labeled. Before digestion, we mixed the product with authentic coat protein isolated from purified phage and thereafter handled them together. Following electrophoresis, the paper was exposed to X-ray film and then stained with ninhydrin to visualize the authentic peptides. Nine peptides appeared in the ninhydrin-stained paper and eight on the autoradiogram (Fig. 1). They were *congruent.*

Thus, our *in vitro* system synthesized a majority of protein that clearly resembled f2 coat protein. Only a few much lighter and differently located spots were on the autoradiogram. Further confirmation came from labeling phage protein *in vivo* with ^{3}H leucine and making products *in vitro* with ^{14}C leucine. Two authentic peptides had ^{3}H leucine and only the same two peptides contained ^{14}C leucine.

CONFUSION SETS IN

We had heard from the grapevine, as had already been mentioned in the original Nirenberg and Matthaei paper, that TMV RNA not only stimulated protein synthesis, but that some of its product was the TMV coat protein. In May of 1962, a paper from Fraenkel-Conrat's laboratory[6] appeared in the *PNAS,* documenting their findings. Six criteria were used to show that product was TMV coat protein. We note here that the properties and complete sequence of the authentic TMV coat protein were already known for direct comparisons. We therefore initiated a further set of controls for our own

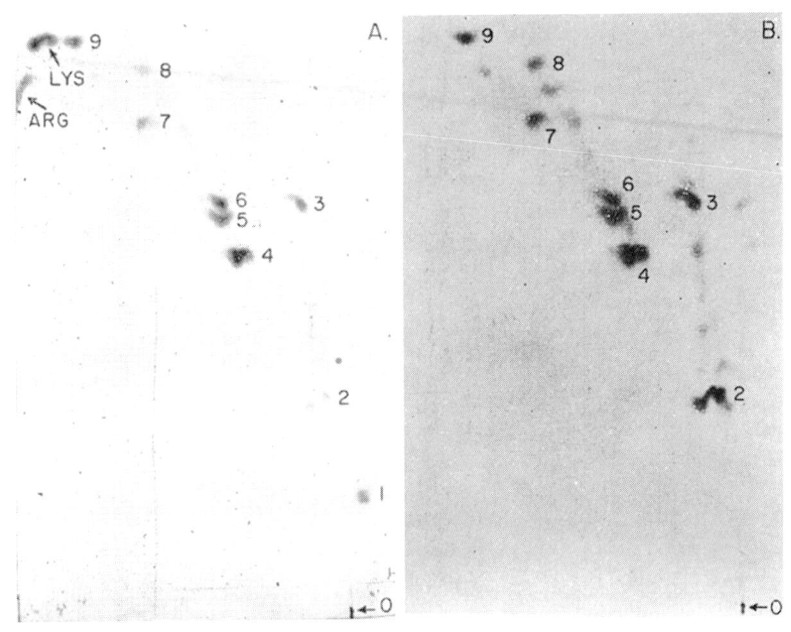

FIGURE 1. (a) Ninhydrin-stained two-dimensional electrophoresis paper and (b) autoradiogram of the tryptic peptides from ^{14}C-lysine- and arginine-labeled product and carrier f2 protein. Arg and Lys represent markers and O indicates the origin. Figure reproduced with permission from Ref. 7.

experiments, attempting to make TMV coat protein. This unfortunately led to one of those polemics for which science is noted.

As reported in a "note added in proof" of our paper on *in vitro* f2 coat synthesis,[7] the following two additional controls had been done. In the first, TMV protein was mixed with *in vitro* product of f2 RNA and fingerprinted. There was no overlap of the peptides, showing that the real authentic peptides did not get contaminated by the free label; even a trace of contamination would have given the autoradiograms we saw. In the second control, f2 coat protein was added to the product from TMV RNA and again there was no overlap of the peptides. Specificity resided in the particular RNA message used. We did not report one further control, that of TMV RNA product with authentic TMV coat, although the result of this was apparent from the position of the peptides in the experiments described above. TMV coat protein has 158 amino acids and with 11 arginines and two lysines, should have given about 14 tryptic peptides were they all soluble. In any event, both the autoradiograms and the ninhydrin stained paper gave ten clear spots, but none of them

Conference delegates at a Bacteriophage meeting in 1954. Indicated are (a) J.D. Watson, (b) N. Zinder, (c) A. Hershey, (d) D. Demerec, (e) S. Brenner and (f) F.H.C. Crick. (Photograph courtesy of the Cold Spring Harbor Laboratory Archive.)

overlapped. Not understanding this result or wishing to detract from the positive results with the RNA phage, we just put this result aside for the moment.

When you think about it as we did then, f2 coat had 129 amino acids (387 nucleotides), which were encoded by some part of the 3000 nucleotides of phage RNA (4%) while TMV was less than half that (158 nucleotides from 6400). In any case, it should have been shocking that the majority of product was phage coat, not that we could not find the TMV coat. But out of blissful ignorance of all regulatory mechanisms at the time, and with full knowledge that the vast majority of the protein made following viral infection was the coat protein, we readily accepted the phage result and fretted over the TMV result, especially as the TMV result was buttressed by data from six independent analytical techniques, in the hands of first rate scientists.

Why was all this important? Again we return to our excitement at that time with the emergence of access to the genetic code, but still with an almost complete ignorance of its structure. We had fulfilled the Monod and Jacob proof of the mRNA hypothesis—we had made a protein by adding only an RNA message to a system that did not contain it or its DNA homologue. Thereby, we had verified that the *in vitro* protein synthesizing system was both real and faithful. Not only was it a faithful one, but taken together with the TMV result (if true), the *message* was the only specific component in protein biosynthesis. The large number of components needed for protein synthesis and any nucleic acid encoded signals were interchangeable between the bacteria *E. coli* and the tobacco plant. Moreover, the code, whatever it was, was *universal.*

INFORMATIONAL MACROMOLECULES—RUTGERS, 1962

As things will happen, there were several symposia in that and the next year to which both Fraenkel-Conrat and I had been invited. The first in September 1962 at Rutgers. It was entitled "Informational Macromolecules."[8] In a way only possible in these early days, almost everybody who was anybody in the young molecular biology was there. For example, the K's alone had H. Kalckar, W Kauzman, H.G. Khorana and A. Kornberg. While proceeding to the "L's" there were O. Lampen, J. Lederberg, and F. Lipmann, among others.

I had sent Fraenkel-Conrat copies of the chromatograms and autoradiograms with a detailed description of what we had done and informed him that I would say that we didn't know whether TMV coat protein was made but with our procedures, we could not find it. Proving negations in science is not generally easy. At the meeting, arguments by the speakers for the fidelity of the *in vitro* incorporation system and for the universality of the code rested more on the TMV experiments than on the f2 experiments. However, Fraenkel-Conrat was in for one hard day.

ROUND ONE

The first section of the meeting on the Genetic Code opened with a talk by H. Wittman (soon to become famous for isolating a large number of proteins from ribosomes) from Tubingen.[8] He was studying aspects of the genetic code by inducing mutations in TMV with nitrous acid. Because this produces mainly cytosine to uracil changes, by noting which amino acids changed to which others, some insight into codons could be obtained. He noted that, in his hands, the treatments generally led to only a single amino-acid change and rarely to two. However, he noted that similar experiments by Tsugita and Fraenkel-Conrat occasionally yielded as many as 16 changes. However, these changes were identical to those of a particular strain of TMV that grows only in certain strains of tobacco. He noted that controls for contamination of the TMV stock used before the mutagen treatment were not reported. (I should note here that a cleaned-up version of the discussion is printed in the volume.) Fraenkel-Conrat's response was: ". . . we tested many controls and the virus did not contain the same mutants before . . . possibly one fell in along the way somewhere . . . we are fallible and do make mistakes . . . and all kinds of things."[8] Wittman then repeated his charge in somewhat greater detail, which as you might expect, settled that discussion.

HEAD TO HEAD

After two speakers, I delivered my paper recounting the results with phage RNA and then noted that we could not find any TMV coat protein product with our techniques.[9] Fraenkel-Conrat then got up and, although not recorded in the discussion, essentially made the famous Chargaff charge about molecular biologists: that I "was practicing biochemistry without a license." To paraphrase, I was using imprecise techniques such as paper electrophoresis rather than the then highly defined peptide "Dowex" resin columns. In the published text, he praises my *E. coli* experiments, but says that he purifies the protein product first, saving only 1% that is the pure coat and, of course, I am naive in thinking my gross procedures could find it. He ends up saying that more work needs to be done and I readily agree. In the next year, our paths cross several times, leading to similar exchanges. It seemed to me that he worried that if he proved me wrong about f2, then he could be wrong about TMV. In fact, within a couple of years, further studies in Fraenkel-Conrat's laboratory failed to reproduce the finding of *in vitro* TMV coat protein synthesis from TMV virion RNA,[9] and reported probable contamination of their virus stocks.[10]

The *in vitro* system was faithful and the code was elucidated and shown to be universal.[11] This latter being one of the most important unifying principles of biology. We would really be in trouble if there were many codes that were hard to interrelate. Two years later, we made f2 coat in a *Euglena gracilis* chloroplast system.[12] Chloroplast protein synthesis is much like prokaryote protein synthesis, but certainly phylogenetically distant from that in *E. coli.*

CONCLUDING REMARKS

Of course, if we had known then what we know now, we could have said, as Francis (Crick) would have done: "Fraenkel-Conrat is wrong for theoretical reasons," and be done with it. The protein synthetic system and the mRNAs of eukaryotes and prokaryotes are quite different.[13] In prokaryotes, mRNAs have ribosome-binding sites known as Shine-Delgarno sequences (polypurines) upstream of the initiation codon, and can be polycistronic. In eukaryotes, the messages are monocistronic and have a 5′ turned-around methyl-guanosine just before the initiating codon. TMV is, in this respect, a eukaryote. With an *in vitro* protein synthesis system derived from eukaryotes, some TMV protein encoded near the 5′ end of the virion RNA is made, but not the coat protein.[14,15] Instead, a special mRNA, which directs coat protein synthesis, is made in plant cells from the 3′ end of the TMV virion RNA. Thus, while f2 virion RNA can direct phage coat protein synthesis, TMV RNA does not directly code for TMV coat. Like everywhere else in biology, one can be wrong in theory for the right reasons, as well as being right in theory for the wrong reasons.

REFERENCES

1. Monod J. and Jacob F. 1961. *Cold Spring Harbor Symp.* **26:** 395–401.
2. Loeb T. and Zinder N.D. 1961. *Proc. Natl. Acad. Sci. U. S. A.* **47:** 282–289.
3. Cooper S. and Zinder N.D. 1962. *Virology* **18:** 415–414.
4. Matthaei J.H. and Nirenberg M. 1961. *Proc. Natl. Acad. Sci. U. S. A.* **47:** 1580–1588.
5. Nirenberg M. and Matthaei J.H. 1961. *Proc. Natl. Acad. Sci. U. S. A.* **47:** 1588–1602.
6. Tsugita A., Fraenkel-Conrat H., Nirenberg M.W., and Matthaei J.H. 1962. *Proc. Natl. Acad. Sci. U. S. A.* **48:** 846–853.
7. Nathans D., Notani G., Schwartz J.H., and Zinder N.D. 1962. *Proc. Natl. Acad. Sci. U. S. A.* **48:** 1424–1431.
8. Vogel H., Bryson V., and Lampen J.O., eds. 1963. *Informational Macromolecules,* Academic Press.
9. Aach H.G., Funatsu G., Nirenberg M.W., and Fraenkel-Conrat H. 1964. *Biochemistry* **3:** 1362–1366.
10. Funatsu G. and Fraenkel-Conrat H. 1964. *Biochemistry* **3:** 1356–1362.
11. On the Genetic Code. 1966. *Cold Spring Harbor Symp. Quant. Biol.* **XXXI**.
12. Schwartz J.H., Eisenstadt J.M., Brawerman G., and Zinder N.D. 1965. *Proc. Natl. Acad. Sci. U. S. A.* **53:** 195–200.
13. Watson J.D. et al. 1987. *The Molecular Biology of the Gene.* Benjamin Cummings.
14. Hunter T., Jackson R., and Zimmern D. 1983. *Nucleic Acids Res.* **11:** 801–808.
15. Hunter T., Hunt T., Knowland J., and Zimmern D. 1976. *Nature* **260:** 759–764.

Early Days of Ribosome Research

MASAYASU NOMURA

University of California,
Irvine, California, USA

The essay covers the early days of ribosome research, including the discovery of messenger RNA and the naming of ribosomes, from the perspective of a young post-doc from Japan working in the USA during these memorable years for ribosome research (1958–1960).*

It was the spring of 1958 when I met Jim Watson for the first time. I had come from Japan to work as a post-doctoral fellow in Sol Spiegelman's laboratory at the University of Illinois. Jim came to the campus to give a series of lectures sponsored by his former mentor, Salvador Luria, then a professor at Illinois working in the same department as Spiegelman. In addition to giving formal lectures to students and faculty members on the campus, Jim spent time in Spiegelman's laboratory discussing with Sol the nature of the ribosomes seen in bacterial extracts by analytical ultracentrifugation, as first observed by Schachman, Pardee and Stanier.[1] The specific question Jim and Sol discussed was whether ribosomes consist of a single species (70S) or of two species (30S and 50S subunits). People who were interested in ribosomes (then variously called "microsomal particles" or "ribonucleoprotein particles of the microsome fraction") had sometimes observed a single peak and sometimes two (or more) peaks in Shlieren patterns obtained with the analytical

*This article represents an edited version of the very first part of an extensive historical article written by Masayasu Nomura entitled "History of Ribosome Research: A Personal Account," which will appear in the book *The Ribosome: Structure, Function, and Evolution* (ed. W.E. Hill, A.E. Dahlberg, R.A. Garrett, P.B. Moore, D. Schlessinger and J.R. Warner), due to be published in 1990 by the American Society for Microbiology.

Reprinted from *Trends in Biochemical Sciences* June 1990, 15(6): 244–247.

ultracentrifuge. Jim and Sol agreed that these differences in the patterns were due to the solution conditions, specifically Mg^{2+} concentrations, used to prepare bacterial extracts and isolate the particles, as suspected from the work of Chao on yeast ribosomes which had just been published.[2] The exciting question was the significance of this unequal subunit structure. Of course, the "subunit structure" of DNA, the double helical structure, was the key to revealing the secret of self-replication of DNA. Ribosomal particles present in various living cells apparently all consisted of two non-identical subunits, small and large. "What does this mean?" Sol asked me one evening, and this question remained with me for a long time.

RIBOSOMES AND PROTEIN SYNTHESIS

The story of ribosomes is of course intimately connected to the story of protein synthesis. So, the history of ribosome research is almost intimately connected to the history of research on protein synthesis. Before coming to the United States as a "post-doc," I was interested in protein synthesis, randomly reading research papers on protein synthesis published in several journals of which a limited selection was accessible to scientists in Japan at that time. From these readings, I started to learn about ribonucleoprotein particles in the microsome fraction (which contained endoplasmic reticulum membrane fragments) in mammalian cells as the possible site of protein synthesis. For example, Zamecnik and his collaborators published several papers using intact animals, tissues and tissue homogenates, that demonstrated that radioactive amino acids administered to animals, or added to *in vitro* systems, appeared first in protein attached to ribonucleoprotein particles in the microsome fractions, followed by gradual incorporation into protein in the soluble protein fraction.[3] However, when I came to the United States in late 1957, protein synthesis in cell-free extracts from bacterial cells had not been clearly demonstrated, even though Zamecnik had already developed a cell-free system from rat liver,[4] and people such as Sol Spiegelman were struggling to develop a similar system from bacteria. Nevertheless, we believed in the unity of fundamental principles in biology and thus in the importance of ribosomal particles seen in bacterial extracts in connection with protein synthesis.

During Jim Watson's visit to Urbana, I arranged to spend three months in his laboratory at Harvard the following summer. At this time I had been in the United States for only six months and was struggling to develop an *in vitro* system in which information transfer from DNA to protein could be studied. However, as a post-doctoral fellow I was eager to learn as many different approaches as possible, and working in Jim Watson's laboratory was certainly appealing.

CHLORAMPHENICOL PARTICLES

At Harvard, Jim Watson, together with Alfred Tissières, had started to isolate and characterize ribosomes from *E. coli.* Other people in the Watson laboratory included David Schlessinger and Charles G. Kurland. They were busy estimating the molecular weights of ribosomes and rRNAs and studying other physicochemical and chemical properties of ribosomes, and Jim and Alfred were just preparing their first paper on the characterization of *E. coli* ribosomes.[5] In terms of my research, Jim suggested that I study the effects of chloramphenicol on ribosomes. This was my formal introduction to ribosome research. I discovered that extracts prepared from *E. coli* treated with chloramphenicol showed patterns very different from normal extracts in the analytical ultracentrifuge, indicating formation of abnormal particles. By repeated differential centrifugation, I isolated the particles, named chloromycetin particles (later called chloramphenicol particles), and showed that they were deficient in proteins.[6,7] Although chloramphenicol particles were once suggested as artefacts,[8] later studies did not agree with this suggestion (see, for examples, Refs 9 and 10) and I now think that the formation of these protein-deficient particles in *E. coli* under these conditions is real, and reflects certain features of the regulation and assembly of ribosomes.

THE HOLY GRAIL

Although my own chloramphenicol particle project was concerned with the subject of ribosome biosynthesis, the mission of the ribosome research of the Watson laboratory and for several other laboratories was the quest for the "Holy Grail" of biology at the time, namely, the mechanism of information transfer from gene to protein. Francis Crick had just formulated the central dogma and specifically proposed that RNAs contained in the ribosomes are the information carriers between DNA and protein. Although Crick's article[11] had not yet been published, the concept of the central dogma was already known and there was a sense of excitement and urgency. In fact, only a few days after my arrival at Harvard (in June 1958), the first Nucleic Acid Gordon Conference was held at New Hampton, New Hampshire, with Crick from England and other major players on this subject participating in the conference. Attending the conference, I had the chance to meet and talk

A get-together of molecular biologists during my three years post-doctoral time in the United States. Francis Crick is at the centre. This photograph was taken in June 1960 by Seymour Benzer. From left: E.K.F. Bautz, A. Kaplan, M. Nomura, F.H.C. Crick, S. Champe and D. Krieg.

with these people, gradually learning ideas and facts in the mainstream of the emerging new biology, molecular biology as we now call it, and the significance of ribosome research in this context.

In addition to the proposed role of rRNAs as the information carrier from gene to protein, Crick also speculated on the structure of ribosomes in his 1958 article on the central dogma.[11] The favored working model was a structure similar to that of spherical viruses with many identical protein subunits. Of course, these early speculations were soon destined to be challenged by the discovery of mRNA and by the work of Jean-Pierre Waller on the heterogeneity of ribosomal proteins (r-proteins).[12,13]

GRADIENTS, NAMES AND TEMPLATES

Although many people working on protein synthesis were involved in the studies of ribosomes, there were only a handful of laboratories working on ribosomes *per se* with a full commitment. Among them, one of the most active was the group at the Carnegie Institution of Washington. Dick Roberts, Roy Britten, Ellis Bolton and Dean Cowie (and others, such as Brian McCarthy, who joined the group later) were intensively studying the kinetics of flow of radioactive amino acids and bases into protein and RNA, respectively, in *E. coli,* devising various techniques and strategies for this purpose. Thus they devised and started to use sucrose density gradient centrifugation extensively to separate ribosomal particles,[14] and this quickly became an essential technique for research on ribosomes. Dick Roberts also organized a session on ribosomes in the Biophysical Society meeting at MIT, Cambridge, in 1958 and published a slim book entitled *Microsomal Particles and Protein Synthesis* which contained papers presented in this symposium.[15]

In the introduction to this book,[15] Roberts discussed the confusion and inadequacy of various phrases such as "microsomal particles" used during the meeting and proposed the name "ribosome" to designate "ribonucleoprotein particles in the size range 20 to 100S," stating that "it has a pleasant sound." Thus, 1958 was a memorable year for the history of ribosome research.

The concept that rRNAs in the ribosomes are the information carriers soon started to encounter experimental observations that were difficult to explain in simple terms. For example, purified *E. coli* 16S and 23S rRNAs appeared to be too homogeneous in size to function as a template for thousands of different *E. coli* proteins with different sizes. Thus, when Arthur Pardee announced the results of his experiments conducted at Pasteur Institute, Paris, that the expression of the β-galactosidase gene (*lacZ*$^+$) takes place immediately after its entry from male bacterial cells to *lacZ*$^-$ female cells in bacterial conjugation (Ref. 16; later published in Ref. 17), some people interpreted the results to mean that metabolically stable rRNAs are not the intermediates and, hence, DNA might be the direct template for protein synthesis (see the discussion in Ref. 16). There was even a suggestion that bacterial ribosomes serve a function different from protein synthesis, and perhaps represent a storage substance. In fact, in contrast to mammalian cell extracts, reliable cell-free protein synthesizing activity had not been demonstrated with *E. coli* extracts until around 1960.[18,19] (However, there was a paper published in 1959 by Zillig's group in Munich, Germany; see Ref. 20.) As late as 1959, the central dogma was still a hypothesis, and the mechanism of information transfer was in a confused state.

DISCOVERY OF mRNA

Like many other investigators at that time, I was much concerned with the phenomenon of information transfer, and carried out experiments to study the question of whether RNA synthesized after T2 phage infection, which would be the intermediate information carrier between the phage DNA and phage specific proteins, is identical to the rRNAs, as the hypothesis of rRNA being the information carrier would predict. A few years earlier, Volkin and Astrachan had discovered a small amount of RNA synthesized in *E. coli* after T2 phage infection using a ^{32}P-phosphate labeling technique, and showed that the base composition of this RNA was similar to that of T2 phage DNA rather than *E. coli* DNA.[21] However, they were much more concerned with the observed quantitative conversion of this RNA to phage DNA and considered a hypothesis that RNA might be a precursor for T2 phage DNA.[22] I remember the discussion I had with Jim Watson on this subject at Harvard in the early summer of 1959, and the actual hard experimental work I did with Benjamin Hall during the following three months in Sol Spiegelman's laboratory at Illinois. We discovered that the "T2-specific RNA" (the name we gave to the mRNA synthesized after T2 infection) was associated with ribosomes in the presence of a high concentration of Mg^{2+}, as judged by electrophoretic analysis, but was released from ribosomes as free RNA upon decreasing the Mg^{2+} concentration. (However, from one sucrose gradient centrifugation analysis, we incorrectly thought that the RNA sedimented mainly as 30S-like particles.) We also found that T2-specific RNA had sedimentation coefficients of about 8–10S and could be physically separated from 16S and 23S rRNAs.[23] However, I left Spiegelman's laboratory without completing the work (to work in Seymour Benzer's laboratory, then at Purdue, on phage genetics), and the experimental results were published without solving the crucial question of whether T2-specific RNA was a new type of RNA functioning together with pre-existing ribosomes or if it was "special" rRNA contained in "special" ribosomes synthesized after phage infection. As is well known, about a year later, brilliant density transfer experiments using heavy isotopes were done by Sydney Brenner, Francois Jacob and Matt Meselson[24] at the California Institute of Technology, Pasadena, proving that the former is correct. The name "messenger RNA" was coined by Jacob and Monod for this new type of RNA.

I should note that in the work on T2-specific RNA,[23] in control uninfected *E. coli* cells labeled with ^{32}P for two minutes, we observed the heterogeneous RNA sedimenting at 8–10S, which was similar to phage-specific mRNA and was distinct from 16S and 23S rRNAs that were also labeled with ^{32}P. This was a clear demonstration of the presence of a new type of RNA (mRNA) distinct from rRNAs in normal cells, but, again, we failed to make a definitive conclusion in this regard. The correct and explicit interpretation was given a year later in the paper[25] published by Watson's group which appeared in Nature side-by-side with the paper by Brenner, Jacob and

Meselson mentioned above. For some time, I regretted that I did not have time to do more experiments nor time to think more seriously about the results so that we could develop some clear concepts of mRNA by ourselves. However, to do so would have been difficult in any event. We were unaware of the new developments, both experimental and conceptual, which were taking place in Cambridge, UK, as well as in Paris (for a detailed historical account of mRNA discovery, see Ref. 26). In addition, many simple basic facts about ribosomes were still lacking in 1959. For example, it had not been firmly established that the observed two major species of rRNAs, 16S and 23S rRNAs in the case of bacterial rRNAs (studied by several workers, especially by Kurland[27] in Watson's laboratory), are really two distinct species, each representing covalently linked unique and homogeneous molecules without any subunit structures. In fact, some workers explicitly suggested the possibility that these high molecular weight rRNAs consisted of discrete subunits of smaller sizes which are non-covalently linked and that the smaller-sized RNA detected by pulse-labeling might be precursors for rRNAs (see, for examples, Refs 28 and 29). Thus, the discovery of the new type of heterogeneous RNA with sizes smaller than rRNAs alone could not distinguish between the two possibilities; one that the new RNA was mRNA and the other that it was rRNA precursors. In fact, after the discovery of mRNA, there was still a period of confusion. Some people studied mRNA assuming that pulse-labeled RNA is mostly mRNA; while others studied the synthesis of ribosomes assuming that it is mostly rRNA precursors. Of course, we now know that pulse-labeled RNA in *E. coli* consists of both stable RNAs (rRNAs and tRNAs) and mRNAs, that their proportions vary depending on growth conditions, and that there are certainly no subunits for *E. coli* rRNAs. Thanks to the progress made during the past decade, our current knowledge of the primary and the secondary structure of rRNAs as well as the structure of rRNA transcription units (operons) is on a very firm basis, and most of the confusion and debate about the results in these early days can now be easily resolved.

The discovery of mRNA clarified the role of ribosomes in protein synthesis and, in addition, helped in the development of reliable cell-free protein synthesizing systems which were needed to study ribosomes. The role of ribosomes as the site of protein synthesis had been firmly established by that time even in bacterial systems. By carrying out pulse–chase experiments using radioactive amino acids, McQuillen, Roberts and Britten demonstrated that in growing *E. coli* the radioactive proteins appeared first in the 70S ribosomes and then chased into other cellular fractions.[30] In addition, incorporation of radioactive amino acids into protein was also demonstrated in cell-free extracts from *E. coli* around that time.[18–20] However, it was Nirenberg and Matthaei's *in vitro* experiments[31] using externally added synthetic as well as natural polynucleotides as mRNA that really opened up the new era in the field of protein synthesis, enabling studies to solve the genetic code as well as studies of the mechanism of protein synthesis *in vitro.*

A PERSONAL PERSPECTIVE

The three years I spent in the USA as a post-doctoral fellow were important in shaping up my subsequent career as a scientist. I was educated in Tokyo during the miserable post-World War II era in Japan. I did not have a good education in basic science. In addition, Japan was still isolated from the rest of the scientific community and most of us were not well informed of what was happening in science in the world. Thus, I feel that I was fortunate to have had the opportunity to work in the mainstream of a new science (molecular biology) in three different laboratories during my post-doctoral years. As I described above, the major problem (the mechanism of information transfer from gene to protein) was there, but it was not necessarily clear what would be the best approaches to solve the problem. The field attracted bright and ambitious people with various backgrounds including biochemists, geneticists and physicists. Any motivated graduate student or post-doctoral fellow could dream of designing some crucial experiments to get definitive information to help solve the problem. Thus, even though I felt a deficiency in my background education and had some problems communicating in English and adapting to the life in the States, the sense of participation in this scientific endeavor with other American scientists made these problems rather minor. I was especially excited by the fact that famous scientists whose names I had known through their important work were willing to talk with someone like me who was an unknown young scientist from the Far East.

In late 1960, after three years in the States as a post-doctoral fellow, I returned to Japan and took an assistant professorship in the Institute for Protein Research at Osaka University. Knowing the system in Japanese universities and institutions, I had applied to NIH for a research grant and received everything I asked for. Thanks to this financial support and the special understanding of the director of the Institute, Professor Shiro Akabori, I was able to organize my own laboratory and carry out research with complete

Back in Japan after my post-doctoral work in the States, with young Jim Watson visiting Japan and two graduate students (September 1961).

independence. Nevertheless, I started to miss the people and the scientific atmo-sphere in the United States. Even though we could make some significant achievements in science at Osaka, I remained there for only three years and then left Japan to join the faculty in the Department of Genetics at the University of Wisconsin, USA. The decision was personal, but may have had some connections to the general situation in Japan at that time. However, a discussion of the situation in academic and industrial research institutions in Japan during that period, though perhaps interesting in connection with the subsequent remarkable achievements in applied science and technologies and comparatively less distinguished achievements in basic science, is beyond the scope of this essay.

REFERENCES

1. Schachman H.D., Pardee A.B., and Stanier R.Y. 1952. *Arch. Biochem. Biophys.* **38:** 245–260.
2. Chao F.C. 1957. *Arch. Biochem. Biophys.* **70:** 426–431.
3. Littlefield J.W., Keller E.B., Gros J., and Zamecnik P.C. 1955. *J. Biol. Chem.* **217:** 111–123.

4. Zamecnik P.C. and Keller E.B. 1954. *J. Biol. Chem.* **209:** 337–354.
5. Tissières A. and Watson J.D. 1958. *Nature* **182:** 778–780.
6. Nomura M. and Watson J.D. 1959. *J. Mol. Biol.* **1:** 204–217.
7. Kurland C.G., Nomura M., and Watson J.D. 1962. *J. Mol. Biol.* **4:** 388–394.
8. Yoshida K. and Osawa S. 1968. *J. Mol. Biol.* **33:** 559–569.
9. Lefkovits I. and Di Girolamo M. 1969. *Biochim. Biophys. Acta* **174:** 561–565.
10. Sykes J., Metcalf E., and Pickering J.D. 1977. *J. Gen. Microbiol.* **91:** 1–16.
11. Crick F.H.C. 1958. in *The Biological Replication of Macromolecules,* pp. 138–163. Symp. Soc. Exp. Biol. XII, Academic Press.
12. Waller J.P. and Harris J.I. 1961. *Proc. Natl. Acad. Sci. USA* **47:** 18–23.
13. Waller J.P. 1964. *J. Mol. Biol.* **10:** 319–336.
14. Britten R.J. and Roberts R.B. 1960. *Science* **131:** 32–33.
15. Roberts R.B., ed. 1958. *Microsomal Particles and Protein Synthesis.* Pergamon Press.
16. Pardee A.B. 1958. *Exp. Cell Res.* (suppl.) **6:** 142–151.
17. Pardee A.B., Jacob F., and Monod J. 1959. *J. Mol. Biol.* **1:** 165–178.
18. Lamborg M.R. and Zamecnik P.C. 1960. *Biochim. Biophys. Acta* **42:** 206–211.
19. Tissières A., Schlessinger O., and Gros F. 1960. *Proc. Natl. Acad. Sci. USA* **46:** 1450–1463.
20. Schachtschabel D. and Zillig W. 1959. *Hoppe-Seyler's Z. Physiol. Chem.* **314:** 262–276.
21. Volkin E. and Astrachan L. 1956. *Virology* **2:** 149–161.
22. Astrachan L. and Volkin E. 1958. *Biochim. Biophys. Acta* **29:** 536–544.
23. Nomura M., Hall B.D. and Spiegelman S. 1960. *J. Mol. Biol.* **2:** 306–326.
24. Brenner S., Jacob F., and Meselson M. 1961. *Nature* **190:** 576–581.
25. Gros F., Hiatt H., Gilbert W., Kurland C.G., Risebrough R.W., and Watson J.D. 1961. *Nature* **190:** 581–585.
26. Judson H.F. 1978. *The Eighth Day of Creation.* Simon and Shuster.
27. Kurland C.G. 1960. *J. Mol. Biol.* **2:** 83–91.
28. Aronson A. and McCarthy B. 1961. *Biophys. J.* **1:** 215–226.
29. McCarthy B.J., Britten R.J., and Roberts R.B. 1962. *Biophys. J.* **2:** 57–82.
30. McQuillen K., Roberts R.B., and Britten R.J. 1959. *Proc. Natl. Acad. Sci. USA* **45:** 1437–1447.
31. Nirenberg M.W. and Matthaei J.H. 1961. *Proc. Natl. Acad. Sci. USA* **47:** 1588–602.

The Discovery of Polyribosomes

JONATHAN R. WARNER*¶ AND PAUL M. KNOPF†

*Albert Einstein College of Medicine, Bronx, New York, USA;
†Brown University, Providence, Rhode Island, USA

The Watson–Crick DNA structure of 1953 led readily to an understanding of how DNA could be replicated. But how DNA could be used to direct the synthesis of specific proteins was hardly obvious. During the late '50s accumulating evidence implicated RNA in the process of translation, and the Central Dogma, "DNA makes RNA and RNA makes protein," became generally accepted. To reveal how the various species of RNA carried out translation was the task for the early '60s.

THE SETTING

By 1962 the major players in protein synthesis had been identified. The ribosome, named just four years previously, was clearly the machine that conducted the reaction (reviewed in Ref. 1). The tRNA (known then as sRNA or "soluble" RNA) was the adaptor—the intermediate between the nucleotide language and the amino acid language.[2,3] And only the previous year had several laboratories identified mRNA as the carrier of information between genes and the ribosome, presumably serving as the template on which amino-acyl tRNAs were assembled.[4,5] Indeed, a poly U template had just been shown to effect the synthesis of polyphenylalanine.[6] The race to decipher the genetic code was heating up.

Yet, how did these components work together to synthesize a real protein? What were their relative proportions during protein synthesis? Howard Dintzis had just demonstrated that in rabbit reticulocytes the globin chain

Reprinted from *Trends in Biochemical Sciences* July 2002, 27(7): 376–380.

was assembled stepwise from the N terminus towards the C terminus.[7] But how did an mRNA, thousands of nucleotides in length, interact with a ribosome of only 150–200 Å?

James Watson's group had presented some provocative data, observing that mRNA being translated by *Escherichia coli* ribosomes sedimented more rapidly than 70S ribosomes.[8] They suggested that "heavy ribosomes [were] sedimenting faster because of associated mRNA," and that "mRNA-complexed ribosomes aggregate more easily than free ribosomes." However, studies of *E. coli* ribosomes were in some confusion because many of the 70S ribosomes dimerized to form 100S complexes in the low salt, high Mg^{2+} ionic conditions that were routinely used. Indeed, it was suggested that the "100S ribosomes, not the 70S ribosomes, may be the principal sites of protein synthesis."[8]

THE COLLABORATION

At this time we were graduate students at MIT in the laboratory of Alex Rich, where we had been sharing a bench for several years. Paul had developed a cell-free translation system (CFS) from rabbit reticulocytes and had succeeded in repeating the Dintzis experiment *in vitro*, showing that protein synthesis occurred in the same step-wise fashion (N → C) as found in intact reticulocytes. However, no initiation or completion of new globin polypeptide chains was detectable.[9] He was writing up his thesis and trying to substantiate his observation that the reticulocyte CFS could translate poly U and poly A. Meanwhile, Jon was trying to analyze the interaction of tRNA with *E. coli* ribosomes, running lots of sucrose gradients with *E. coli* extracts. But progress was slow, and he jumped at the opportunity when Jim Darnell, then a new faculty member at MIT whom Jon had just met at a softball game, asked if he would be willing to investigate the ability of poliovirus RNA to serve as mRNA in the *E. coli* extract. Poliovirus RNA did bind to ribosomes and stimulated, albeit weakly, the incorporation of labeled amino acids, some of which immunoprecipitated with polio-specific antibodies.[10] Although it was recognized that bacterial ribosomes were smaller than mammalian ribosomes, at that time there was no suggestion that the mechanism of protein synthesis would differ in the two systems. Nevertheless, the low incorporation suggested that the poliovirus RNA might be a more effective mRNA in a mammalian extract.

So, we struck a deal. Paul would help Jon by translating poliovirus RNA in a reticulocyte CFS, and Jon would help Paul by running sucrose gradients of poly U bound to reticulocyte ribosomes. But a problem arose. The ribosomes kept sedimenting too far, often into the pellet at the bottom of the tube, even without poliovirus RNA or poly U. As it was easier to follow ribosomes by radioactivity than by ultraviolet absorbance, we decided to label the nascent globin chains on the ribosomes by incubating the reticulocytes briefly with ^{14}C amino acids. The cells were then lysed osmotically, and their ribosomes were prepared by two cycles of sedimentation and analyzed on a 5–20% sucrose gradient. Much to our surprise we found a complex pattern of several peaks (Fig. 1a). Furthermore, the radioactivity, representing growing globin chains, was not uniformly distributed among the ribosomes, but seemed more abundant in the rapidly sedimenting peaks. Because of reports that active ribosomes tended to "aggregate," we were concerned that the active ribosomes had stuck together during the initial pelleting. An equally important influence in our thinking, however, was the vocal presence down the hall of Cy Levinthal, who had recently shown that long DNA molecules were subject to breakage by even gentle shear forces.[11]

Was the method of ribosome preparation leading to aggregation or was it breaking fragile structures? We decided to determine the arrangement of the ribosomes by direct analysis of a reticulocyte lysate, with only a brief centrifugation to remove cell membranes. A new problem arose: the lysate, with its high concentration of hemoglobin, would not form a stable boundary at the interface with the 5% sucrose then used at the top of gradients. Replacing the 5–20% sucrose gradient with a 15–30% sucrose gradient solved the problem while maintaining a linear rate of sedimentation. Figure 1b is our plot of the first sucrose gradient analysis of reticulocyte lysates that had been briefly labeled with ^{14}C amino acids. It is difficult to describe the rush of excitement as we plotted the CPM (counts per min) from the strip of paper that the counter had been generating overnight. The reticulocyte had ribosomes that seemed to be inactive, and sedimented as single 80S ribosomes, but half the ribosomes were in much larger structures—these were the active ones! Both polio and poly U were forgotten as we rushed to analyze just what these structures were.

The next few weeks passed in a flurry of experiments in which we reproduced this result and determined the nature of the larger structures. Electron microscopy (EM) studies of mammalian cells had often revealed clusters of ribosomes on the endoplasmic reticulum.[12] Our experiment showing that the larger structures were insensitive to 0.5% deoxycholate and to 0.5 M NaCl was, therefore, important as it proved that these structures were not held

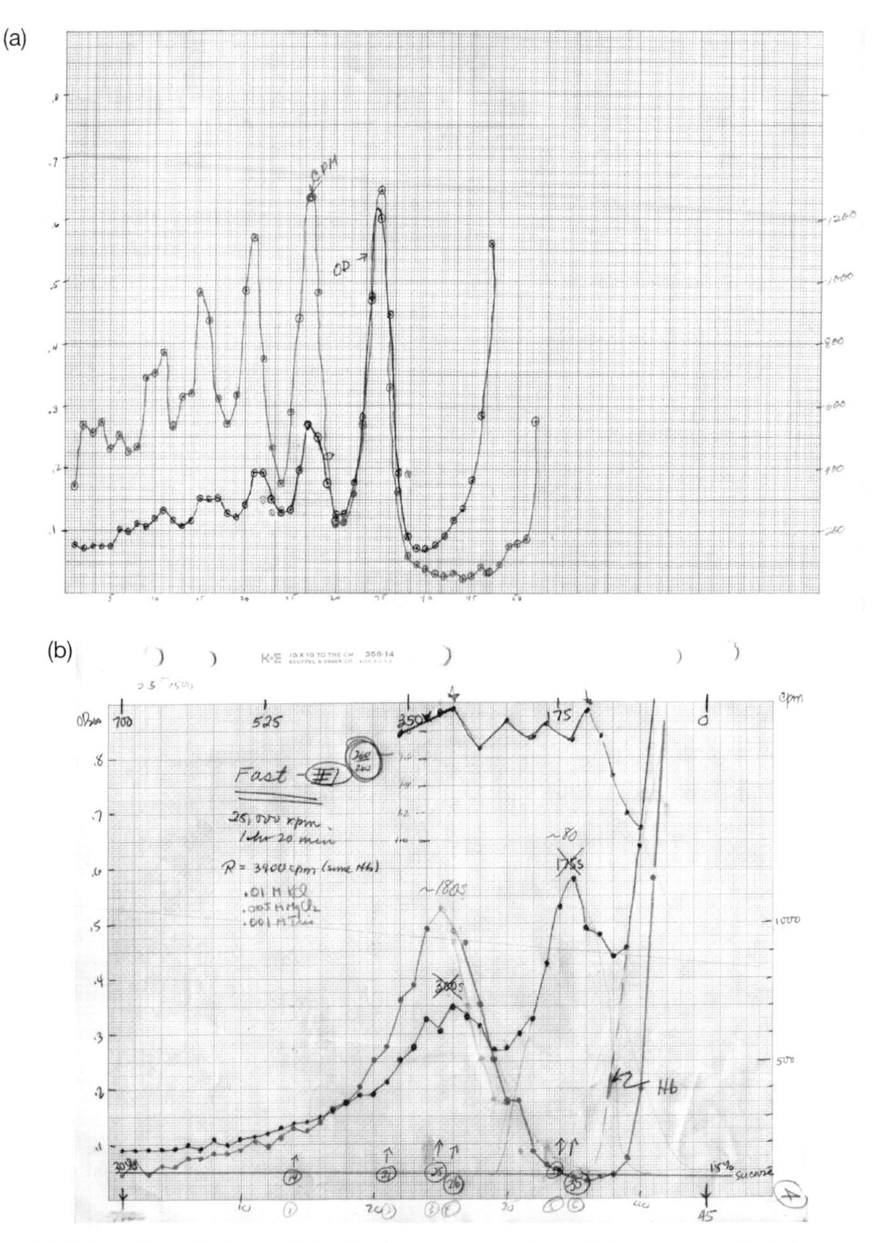

FIGURE 1. (a) Analysis on a 5–20% w/w sucrose gradient of ribosomes purified, by two cycles of centrifugation and resuspension, from reticulocytes that had been labeled for 1.5 min with ^{14}C amino acids (24 July 1962). The *x*-axis represents the 52 fractions from the sucrose gradient: sedimentation from right to left. OD_{260} (left scale); CPM (right scale). (b) Our first analysis on a 15–30% w/w sucrose gradient of the clarified lysate of reticulocytes that had been labeled for 0.5 min with ^{14}C amino acids (8 August 1962). (Note the original, short-lived confusion as to the S values of the peaks.) Sedimentation as in (a). Abbreviations: CPM, counts per min; OD, optical density.

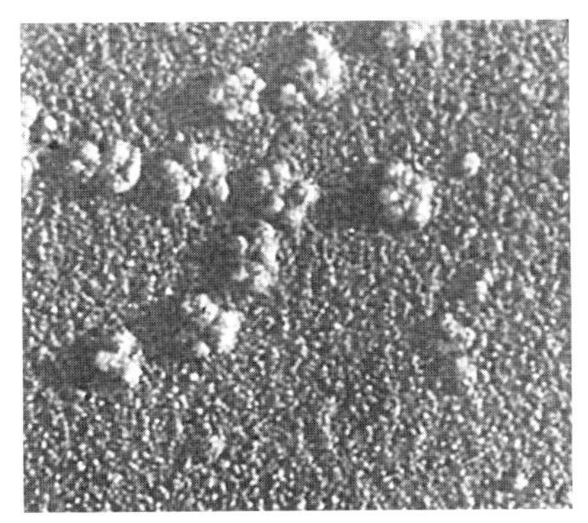

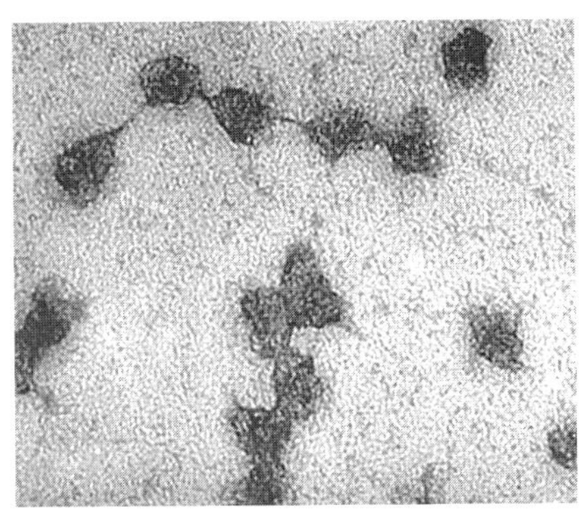

FIGURE 2. (a) Negatively stained[15] and (b) positively stained[13] samples from the polysome peak of a gradient such as that of Fig. 1b. In Fig. 2b, the mRNA connecting the ribosomes is clearly visible.

together by membranes. In addition, the large structures were exquisitely sensitive to RNAse but unaffected by DNAse. They were resistant to lowering of the Mg^{2+} concentration (an obsession at the time, as Mg ions were recognized as being key to the association–dissociation of ribosomal subunits[4]) but destroyed by chelating Mg^{2+} ions with EDTA. As we had worried, the structures were sensitive to shear: they were reduced to 80S ribosomes by grinding with alumina (the usual method of preparing bacterial ribosomes at the time) and were partially degraded by repeated sedimentation and resuspension.

The intermediates apparent as partial breakdown products (Fig. 1a) suggested that the structures in the main peak of Fig. 1b were composed of five ribosomes. Indeed, EM analysis of the fractions from a gradient as in Fig. 1b revealed largely penta-ribosome structures in the peak fractions (Fig. 2a), with hexa- and tetra-ribosome structures in the neighboring fractions (this analysis was conducted in collaboration with Cecil Hall's laboratory, where we simply spotted a drop from each fraction on a grid, sucked off the sucrose solution and negatively stained the sample).

Although no individual mRNA had been isolated at that time, one could calculate that ~450 nucleotides were required to encode a globin chain. In a stretched configuration of 3.4 Å per nucleotide, the 1500 Å globin mRNA could accommodate five or six 200-Å ribosomes. This was subsequently shown to be the case by the exquisite electron micrographs of Henry Slayter[13] (Fig. 2b).

From the moment we saw Fig. 1b, we knew that we were onto a fundamentally new way of thinking about protein synthesis, but it was Alex Rich who was able to put the model in a clear perspective. The rapidly sedimenting component comprised several ribosomes associated with a single mRNA, each ribosome presumably scanning the mRNA from one end to the other as it synthesized a complete polypeptide. What the microscopists had been seeing for years finally became clear.[12] With the advice of Richard Roberts, who had coined the name "ribosome" in 1958, we termed the structure a "polyribosome," or "polysome" for short. Although in retrospect this concept seems obvious, at the time groups that observed similar phenomena had different ideas; for example, ". . . the ability of the heavy ribosome particles to incorporate amino acid may be dependent not on their content of 'messenger' RNA alone but on some other factor which is a determinant of the stability of the heavy ribosomes."[14]

GOING PUBLIC

It was an exciting time in a graduate student's life, and Jon gave his first public talk at a small meeting organized by MIT in October 1962. Unfortunately, as the meeting coincided with the Cuban missile crisis, the audience was rather distracted. Paul had left for a post-doc in Cambridge with Francis Crick in September, and enjoyed a more attentive audience for his presentation on polysomes at the Medical Research Council, fortunately just before the excitement caused by the announcement of Crick's Nobel prize.

Alex and Jon wrote up the work and sent the paper off to John Edsall at Harvard for submission to *Proc. Natl. Acad. Sci. USA*. After some grumbling from reviewers about who had originated the idea of polyribosomes, the paper was published in January 1963.[15] This was clearly an observation whose time had come; independently, Alfred Gierer in Tubingen, also using reticulocytes,[16] and Hans Noll in Pittsburgh, using rat liver,[17] had reached the conclusion that protein synthesis was conducted on structures with multiple ribosomes scanning a single mRNA. Although Noll coined the term "ergosome" (working ribosome) to describe them, "polysome" stuck.

It is interesting that, at a time when the attention of molecular biologists was fixed on *E. coli* and its phages, the most compelling evidence for polyribosomes came from mammalian systems. In part, this was because of the harsh methods needed to open bacterial cells. But another important element was the separation of transcription from translation that occurs in eukaryotic but not prokaryotic cells. Dealing with the cytoplasm and, in the case of reticulocytes, only the cytoplasm, made a much clearer analysis possible.

Indeed, the subsequent studies of polyribosome function were conducted almost exclusively in mammalian cells (see Refs. 18–20 for examples). One could argue that the discovery of polyribosomes was one (of several) turning points that led molecular biologists to realize that mammalian cells were a suitable subject for study.

In Cambridge, Paul and Hildegard Lamfrom continued to develop the reticulocyte CFS, using the membrane-free lysate without further fractionation. They showed that its high activity reflected the recruitment of new ribosomes to the globin mRNA, with completed globin chains being continuously released from the active ribosomes.[21] The lack of translation initiation seen in Paul's thesis work had been a consequence of mRNA fragmentation in the fractionated CFS.

HOW THE PIECES FIT

Whereas the concept of polysomes provided a picture of the ribosome flowing along an mRNA molecule, the way in which this occurred remained obscure. In particular, how many polypeptides were made by a ribosome? How many tRNA molecules could a ribosome accommodate? Although J.-P. Waller had already shown that the protein complement of a ribosome was extremely complex,[22] unlike that of a virus, many still thought of the ribosome as a quasi symmetrical structure that might have several active sites (see example in Ref. 23).

At this point it was known that the growing polypeptide chain is covalently attached to a tRNA molecule,[24] and that the incoming amino acid is attached to another tRNA. Wally Gilbert had found that a single molecule of tRNA bound to a ribosome, and he proposed a model by which the growing polypeptide chain was passed off from one tRNA to another.[25] However, because it was difficult to determine the fraction of active ribosomes in these preparations, the numerology was somewhat uncertain. The reticulocyte system, in which the active ribosomes could be isolated, offered the possibility of solving this problem.

Jon attempted first to determine the number of growing polypeptide chains by labeling reticulocytes with ^{14}C leucine or phenylalanine, and measuring the specific activity of the polyribosomes and of the soluble amino acid pool. Knowing the number and location of the leucine and phenylalanine residues in the α and β globin chains, we could calculate the number of growing chains per ribosome. In numerous experiments, the value came out to be ~0.8. Use of a convenient "fudge factor" (based on the demonstration that the rate of translation appeared slower at the beginning of translation[26]) raised that value to ~0.9 growing chains per ribosome, close enough to 1.0 to conclude that each ribosome carried a single growing chain, a gratifyingly simple number.[27]

With a single growing chain, how many tRNA molecules were associated with each ribosome? Although the reticulocyte, without a nucleus, does not transcribe RNA, it does repair the 3′-CCA portion of its tRNA, a process that was being studied in the neighboring laboratory by Ed Herbert.[28] We reasoned that by providing reticulocytes with ^{3}H adenine, we might label the tRNAs sufficiently to detect them on functioning ribosomes. Although the labeling was not great, we could measure the tRNAs on polyribosomes, though each experiment tied up the scintillation counter for a week. By determining the specific activity of the tRNA fraction and knowing the molecular weight ratio of the ribosome to the tRNA, we calculated that each active ribosome had two molecules of tRNA, whereas the inactive single ribosomes had only one. The data were robust, as the specific activity was constant throughout the polysome peak (Fig. 3).

Thus, we concluded that an active ribosome has not one but two sites for tRNA. One holds the growing chain, which Alex and Jon termed the "P" site for peptidyl tRNA, and the other, which we termed the "A" site, holds the incoming amino-acyl tRNA.[29] This seemed to make sense as it permitted the two tRNAs to reside on the ribosome simultaneously while the peptide bond was formed. A similar conclusion was also reached by the Schweet group, based on dissecting, *in vitro*, the steps of peptide bond formation by reticulocyte ribosomes.[30]

In retrospect, however, this two-site model suggests that the formation of the peptide bond is the rate limiting step, whereas we would expect it to be the selection of the amino-acyl tRNA. This discrepancy was resolved a few years later, when Knut Nierhaus showed that the ribosome actually has three sites, the additional one being the "E" or exit site, which retains the spent tRNA until the A site is filled.[31] Consequently, a translating ribosome has exactly two tRNAs at all times.

POSTSCRIPT

Thus, in little more than a year and a half our views of protein synthesis had evolved from a vague idea of the triple interaction of mRNA, tRNA and

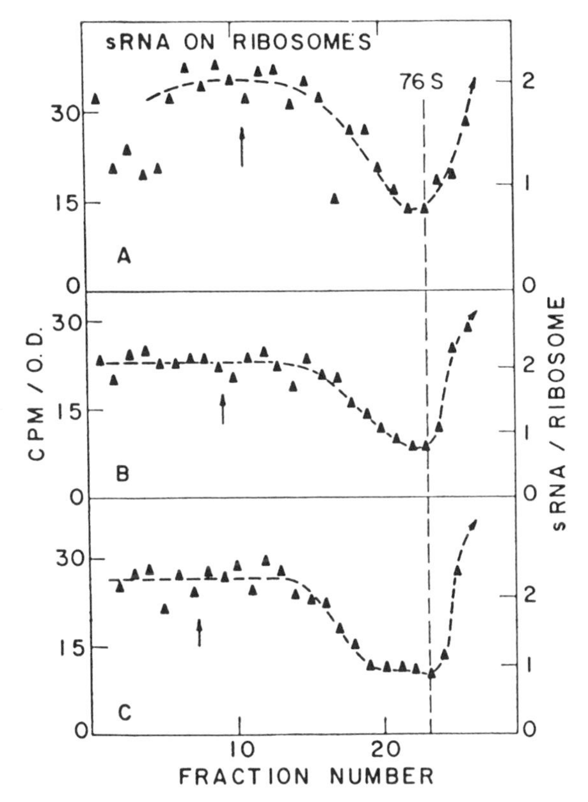

FIGURE 3. Reproduction of Figure 3 from Ref. 29 summarizing three independent experiments showing that active ribosomes have two tRNAs. The specific activity (CPM in tRNA/OD_{260} of ribosomes) is on the left scale, and the number of tRNAs per ribosome on the right scale (calculated from the specific activity of the tRNA). The vertical arrows indicate the polysome peak. Abbreviations: CPM, counts per min; OD, optical density. Reproduced, with permission, from Ref. 29; copyright 1963 National Academy of Sciences, U.S.A.

ribosomes, to a model very close to the current one: ribosomes flow from one end of an mRNA to another, and each has a site for a tRNA attached to the growing polypeptide and a site for the incoming tRNA attached to an amino acid.

For each of us, this experience was the coming-of-age as a scientist because we had the thrill of learning something that *no-one* else knew, the essence of conducting science, coupled with the responsibility to see that we got it right. It was also a learning experience. It taught us the importance of scraping the pellets from the bottom of tubes, that is, of keeping our eyes open. And it taught us the importance of colleagues and of comfortable communication between faculty and students, as not only Alex Rich, but also Howard Dintzis, Jim Darnell, Cy Levinthal, Cecil Hall, Henry Slayter and Ed Herbert, all played a direct role in our thinking and our experiments.

REFERENCES

1. Siekevitz P. and Zamecnik P. 1981. Ribosomes and protein synthesis. *J. Cell Biol.* **91:** 53s–65s.
2. Hoagland M.B. et al. 1958. A soluble ribonucleic acid intermediate in protein synthesis. *J. Biol. Chem.* **231:** 241–257.
3. Crick F.H.C. 1958. On protein synthesis. *Soc. Exptl. Biol. Symposium* **XII:** 138–163.
4. Brenner S. et al. 1961. An unstable intermediate carrying information from genes to ribosomes for protein synthesis. *Nature* **190:** 576–581.
5. Gros F. et al. 1961. Unstable ribonucleic acid revealed by pulse-labelling of *Escherichia coli. Nature* **190:** 581–585.
6. Nirenberg M. and Matthai J.H. 1961. The dependence of cell-free protein synthesis in *E. coli* upon naturally occurring or synthetic polynucleotides. *Proc. Natl. Acad. Sci. U. S. A.* **47:** 1588–1602.
7. Dintzis H.M. 1961. Assembly of the peptide chains of hemoglobin. *Proc. Natl. Acad. Sci. U. S. A.* **47:** 247–261.
8. Risebrough R.W. et al. 1962. Messenger RNA attachment to active ribosomes. *Proc. Natl. Acad. Sci. U. S. A.* **48:** 430–436.
9. Knopf P.M. and Dintzis H.M. 1965. Hemoglobin synthesis in a cell-free system. *Biochemistry* **4:** 1427–1434.
10. Warner J.R. et al. 1963. The interaction of poliovirus RNA with *Escherichia coli* ribosomes. *Virology* **19:** 393–399.
11. Davison P.F. et al. 1961. The structural unit of the DNA of bacteriophage T4. *Proc. Natl. Acad. Sci. U. S. A.* **47:** 1123–1129.
12. Palade G.E. and Siekevitz P. 1956. Pancreatic microsomes; an integrated morphological and biochemical study. *J. Biophys. Biochem. Cytol.* **2:** 671–689.
13. Slayter H.S. et al. 1963. The visualization of polyribosomal structure. *J. Mol. Biol.* **7:** 652–657.
14. Marks P.A. et al. 1962. Protein synthesis in erythroid cells, I. Reticulocyte ribosomes active in stimulating amino acid incorporation. *Proc. Natl. Acad. Sci. U. S. A.* **48:** 2163–2171.
15. Warner J.R. et al. 1963. A multiple ribosomal structure in protein synthesis. *Proc. Natl. Acad. Sci. U. S. A.* **49:** 122–129.
16. Gierer A. 1963. Function of aggregated reticulocyte ribosomes in protein synthesis. *J. Mol. Biol.* **6:** 148–157.
17. Wettstein F.O. et al. 1963. Ribosomal aggregate engaged in protein synthesis; characterization of the ergosome. *Nature* **197:** 430–435.

18. Penman S. et al. 1963. Polyribosomes in normal and poliovirus-infected HeLa cells and their relationship to messenger-RNA. *Proc. Natl. Acad. Sci. U. S. A.* **49:** 654–661.
19. Goodman H.M. and Rich A. 1963. Mechanism of polyribosomal action during protein synthesis. *Nature* **199:** 318–322.
20. Noll H. et al. 1963. Ribosomal aggregates engaged in protein synthesis: ergosome breakdown and messenger ribonucleic acid transport. *Nature* **198:** 632–638.
21. Lamfrom H. and Knopf P.M. 1964. Initiation of hemoglobin synthesis in cell-free systems. *J. Mol. Biol.* **9:** 558–575.
22. Waller J.-P. and Harris J.I. 1961. Studies on the composition of the protein from *Escherichia coli* ribosomes. *Proc. Natl. Acad. Sci. U. S. A.* **47:** 18–23.
23. Davis B.D. 1963. Discussion in *Cold Spring Harb. Symp. Quant. Biol.* **XXVIII:** 294–296.
24. Gilbert W. 1963. Polypeptide synthesis in *Escherichia coli* II. *J. Mol. Biol.* **6:** 389–403.
25. Cannon M. et al. 1963. The binding of sRNA by *Escherichia coli* ribosomes. *J. Mol. Biol.* **7:** 360–370.
26. Naughton M.A. and Dintzis H.M. 1962. Sequential synthesis of the peptide chains of hemoglobin. *Proc. Natl. Acad. Sci. U. S. A.* **47:** 1822–1830.
27. Warner J.R. and Rich A. 1964. The number of growing polypeptide chains on reticulocyte polyribosomes. *J. Mol. Biol.* **10:** 202–211.
28. Holt C.E. et al. 1966. Turnover of terminal nucleotides of soluble ribonucleic acid in intact reticulocytes. *J. Biol. Chem.* **241:** 1819–1829.
29. Warner J.R. and Rich A. 1964. The number of soluble RNA molecules on reticulocyte polyribosomes. *Proc. Natl. Acad. Sci. U. S. A.* **51:** 1134–1141.
30. Arlinghaus R. et al. 1964. Mechanism of peptide bond formation in polypeptide synthesis. *Proc. Natl. Acad. Sci. U. S. A.* **51:** 1291–1299.
31. Rheinberger H.J. et al. 1981. Three tRNA binding sites on *Escherichia coli* ribosomes. *Proc. Natl. Acad. Sci. U. S. A.* **78:** 5310–5314.

Peptidyl Transfer, the Monro Era

B. EDWARD H. MADEN

University of Liverpool, Liverpool, UK

The peptide bond-forming reaction of protein synthesis, the peptidyl transfer reaction, takes place in a region of the 50S ribosomal subunit that consists entirely of RNA, the peptidyl transferase centre. Basic to the present knowledge of peptidyl transfer was the discovery by Robin Monro and his colleagues in the 1960s that the reaction is catalyzed by the 50S ribosome. The Monro experiments, and the historical context in which they were conceived, are described in this personal recollection. Monro's "fragment reaction," the ribosome catalyzed reaction of a fragment of formylmethionyl-tRNA with puromycin, remains in use in work on peptidyl transfer.

Peptidyl transfer is the fundamental reaction of protein synthesis, whereby amino acid residues become linked through peptide bonds. It occurs on the large ribosomal subunit, known as the 50S subunit in prokaryotes (see Box 1, note 1), in a site called the peptidyl transferase centre. Mounting evidence over several years has pointed to the direct involvement of rRNA in the peptidyl transferase centre. It is now manifest from structural work on the ribosome[1–3] that the centre consists entirely of rRNA. Given the central role of peptidyl transfer in biology, it is of interest to review developments that have led to our current knowledge of the reaction.

The foundations were laid in the 1960s, especially by Robin Monro and his colleagues at the Medical Research Council Laboratory for Molecular Biology (LMB) in Cambridge, UK. I was a research student under Robin's supervision during that exciting era, and the major emphasis of this article will be the work in the Monro Laboratory.

Reprinted from *Trends in Biochemical Sciences* November 2003, 28(11): 619–624.

Box 1

1. Most of the work described here was done (or first done) with ribosomes from prokaryotic sources. Prokaryotic ribosomal subunits are called 50S and 30S; the complete ribosome is called 70S. Prokaryotic nomenclature is used throughout the article. tRNA was formerly called SRNA or sRNA, as in titles of some pre-1967 papers in the reference list.
2. Puromycin lacks the polynucleotide chain of tRNA, the amino acid is linked to the 3´ position of the nucleoside by an amide bond, and the amino acid is a modified tyrosine, p-methoxyphenylalanine. Functional aspects of these differences are outlined in note 5.
3. It was reported[19] that polylysyl-tRNA grew to only 12–15 residues. By contrast, polyphe-tRNA grew to ~40 residues.[9] Moreover, ribosome-bound short oligolysyl-tRNAs, especially dilysyl and trilysyl-tRNA, were reactive towards puromycin, but longer oligolysyl-tRNAs were unreactive. Thus, failure of oligolysyl-tRNA to grow beyond a certain length correlated with failure to react with puromycin.[19] The unreactivity of longer oligolysyl-tRNAs was unexplained at the time. I now speculate that when the oligolysyl chain attains a crucial length, its high positive charge inhibits further chain growth through an electrostatic effect upon some step(s) in the elongation cycle.
4. The overall free energy change in normal peptide bond formation in protein synthesis is that of hydrolysis of aa-tRNA minus formation of the new peptide bond (peptidyl-$tRNA_n$ plus aa-$tRNA_{n+1}$ yields peptidyl-$tRNA_{n+1}$ plus deacylated tRNA) [21]. Because the free energy of hydrolysis of the aa-tRNA ester bond is approx. -31 $kJ.mol^{-1}$ and that of the peptide bond is roughly -12 $kJ.mol^{-1}$, normal peptide bond formation is exergonic by about -19 $kJ.mol^{-1}$. aa-tRNA synthesis is driven by consumption of two high-energy phosphates because in the reaction, PP_i is split from ATP, and this is subsequently hydrolyzed to $2P_i$.
5. Figures depicting the P- and A- sites and the more recently recognized E- (exit) site are in standard Biochemistry textbooks. In translocation, the "n + 1" peptidyl-tRNA and its associated mRNA codon are moved from the A-site to the P-site, with concomitant movement of deacylated tRNA from the P- to E- site. Translocation is mediated by elongation factor (EF)-G with hydrolysis of one GTP to GDP and P_i. Entry of the aminoacyl end of aa-tRNA into the A-site of peptidyl transferase on the 50S subunit is preceded by decoding in the 30S A-site. In decoding, there is high accuracy in aa-tRNA selection, involving EF-T_u, ribosomal sensing of correct codon–anticodon interaction, and hydrolysis of GTP to GDP and P_i (for review, see Ref. 39). Thus, including amino acid activation, four high-energy phosphates are consumed per round of the elongation cycle.) Differences between peptidyl transfer to puromycin instead of to aa-tRNA are as follows. Puromycin's small size and lack of polynucleotide enable it to slip unmonitored into the A-site of the 50S subunit. Studies with modified puromycins showed that a phe-like amino acid is necessary for maximum activ-

ity[14]; presumably, in the absence of the tRNA polynucleotide, the amino acid contributes importantly to binding to the 50S A-site. Whereas the amino acid residue in aa-tRNA undergoes rapid acyl migration between the 2´ and 3´ positions of the terminal adenosine, that in puromycin is fixed by the amide bond to the 3´ position of the nucleoside. A modified puromycin with a 2´ linkage was inactive[14]; this provided the first indication that the 3´ form of aa-tRNA is used in peptidyl transfer. (See Ref. 30 for discussion of acyl migration in aa-tRNA.) Lastly, structural features of puromycin, particularly the amide bond, might modify the energetics of peptidyl transfer to this molecule compared with the reaction with aa-tRNA.

6. As I recall, Robin Monro originally added ethanol with the intention of terminating the fragment reaction, but he obtained erratic reaction kinetics! This led him to discover that addition of ethanol started the reaction. It was not possible to eliminate the requirement, and ethanol (or a related substance such as methanol) has remained a standard component of the reaction mix. Robin further showed that the 50S subunit, without the 30S subunit or mRNA or supernatant proteins, catalyzes the synthesis of short peptidyl-tRNAs from fmet-$tRNA_F$ and aa-tRNAs in the presence of methanol.[40] A simple explanation, but not necessarily the only one, might be that the alcohol additive loosens the 50S structure without completely denaturing it, facilitating entry of substrates.

PROTEIN SYNTHESIS IN THE EARLY 1960s

The discovery of ribosomes as the sites of protein synthesis and of tRNAs as the carriers of activated amino acids took place over a few years up to 1960. Then came the *annus mirabilis* of 1961. On the informational side of protein synthesis, mRNA was discovered,[4] genetic experiments indicated a triplet code[5] and polyuridylic acid (poly-U), the first artificial mRNA to be tested in a cell-free protein-synthesizing system, was found to direct the synthesis of polyphenylalanine (polyphe).[6] On the assembly side, polypeptide chain growth was shown to proceed from N-terminus to C-terminus.[7] Sydney Brenner concluded his Cold Spring Harbor Symposium paper on mRNA[4] with the question "how does it all work?" With charismatic brevity, he had articulated the thoughts of a growing body of researchers. Soon, intense efforts were going into deciphering the genetic code and unraveling the mechanistic details of protein synthesis.

A crucial issue was the mode of attachment of the growing peptide to the ribosome. Walter Gilbert, who was then making a famous transition from physics to molecular biology, used an *Escherichia coli* cell-free protein-synthesizing system programmed with poly-U to investigate the ribosomal steps in protein synthesis.[8,9] He established that the nascent peptide is bound primarily to the 50S subunit, and is covalently linked to tRNA through a bond that resembles the ester bond in aa-tRNA but is relatively more stable.[9] When ribosomes carrying peptidyl-tRNA were incubated with puromycin, the peptidyl-tRNA bond was broken.[9] Mark Bretscher established that the nascent peptide is attached to a hydroxyl group of the terminal adenosine of tRNA, as in aa-tRNA.[10]

Polypeptide chain growth was also known to require GTP and protein factors from the postribosomal supernatant but it was not known what these components did in the overall process. It was suspected that special mechanisms existed for polypeptide chain initiation and termination but no details were available. There were no sequence data on any of the participating macromolecules until the mid-1960s. Three-dimensional structural data did not come until much later, starting with tRNA.

PUROMYCIN

Puromycin, an antibiotic produced by *Streptomyces alboniger*, has had a special place in the elucidation of peptidyl transfer. It resembles the aminoacyl end of aa-tRNA, although with some differences (Figure 1; Box 1, note 2). It was found to inhibit protein synthesis subsequent to aminoacylation of tRNA[11] and to release nascent peptides from ribosomes.[12,13] Release was accompanied by blocking of the reactivity of puromycin's primary amino group, and radioactivity from puromycin labeled in its amino acid moiety appeared in the released peptides.[13] Allen and Zamecnik made the following proposal[13]:

> An attractive possibility would be a nucleophilic substitution of the amino group of the *p*-methoxyphenylalanyl moiety of puromycin on the C-terminal acyl group of the growing polypeptide chain with displacement of the soluble RNA to which the peptide chain is considered to be esterified.

(The attachment of the nascent peptide to tRNA by an ester bond had not yet been proven.)

Nathans and Neidle reported the effects of several analogues and isomers of puromycin upon protein synthesis, and were led to consider the enzymic nature of puromycin's site of action[14]:

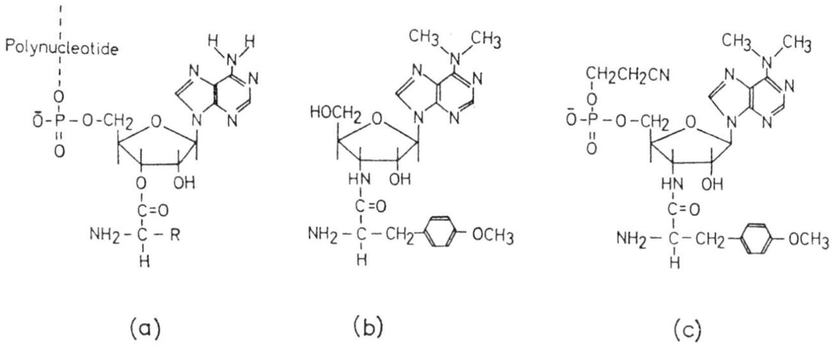

FIGURE 1. The structures of (a) the 3′ end of aminoacyl-tRNA, (b) puromycin, (c) puromycin cyanoethylphosphate. Reproduced, with permission, from Ref. 21.

> . . . the specific structural requirements for inhibition suggest that puromycin acts by binding to the active centre of an enzyme or to a ribonucleoprotein involved in the last steps of protein synthesis. In considering the reactions which occur, the possibility that puromycin interacts with the peptide bond-forming enzyme is an attractive hypothesis.

Nathans also showed that puromycin is incorporated C-terminally into released peptides,[15] as expected if the antibiotic substitutes for aa-tRNA, and that puromycin-mediated release of peptides is inhibited in an *E. coli* system by the bacterial protein synthesis inhibitor chloramphenicol.[15]

RIBOSOME-CATALYZED PEPTIDYL TRANSFER

This work pointed the way to the use of puromycin as a model substrate for the peptide bond-forming reaction of protein synthesis, and particularly for investigating the nature of the enzyme. Robin Monro had been a postdoctoral scientist in Fritz Lipmann's laboratory at the same time as Nathans, and had worked on the early characterization of postribosomal supernatant factors that take part in the elongation reactions of protein synthesis[16] (now called elongation factors, or EF). Two protein factors were separated, and were provisionally called A and B. The authors drew provisional inferences as to their possible functions[16]:

> . . . we are very tentatively inclined to identify the A fraction with the amino acid polymerase. The B fraction, on the other hand, appears to be linked to the GTP effect, a chemical definition of which still needs to be supplied.

Soon, Robin was to establish that the "amino acid polymerase" (the peptide bond-forming enzyme) is neither of the supernatant proteins but is part of the ribosome. Robin joined the new LMB laboratory in Cambridge shortly after it opened in 1962. With him from Lipmann's laboratory came Rob Traut, who was continuing as a postdoctoral researcher at LMB. Together, they carried out the first of the series of studies that were later to earn the accolade "the Monro era,"[17] in recognition of Robin's leading role for several years in analyzing the peptide bond-forming reaction.

"The Puromycin Reaction and Its Relation to Protein Synthesis"

The essence of the Traut and Monro study[18] was to uncouple the puromycin reaction from the other steps of protein synthesis. In previous work, puromycin had been added either to living cells or to cell-free extracts containing all of the components that were needed to synthesize protein starting from aa-tRNA or from free amino acids. The cell-free systems contained, in addition to ribosomes and mRNA, the post-ribosomal supernatant proteins, GTP, the components for amino acid activation (if needed) and any other unresolved entities. The question was, which ingredients were required for the actual peptide bond-forming step?

Traut and Monro addressed this question by *first* incubating ribosomes in a complete protein synthesizing system programmed by poly-U, *then* washing the ribosomes by centrifugation and *then* adding puromycin. The synthesized polyphe-tRNA (radio-labeled in the polyphe moiety) remained attached to the ribosomes through the washing procedures. Puromycin-mediated release of polyphe from tRNA was measured by a centrifugation assay.

The main findings were as follows. A proportion of polyphe (typically ~40%) was released rapidly by puromycin in the absence of added supernatant proteins and GTP. 50S ribosomal subunits bearing polyphe-tRNA were reactive. The reaction was partly inhibited by chloramphenicol and completely inhibited by EDTA, which disrupts ribosome structure by chelating Mg^{2+}.

Although a proportion of polyphe was released in the absence of added supernatant and GTP, addition of these components increased the amount of polyphe released. It was concluded that the ribosome-polyphe-tRNA complexes existed in two states, only one of which was reactive towards puromycin. The authors related this interpretation to a "two-site" model of the ribosome, in which peptidyl-tRNA could be in either a puromycin-sensitive or a puromycin-insensitive site. They proposed that a supernatant protein and GTP mediated translocation from the latter to the former.

Polylysine and Puromycin

Although it was known[13,15] that puromycin is linked C-terminally to released peptides by a bond with the characteristics of the peptide bond, it was uncertain whether *all* or only *some* of the released peptides were attached to puromycin.[15] Clarification was crucial if the puromycin reaction was to serve as a straightforward model of peptide bond formation but was difficult with existing experimental systems because the released peptides were mixtures of sequences and/or sizes.[13,15,18]

Polyadenylic acid (poly-A)-directed polylysine synthesis proved amenable in this regard.[19] The products of poly-A directed cell-free protein synthesis are oligolysyl-tRNAs in the size range lys_{2-12} (see Box 1, note 3). Oligolysines and oligolysyl puromycin products can be resolved by column chromatography. The lysine peptides were shown to be attached through their carboxyl groups by peptide linkage to puromycin, with one puromycin residue per released peptide. Quantification was established by experiments using ^{3}H lysine and a puromycin derivative bearing a ^{32}P-labeled 5′-cyanoethylphosphate group, synthesized by Mike Blackburn who was then in the Cambridge University Chemistry Laboratory.[19]

Incorporation of puromycin cyanoethylphosphate into the lysine peptides also proved that the entire puromycin molecule, including the nucleoside moiety, is transferred in the peptide bond-forming reaction. The earlier experiments, which had used puromycin labeled in its amino acid group,[13,15] had not reported directly on the fate of the nucleoside. The puromycin cyanoethylphosphate experiments thereby strengthened the evidence that the puromycin reaction indeed resembles the naturally occurring reaction with aa-tRNA.

A Non-hydrolyzable GTP Analog

With a view to further investigating the role of GTP in protein synthesis, John Hershey, who was also then in the Chemistry Laboratory at Cambridge, synthesized a GTP analog with a non-hydrolyzable methylene bridge between the β and γ phosphorus atoms. The compound, 5′-guanylyl-methylenediphosphonate (GMP-PCP), was shown by Hershey and Monro to be a competitive inhibitor of GTP in protein synthesis.[20] The analog did not inhibit the puromycin reaction assayed in the polyphe system,[21] supporting the inference[18] that GTP is not required for the actual peptide bond-forming step.

An Offer I Did Not Refuse

I came to work with Robin by a circuitous route. I had trained in medicine (1954–60) and had done two years of junior hospital appointments (1960–62). As a medical student I had become interested in the then new molecular biology, which had not yet entered the curriculum. During one of my hospital appointments, a surgical chief encouraged me to apply for a Medical Research Council research studentship. This was awarded and was to be held at the Department of Radiotherapeutics in Cambridge. Research interests in the Department included measurements on nucleic acid synthesis in cultured cells by microscopical techniques.

Unknown to me when I applied, the research wing of Radiotherapeutics was located in the new LMB building. The small laboratory in which I started my research was next to the LMB seminar room. The library was one floor above. These circumstances had both a stimulating and a distracting effect. On the stimulatory side, my knowledge grew rapidly as I attended numerous seminars and read avidly. Sydney Brenner gave a memorable review seminar on the Gilbert papers cited above.[8,9] He also gave or hosted weekly "open house" evening seminars in his rooms at King's College, aimed at younger molecular biologists. On the distractive side, my experiments, which involved estimating RNA synthesis in cultured cells by counting autoradiography grains through a microscope, seemed like seeing "through a glass darkly" compared with work along the corridor on the genetic code and the mechanism of protein synthesis. I wanted to get to grips with ribosomes.

In time, I struck up an acquaintanceship with Robin. We had met previously when he had been a research student and I a pre-clinical student in the same Cambridge college. When we met again, Robin and Rob Traut were writing their paper on the puromycin reaction.[18] Robin offered me the opportunity to do some experiments with a modified, higher-throughput assay that he had developed for the polyphe-puromycin reaction. He gave me excellent supervision, and Rob also gave me guidance, particularly in the centrifugation of ribosomes. I found immediately that I was doing what really interested me. Unofficially at first, and then officially, my PhD project became the polyphe-puromycin reaction under Robin's supervision.

I concentrated on the reaction that occurred in the absence of added supernatant proteins and GTP. Aspects of the initial work[18] were amplified,[22] and I described the ionic and pH characteristics of the reaction in some detail.[23] The pH profile suggested participation of a functional group with a pK of around 7.4. Robin and I recognized in our paper[23] that the response could be due either to the state of ionization of the nucleophilic amino group of puromycin or to that of a functional group within the peptidyl transferase enzyme.

A Visit to Warsaw, and Naming the Enzyme

The naming of the enzyme as peptidyl transferase was linked to preparations for the 1966 Federation of European Biochemical Societies Meeting in Warsaw. Robin was an invited speaker and I also went to the Meeting to give a short talk. The least expensive accommodation, and therefore attractive to research students, was the "Hotel Tramp." Fortunately, this turned out to be a Youth Hostel-type establishment and not a refuge for down-and-outs. Of six students who shared one room, four of us later became UK professors, so the experience was evidently a formative one!

The invited molecular biology contributions were to be published in book form. Robin wanted to bring the peptide bond-forming reaction within the compass of standard enzymology and nomenclature. He revisited his alma mater, the Cambridge Biochemistry Department, to confer with Malcolm Dixon (of Ref. 24). From their discussion, the term "synthetase" was deemed inappropriate because all of the evidence, including that from GMP-PCP, indicated that the reaction did not directly consume a nucleoside triphosphate. Instead, the reaction appeared to be an energetically favorable group transfer of the nascent peptide from peptidyl-tRNA to aa-tRNA, and the enzyme became called peptidyl transferase.[21] (See Box 1, notes 4–5, for energetic aspects of elongation.) What was most notable about the enzyme was that it appeared to be an integral component of the 50S ribosome.[21]

The "Fragment Reaction" and Its Conception

There remained a possible objection to the conclusions just described. Precharging ribosomes with polyphe-tRNA required supernatant proteins and GTP. It was imaginable that during pre-charging, a molecule of supernatant protein and one of GTP became firmly bound in a protected form, and then participated in the subsequent puromycin reaction. Various data rendered this possibility unlikely[21] but it could not be absolutely ruled out with the polyphe system.

Any such objection was eliminated by means of the "fragment reaction." The fragment reaction was developed in the context of the two-site model of the ribosome and the discovery of the mechanism of initiation of protein synthesis.

It was obvious that for peptidyl transfer to occur, there must be two adjacent sites on the ribosome, one for peptidyl-tRNA and one for the incoming aa-tRNA; the polyphe-puromycin data were interpreted in relation to the two-site model, as mentioned above.[18] The two sites are now called the P- and A-sites. Peptidyl transfer converts the aa-tRNA into the new "n +1" peptidyl-tRNA, and this must be translocated back to the P-site to allow binding of another aa-tRNA and repetition of the cycle (see Box 1, note 5).

In the mid-1960s, formylmethionyl-tRNA (fmet-tRNA$_F$) was discovered and its role in the initiation of protein synthesis in *E. coli* was established. fmet-tRNA$_F$, unlike other aa-tRNAs, binds to the P-site. This conclusion came from the finding that fmet-tRNA$_F$, bound to ribosomes under the direction of the codon AUG, reacts with puromycin, yielding fmet-puromycin.[25] Other ribosome-bound aa-tRNAs are unreactive to puromycin because they and puromycin compete for A-site occupancy. Entry of fmet-tRNA$_F$ into the P site enables it to fulfill its initiator role by being in place to react with the next aa-tRNA.

No low molecular weight P-site substrate had yet been developed. Monro and Marcker recognized the potential usefulness of such a substrate. This led them to test the activity of a fragment of fmet-tRNA$_F$ generated by digestion of the molecule with T$_1$ ribonuclease. The fragment was the terminal hexanucleotide with its formylmethionine attached, CAACCA-met-f. The fragment reacted with puromycin in the presence of ribosomes, suitable ions and, surprisingly, ethanol, generating fmet-puromycin.[26] (See Box 1, note 6, for a comment on the role of alcohol.) The fragment reaction afforded an assay for peptidyl transfer in which the ribosomes had not been exposed *in vitro* to those other components of protein synthesis (mRNA, supernatant proteins, GTP) that were needed for generating the peptidyl-tRNA substrate in the polyphe system.[18] By means of the reaction, Monro conclusively established that peptidyl transfer is catalyzed by the 50S ribosomal subunit and does not involve supernatant protein or GTP.[26,27] The reaction also enabled the testing of several antibiotics for their effects on peptidyl transfer.[28]

The fragment reaction was the culmination of Robin's work in resolving peptidyl transfer from the other reactions of protein synthesis. After developing the reaction in 1966, he published the first three papers[26–28] in rapid succession in 1967, indicating the importance he attached to the reaction and the speed with which he identified and performed crucial experiments (as sole scientist or in small collaborations). He invited me to join him in the fragment experiments during the latter part of 1966 but, in contrast to the earlier invitation, I had to decline, for I had limited time in which to conclude the polyphe work and to write and defend my PhD before departing to a postdoctoral position with Jim Darnell and Jon Warner in New York.[29] Robin and I combined the reporting of our ionic and pH data from the polyphe and fragment reactions,[23] noting the similar pH responses in the two systems.

Ribosome to Ribozyme

Peptidyl transfer was now attracting growing interest. Work in several laboratories, including Monro's, addressed two main fronts: further definition of the substrate specificity, and the nature of the enzyme. Peptidyl transferase was found under suitable experimental conditions to be somewhat flexible with regard to substrates and to catalyze the formation of various peptide-related bonds such as esters, and was implicated in the cleavage of peptidyl-tRNA in polypeptide chain termination. (Primary literature on alternative substrates and catalytic versatility is reviewed in Refs. 30,31.) Regarding the nature of the enzyme, it was expected that peptidyl transferase would be found among the 50S ribosomal proteins. However, when 50S ribosomes were partly stripped of proteins, transferase activity remained with the ribosomal core. After 50S inactivation by further stripping, activity was restored by adding back some proteins but was absent in proteins in their detached state.[32] In this and further such work, the transferase eluded identification as any specific protein.

Francis Crick[33] and Leslie Orgel,[34] in considering the evolutionary origins of the genetic code and the protein synthetic apparatus, argued that the primitive ribosome might have consisted entirely of RNA. It is interesting to revisit the relevant passages:

> If indeed rRNA and tRNA were essential parts of the primitive machinery, one naturally asks how much protein, if any, was then needed. It is tempting to wonder if the primitive ribosome could have been made *entirely* of RNA. Some parts of the structure, for example the presumed polymerase, might now be protein, having been replaced because a protein could do the job with greater precision.[33]

> What proteins, if any, seem indispensable for a primitive ribosome? The fact that RNA is the major component of ribosomes suggests that it might be both primitive and difficult to replace. It is interesting that unlike tRNA it has been perfected with proteins . . . .[34]

Before 1980, all known enzymes were proteins. Crick evidently thought it likely that in present day ribosomes, the "presumed polymerase" (i.e. peptidyl transferase) is a protein. The possibility that the peptidyl transferase of present day ribosomes is part of rRNA began to emerge in the 1980s with the discovery of ribozymes. Moreover, methods that enabled the mapping of ribosomal binding sites for components and inhibitors of protein synthesis were pinpointing specific bases in rRNA. A region of 23S rRNA containing several bases that interact with substrates and inhibitors of peptidyl transfer[17] became known as the "peptidyl transferase ring."

Harry Noller and colleagues reported experiments[35] that surpassed earlier ones[32] in stripping 23S rRNA of ribosomal proteins and yet retaining catalytic activity in the fragment reaction. However, a few percent of proteins resisted extraction, so it was still uncertain whether 23S rRNA was the catalytic agent.[35]

The determination of ribosome structure at atomic resolution at last revealed "what fits where" in the machinery of protein synthesis. The peptidyl transferase centre is within a region of the 50S ribosome that is bare of ribosomal proteins,[2,3] clearly implicating 23S rRNA in catalysis of the reaction. The status of work on the role of the active site of 23S RNA in catalysis has been reviewed in this journal.[36]

To conclude on a general note, it is difficult to imagine an Origins-of-Life scenario whereby proteins of defined amino acid sequence could have evolved in the absence of template-directed protein synthesis. (See Ref. 34 for a discussion of the difficulties.) Steps whereby a primitive system of protein synthesis might have evolved from an RNA World are discussed in Refs. 37,38. The direct involvement of rRNA in present-day ribosome-catalyzed peptidyl transfer strongly implies that this mechanism for making peptide bonds has persisted unchanged in its essentials from that remote era of the Earth's past.

ACKNOWLEDGEMENTS

Almost 40 years on, I remain deeply grateful to Robin Monro for guiding me into molecular biology by way of this most fundamental of biosynthetic reactions. I dedicate this article to him. I thank Sybil Maden for much help with the preparation of the manuscript.

REFERENCES

1. Ban N. et al. 2000. The complete atomic structure of the large ribosomal subunit at 2.4A resolution. *Science* **289:** 905–920.
2. Nissen P. et al. 2000. The structural basis of ribosome activity in peptide bond synthesis. *Science* **289:** 920–930.
3. Yusopov M. et al. 2001. Crystal structure of the ribosome at 5.5 A resolution. *Science* **292:** 883–896.
4. Brenner S. 1961. RNA, ribosomes, and protein synthesis. *Cold Spring Harb. Symp. Quant. Biol.* **XXVI:** 101–109.
5. Crick F.H.C. et al. 1961. General nature of the genetic code for proteins. *Nature* **192:** 1227–1232.
6. Nirenberg M.W. and Matthaei J.H. 1961. The dependence of cell-free protein synthesis in *E. coli* upon naturally occurring or synthetic polyribonucleotides. *Proc. Natl. Acad. Sci. U. S. A.* **41:** 1588–1602.
7. Dintzis H.M. 1961. Assembly of the peptide chains of hemoglobin. *Proc. Natl. Acad. Sci. U. S. A.* **41:** 247–261.
8. Gilbert W. 1963. Polypeptide synthesis in *Escherichia coli I.* Ribosomes and the active complex. *J. Mol. Biol.* **6:** 374–388.
9. Gilbert W. 1963. Polypeptide synthesis in *Escherichia coli II.* The polypeptide chain and S-RNA. *J. Mol. Biol.* **6:** 389–403.
10. Bretscher M.S. 1963. Chemical nature of the S-RNA polypeptide complex. *J. Mol. Biol.* **7:** 446–449.
11. Yarmolinski M. and de la Haba G.L. 1959. Inhibition by puromycin of amino acid incorporation into protein. *Proc. Natl. Acad. Sci. U. S. A.* **45:** 1721–1729.
12. Morris A.J. and Schweet R.S. 1961. Release of soluble protein from reticulocyte ribosomes. *Biochim. Biophys. Acta* **47:** 415–416.
13. Allen D.W. and Zamecnik P.C. 1962. The effect of puromycin on rabbit reticulocyte ribosomes. *Biochim. Biophys. Acta* **55:** 865–874.
14. Nathans D. and Neidle A. 1963. Structural requirements for puromycin inhibition of protein synthesis. *Nature* **197:** 1076–1077.
15. Nathans D. 1964. Puromycin inhibition of protein synthesis: incorporation of puromycin into peptide chains. *Proc. Natl. Acad. Sci. U. S. A.* **51:** 585–592.
16. Allende J.E. et al. 1964. Resolution of the *E. coli* amino acyl sRNA transfer factor into two complementary fractions. *Proc. Natl. Acad. Sci. U. S. A.* **51:** 1211–1216.
17. Noller H.F. 1993. Peptidyl transferase: protein, ribonucleoprotein or RNA? *J. Bacteriol.* **175:** 5297–5300.
18. Traut R.R. and Monro R.E. 1964. The puromycin reaction and its relation to protein synthesis. *J. Mol. Biol.* **10:** 63–72.
19. Smith J.D. et al. 1965. Action of puromycin in polyadenylic acid-directed polylysine synthesis. *J. Mol. Biol.* **13:** 617–628.

20. Hershey J.W.B. and Monro R.E. 1966. A competitive inhibitor of the GTP reaction in protein synthesis. *J. Mol. Biol.* **18:** 68–76.
21. Monro R.E. et al. 1967. The mechanism of peptide bond formation in protein synthesis. In *Genetic Elements, Properties and Function* (Ed. D. Shugar), pp. 178–203, Academic Press, London.
22. Maden B.E.H. et al. 1968. Ribosome-catalysed peptidyl transfer: the polyphenylalanine system. *J. Mol. Biol.* **35:** 333–345.
23. Maden B.E.H. and Monro R.E. 1968. Ribosome-catalysed peptidyl transfer: effects of cations and pH value. *Eur. J. Biochem.* **6:** 309–316.
24. Dixon M. and Webb E.C. 1964. *Enzymes,* 2nd edn, Longmans, London.
25. Bretscher M.S. and Marcker K.A. 1966. Polypeptidyl-sRibonucleic acid and amino-acyl-sRibonucleic acid binding sites on ribosomes. *Nature* **211:** 380–384.
26. Monro R.E. and Marcker K.A. 1967. Ribosome-catalysed reaction of puromycin with a formylmethionine-containing oligonucleotide. *J. Mol. Biol.* **25:** 347–350.
27. Monro R.E. 1967. Catalysis of peptide bond formation by 50S ribosomal subunits from *Escherichia coli. J. Mol. Biol.* **26:** 147–151.
28. Monro R.E. and Vazquez D. 1967. Ribosome catalysed peptidyl transfer: effects of some inhibitors of protein synthesis. *J. Mol. Biol.* **28:** 161–165.
29. Maden B.E.H. 1998. Eukaryotic rRNA methylation: the calm before the Sno storm. *Trends Biochem. Sci.* **23:** 447–450.
30. Chladek S. and Sprinzl M. 1985. The 3′-end of tRNA and its role in protein biosynthesis. *Angew. Chem. Int. Ed. Engl.* **24:** 371–391.
31. Lieberman K.R. and Dahlberg A.E. 1995. Ribosome-catalyzed peptide-bond formation. *Prog. Nucleic Acid Res. Mol. Biol.* **50:** 1–23.
32. Staehelin T. et al. 1969. On the catalytic center of peptidyl transfer: a part of the 50S ribosome structure. *Cold Spring Harb. Symp. Quant. Biol.* **XXXIV:** 39–48.
33. Crick F.H.C. 1968. The origin of the genetic code. *J. Mol. Biol.* **38:** 367–379.
34. Orgel L.E. 1968. Evolution of the genetic apparatus. *J. Mol. Biol.* **38:** 381–393.
35. Noller H.F. et al. 1992. Unusual resistance of peptidyl transferase to protein extraction procedures. *Science* **256:** 1416–1419.
36. Steitz T.A. and Moore P.B. 2003. RNA, the first macromolecular catalyst: the ribosome is a ribozyme. *Trends Biochem. Sci.* **28:** 411–418.
37. Noller H.F. 1999. On the origin of the ribosome: coevolution of subdomains of tRNA and rRNA. In *The RNA World,* 2nd edn, (ed. R.F. Gesteland et al.), pp. 197–209, Cold Spring Harbor Press, New York.
38. Yarus M. 2001. On translation by RNAs alone. *Cold Spring Harb. Symp. Quant. Biol.* **66:** 207–215.
39. Ogle J.M. et al. 2003. Insights into the decoding mechanism from recent ribosome structures. *Trends Biochem. Sci.* **28:** 259–266.
40. Monro R.E. 1969. Protein synthesis: uncoupling of polymerisation from template control. *Nature* **223:** 903–905.

Forty Years under the Central Dogma

DENIS THIEFFRY* AND SAHOTRA SARKAR*†

Max Planck Institute for the History of Science, Berlin, Germany; †University of Texas, Austin, Texas, USA

> The Central Dogma. This states that once "information" has passed into protein it cannot get out again. In more detail, the transfer of information from nucleic acid to protein may be possible, but transfer from protein to protein, or from protein to nucleic acid is impossible. Information means here the precise determination of sequence, either of bases in the nucleic acid or of amino-acid residues in the protein.[1]

The quotation above is from a seminal paper, "On Protein Synthesis," presented by Francis Crick at the 1957 annual meeting of the Society of Experimental Biology and published in 1958.[1] In this paper, Crick listed the standard set of 20 amino acid residues for the first time; argued that "the specificity of a piece of nucleic acid is expressed solely by the sequence of its bases, and that this sequence is a (simple) code for the amino acid sequence of a particular protein"; argued that the three-dimensional conformation of a protein must be determined by its amino acid sequence; pointed out that protein synthesis must be sequential; and presented his hypothesis of "adaptor" molecules mediating protein formation at the ribosome. Most importantly, Crick formulated what he called the "Central Dogma," which is the focus of this article.

The Central Dogma had and still has an influential place in molecular biology. It has been a constant point of reference that served as a distinctive icon for proponents of the new molecular biology of the late 1950s and 1960s. However, it has also provided a target for many criticisms of the molecular approach. In this article, we give a short overview of the history of the Central Dogma, in the hope of stimulating further reflection and analysis.

Reprinted from *Trends in Biochemical Sciences* August 1998, 23(8): 312–316.

INTRODUCTION OF THE CONCEPT OF INFORMATION INTO BIOLOGY

The concept of information, introduced into molecular biology only five years earlier, stands at the core of Crick's view of protein synthesis. In 1953, Ephrussi, Leopold, Watson and Weigle[2] suggested that the term "interbacterial information" be introduced in order to allow navigation through what they perceived as a terminological morass into which bacterial genetics had entered. The objects of concern were terms such as transformation and transduction, which had come into vogue during the preceding decade. This was the first printed use of the word information in what became molecular biology. Although, Ephrussi et al.[2] did not define the term, its use spread almost immediately. In 1956, Mazia[3] argued that the role of RNA was to carry information from nuclear DNA to the cytoplasm, for protein synthesis. Spiegelman argued that only RNA and DNA had sufficient informational complexity to serve as templates for protein formation,[4] and Lederberg noted perceptively that "information" was what "specificity" was "called nowadays."[5] It remained for Crick, in 1958, to incorporate Lederberg's observation into an explicit definition of information as the "specification" of sequence.[1]

Crick's scheme for protein synthesis puts DNA at the centre of attention. According to this scheme, DNA or, more generally, nucleic acids are the source of a unidirectional information flow ultimately specifying proteins. This linear scheme contrasts with former circular schemes for protein synthesis, where proteins specify proteins ad infinitum.[6] Earlier in the 1950s, some biologists had already proposed a way out of the circle by positing a role for nucleic acids as templates for protein synthesis.[7] In this context, Crick's move capped the development of a gene-centred view of biological processes that had been gradually emerging since the 1920s.

INTERPRETATIONS OF THE CENTRAL DOGMA

The most obvious interpretation of Crick's original (1958) formulation of the Central Dogma is in negative terms. The Central Dogma only forbids a few types of information transfer, namely, from proteins to proteins and from proteins to nucleic acids. However, after its rapid adoption by most of the biologists interested in protein synthesis, it was most often interpreted or reformulated in a more restrictive way, constricting the flow of information from DNA to RNA and from RNA to protein (Fig. 1).[8]

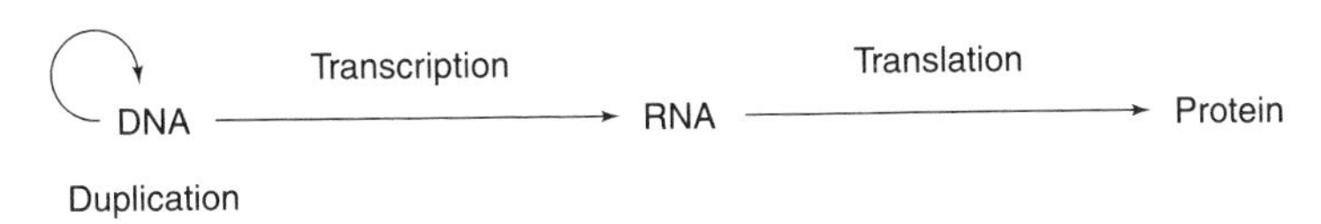

FIGURE 1. The Central Dogma as envisioned by Watson in 1965. "We should first look at the evidence that DNA itself is not the direct template that orders amino acid sequences. Instead, the genetic information of DNA is transferred to another class of molecules, which then serve as the protein templates. These intermediate templates are molecules of ribonucleic acid (RNA) . . . Their relation to DNA and protein is usually summarised by the formula (often called the central dogma). [T]he arrows indicate the direction of transfer of the genetic information. The arrow encircling DNA signifies that it is the template for its self-replication; the arrow between DNA and RNA indicates that all cellular RNA molecules are made on DNA templates. Correspondingly, all protein sequences are determined by RNA templates. Most importantly, both these latter arrows are unidirectional, that is, RNA sequences are never copied on protein templates; likewise, RNA never acts as a template for DNA."[8] Figure reproduced, with permission, from Ref. 8.

According to Watson's autobiography, he had already derived this "formula" (Fig. 1) in 1952.[9] In fact, such schemes were commonly entertained during the early 1950s, at least among the biologists interested in protein synthesis. This is attested to, for example, by a figure published by Jean Brachet[10] in 1952, and by the English summary of a paper published in French by Boivin and Vendrely[11] in *Experientia* in 1947. Much more restrictive than Crick's original statement, Watson's formula was immediately confronted with a series of possible exceptions, some of which are mentioned below. Crick, meanwhile, remained rather cautious in his interpretation of the Central Dogma. On several occasions, he felt it necessary to come back to his original idea and explicate what he thought to be its correct interpretation. For example, in 1970, Crick[12] devoted a paper specifically to the Central Dogma, including a diagram reportedly conceived (but not published) in 1958.

In Crick's diagram (Fig. 2), the looped arrows represent the self-template properties of DNA and RNA, whereas the other solid arrows represent the same unidirectional flow of information that was represented in Watson's 1965 formula. Crick also explicitly mentions the possibility of a direct flow of information from DNA to proteins, as well as of a reverse flow of information from RNA to DNA. As in the original 1958 formulation, the only

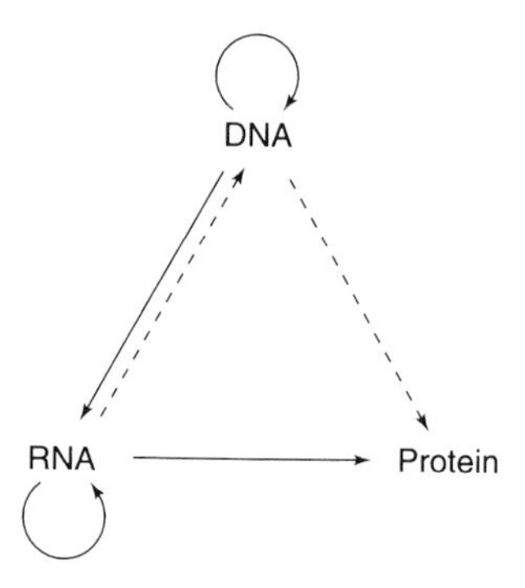

FIGURE 2. The situation envisioned in 1958. Crick described the figure using the following words. "The arrows show the situation as it appeared in 1958. Solid arrows represent probable transfers, dotted arrows possible transfers. The absent arrows represent the impossible transfers postulated by the central dogma. They are the three possible arrows starting from proteins."[12] Figure reproduced, with permission, from Ref. 12.

information flows that are forbidden are those from protein to DNA or RNA, and from protein to protein. Crick wrote the following.

> The discovery of just one type of present-day cell which could carry out any of the three unknown transfers would shake the whole intellectual basis of molecular biology, and it is for this reason that the Central Dogma is as important today as when it was first proposed.[12]

According to Crick, three types of transfer could be distinguished on the basis of the available data: "unlikely" transfers from proteins to nucleic acids; "possible" transfers from RNA to DNA and from DNA to proteins; and finally "known" transfers (e.g. from DNA to RNA to proteins, plus auto-transfers involving DNA or RNA). Strikingly, Crick emphasized auto-transfers involving RNA, over RNA to DNA transfers. Although it was clear by 1958 that some viruses used RNA as a carrier of genetic information, clear indications of the existence of RNA duplication had to wait until the early 1960s.

It is necessary to consider Crick's remarks in the light of his own conception of the respective roles of theory and experiment. In his view, the role of theory is primarily heuristic. In 1958, Crick prefaced the formulation of both the Sequence Hypothesis and the Central Dogma with the following paragraph.

> My own thinking (and that of my colleagues) is based on two general principles, which I shall call the Sequence Hypothesis and the Central Dogma. The direct evidence for both of them is negligible, but I have found them to be of great help in getting to grips with these very complex problems. I present them here in the hope that others can make similar use of them. Their speculative nature is emphasised by their names. It is an instructive exercise to attempt to build a useful theory without using them. One generally ends in the wilderness,[1]

Similar arguments can been found in other papers by Crick—for instance, in his 1963 review of the "coding problem."[13] In an interview with H.F. Judson in 1975, Crick emphasized the original speculative dimension of the Central Dogma.

> My mind was, that a dogma was an idea for which there was no reasonable evidence . . . I just didn't know what dogma meant. And I could just as well have called it the Central Hypothesis . . . Dogma was just a catch phrase. And of course one has paid for this terribly, because people have resented the use of the term dogma, you see, and if it had been Central Hypothesis nobody would have turned a hair.[14]

THE CENTRAL DOGMA'S IMPACT IN THE EARLY 1960s

Even a rudimentary citation analysis of Crick's 1958 paper gives some insight into its influence. In the early 1960s, the paper was quite widely cited but never reached levels as high as, for example, Watson and Crick's 1953 DNA-double-helix paper,[15] or Jacob and Monod's 1961 operon paper[16]—both of which had hundreds of citations per year for more than a decade. However, a closer look at the citations of Crick's paper reveals that several established biologists, and almost all of the rising stars of the new biology, cited the 1958 paper on at least one occasion (e.g. G.W. Beadle, S. Benzer, S. Brenner, J.N. Davidson, M.B. Hoagland, F. Jacob, S.E. Luria, J. Monod, S. Ochoa, G. Pontecorvo, M.F. Singer, F. Vogel, J.D. Watson, C. Yanofsky, M. Ycas and G.E. Zubay, among many others).

The extent of this influence can be attributed to the fact that the Central Dogma, together with the notion of base-pair complementarity in DNA, provided a theoretical framework for molecular biologists interested in protein synthesis and mechanisms of gene expression. Consequently, the main focus of much of that research shifted from proteins to DNA. Even though most cellular functions were still thought to be carried out by proteins (enzymes), the central problem addressed by the new molecular biologists became that of understanding how hereditary information encoded in DNA is translated into specific enzymatic configurations. A thorough answer to this question, however, had to await the characterization of the different molecules involved in the processing of genetic information (messenger and transfer RNAs, and RNA polymerases). In short, the establishment of the molecular mechanisms that account for the information processing envisioned by Crick proved to be much more complex than expected and required the contributions of various independent lines of research.[17]

During the 1960s, the efforts of bacterial geneticists, biophysicists and biochemists led progressively to a comprehensive molecular picture of protein synthesis. The emerging community of molecular biologists succeeded in spreading their concept of gene expression in informational terms—that is, in terms of information transfer, transcription and translation. The Central Dogma was easily stated, explained and taught, and offered a clear landmark by comparison with the complex and sometimes confusing picture of the 1950s. Watson's 1965 textbook[8] and its successive editions were critically influential in the spread of the informational molecular gospel.

CHALLENGES TO THE CENTRAL DOGMA

From the time of its first formulation, however, experimental results threatened the Central Dogma, at least in the case of its most restrictive definitions (e.g. Watson's definition). In the 1950s, several groups working on the tobacco mosaic virus (TMV) questioned DNA's monopoly in carrying hereditary specifications. In 1956, a series of experimental results were converging towards RNA as the genetic material in TMV.[18–20] Thus, it was possible to bypass DNA and, therefore, the first step of Watson's scheme. RNA viruses, as they are now called, provided still more surprises. In 1970, two groups independently reported the characterization of an RNA-dependent DNA polymerase isolated from Rous sarcoma viruses.[21,22] Confirming earlier suspicions, these results clearly demonstrated the possibility of a flow of information from RNA to DNA. These findings prompted Crick[12] to write his 1970 piece for *Nature*, in which he explicitly showed how the new facts fitted into his scheme.

Bolder criticism of the general relevance of the Central Dogma was also made. For example, Barry Commoner[23,24] published several papers directly questioning the exclusive role of nucleic acids in inheritance. Relying on several sets of experimental results that indicated that compounds other than DNA (e.g. DNA polymerase and aminoacyl-tRNA synthetase) affect the result of transcription, Commoner replaced the unidirectional flow of information prescribed by the Central Dogma with a more complex scheme that explicitly includes feedback from proteins to DNA and RNA (Fig. 3). Though widely disseminated, Commoner's point of view found little support, even among those responsible for the observations on which the scheme was based.

During the 1970s, molecular biologists began to work on eukaryotes (rather than almost exclusively on prokaryotic systems). A number of surprises resulted. One of the most striking was the recognition (from 1977 onwards) of the widespread occurrence of non-coding sequences in genes, which led to the distinction between introns and exons being made, and to the discovery of splicing mechanisms.[25,26] Moreover, alternative or differential splicing (i.e. the production of different mature mRNAs from the same

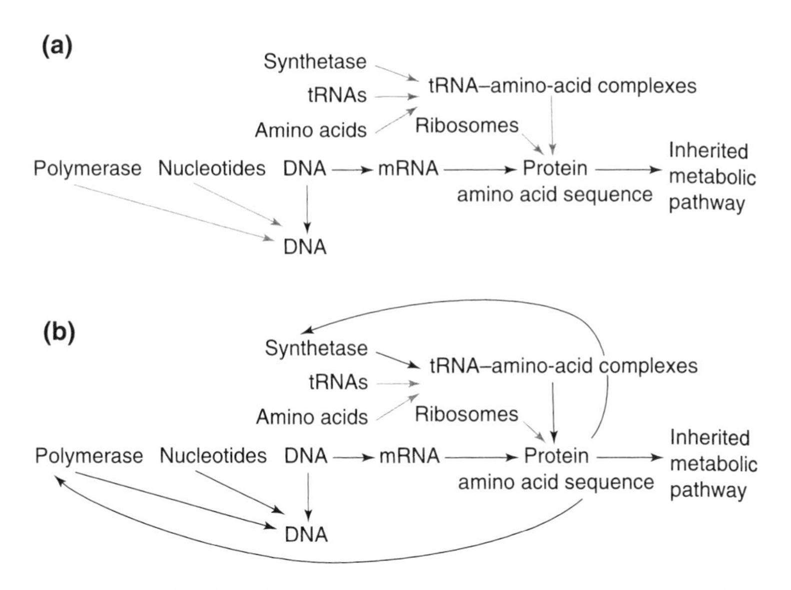

FIGURE 3. (a) Transfer of biochemical specificity in the system proposed by the theory of the DNA code.[23] (b) Modification accounting for the evidence that part of the biochemical specificity of DNA originates in the polymerase enzyme that catalyses DNA synthesis, and possibly in the relative availability of the several requisite nucleotides. Also shown is the effect of the contribution of specificity originating in amino-acid-tRNA synthetase, an enzyme that is essential to protein synthesis. Figure adapted, with permission, from Ref. 23.

transcript) was discovered,[27] and a regulatory role for splicing was suggested by Gilbert[28] in 1978. With respect to the Central Dogma, this discovery has at least two limiting consequences: (1) the specification of protein sequences must thus be considered as discontinuously encoded in DNA; and (2) additional feedback from regulatory proteins is needed to ensure proper RNA splicing. Together, these two facts preclude the possibility of inferring the protein sequence solely from the DNA sequence, an idea that was a key consequence of Crick's Central Dogma and the Sequence Hypothesis.

At the end of the 1980s, several types of mRNA editing were reported, first in the genomes of the mitochondria of some unicellular eukaryotes (paramecia, trypanosomes), but soon also in some metazoan cell types.[29] These editing processes include deamination of cytosine to yield uracil, in mRNA, and the reverse process. Even more unusual behaviours have been observed involving mitochondrial RNAs in which bases can be deleted or inserted. This led to observations interpreted as indicating the formation of proteins for which there are not genes. In an extreme case, in the human parasite *Trypanosoma brucei*, as many as 551 uracils are inserted throughout the transcript that codes for NADH dehydrogenase subunit 7, and 88 are deleted.[30] When looking at the DNA segment that encodes a primary transcript, it is impossible to predict the sequence of amino acid residues in the final protein with complete certainty.

Still another challenge is provided by an infectious neurological disease first found in sheep and called scrapie. As early as the 1960s, the scrapie agent was shown to behave rather strangely. Smaller than all known viruses, it resisted various stringent treatments known to destroy nucleic acids or to inactivate most proteins. As repeated attempts to characterize any nucleic acid associated with this disease failed, several unorthodox explanations were proposed. At first not taken seriously by most researchers, the most controversial of them involved a protein-only agent, christened the prion, which was proposed to either induce the expression of a normally silent host gene or catalyse post-translational changes in a normal cell protein.[31] This hypothesis soon received support when a host gene coding for a protein that matches the sequence of the pathogenic agent was isolated. Today, debates on the very nature and mechanisms of prion transmission and expression still go on, but most researchers now accept the idea of a protein-only infectious agent. Clearly, this idea implies the occurrence of a protein–protein transfer of some kind of specificity.[32]

Finally, perhaps the most controversial challenge to the Central Dogma is provided by recent evidence for epigenetic inheritance in a wide variety of eukaryotic species.[33] Reportedly, part of the inherited specifications—what normally counts as information—are not encoded in the DNA sequence but, rather, are specified by other mechanisms, such as patterns of methylation of the DNA. Methylation modifies the cytosine bases, and excess methylation often leads to transcriptional inactivity. In some cases, these patterns of methylation can be inherited and thus constitute a carrier of transgenerational information other than the sequence itself.

In a recent answer to a question addressing the relevance of these challenges, Crick stated that he still believes in the value of the Central Dogma today (F.H.C. Crick, pers. commun.). However, he also acknowledges the existence of various exceptions, most of which he regards as minor. For him, the most significant exception is RNA editing. Still, according to Crick, simplifications of the Central Dogma in terms such as "DNA makes RNA, and RNA makes protein" were clearly inadequate from the beginning.

DISCUSSION

Forty years after its initial formulation by Crick, the Central Dogma is still generally regarded as one of the keystones of molecular biology. However, popular textbooks and histories frequently overlook the contemporaneous controversies, conveying a success story rather than a story that includes numerous criticisms and corrections. Even though the history of molecular biology has been enriched recently by many new accounts (see Box 1), further examination of the history of the Central Dogma and its subsequent role in the development of molecular biology is necessary.

Beyond the Central Dogma's influence that we have noted here and heuristic value within Crick's own research programme, one question that remains open is that of the Central Dogma's heuristic role in the general development of molecular biology. Many of the discoveries that have followed the formulation of the Central Dogma (e.g. the characterization of transfer RNAs, the uncovering of the mechanisms of gene regulation and even the deciphering of the genetic code) came out of research programmes that had little or no connection to Crick's. Moreover, despite extensive efforts, the initial hope of a resolution of protein three-dimensional (tertiary) structure and function on the basis of the amino acid sequence has so far remained largely unrealized.

Box 1. Selected sources for the history of molecular biology

Books

Brock T.D. 1990. *The Emergence of Bacterial Genetics,* Cold Spring Harbor Laboratory Press.

Debru C. 1987. *Philosophie Moleculaire. Monod, Wyman, Changeux,* Vrin.

Jablonka E. and Lamb M.J. 1995. *Epigenetic Inheritance and Evolution: The Lamarckian Dimension,* Oxford University Press.

Kay L.E. *Who Wrote the Book of* Life? *A History of the Genetic Code,* Stanford University Press (in press).

Keller E.F. 1995. *Refiguring Life: Metaphors of Twentieth Century Biology,* Columbia University Press.

Judson H.F. 1979. *The Eighth Day of Creation,* Simon & Schuster (Expanded edition: Cold Spring Laboratory Press, 1996).

Latour B. and Woolgar S. 1979. *Laboratory Life,* SAGE Publications.

Morange M. 1994. *Histoire de la Biologie Moleculaire,* La Decouverte. (English edition: *The History of Molecular Biology,* Harvard University Press, 1998.)

Olby R. 1974. *The Path to the Double Helix,* Macmillan Press. (New edition: Dover Publications, 1994.)

Portugal F.H. and Cohen J.S. 1977. *A Century of DNA,* MIT Press.

Rheinberger H.-J. 1997. *Toward a History of Epistemic Things. Synthesizing Proteins in the Test Tube,* Stanford University Press.

Sapp J. 1987. *Beyond the Gene. Cytoplasmic Inheritence and the Struggle for Authority in Genetics,* Oxford University Press.

Stent G. 1971. *Molecular Genetics. An Introductory Narrative,* Freeman.

Collections of essays

Bunan R.M. and Thieffry D., eds. 1997. *Research Programs of the "Rouge-Cloitre group."* Special issue of *Hist. Philos. Life Sci.* **19:** 1–144.

Cairns J., Stent G.S. and Watson J.D., eds. 1966. *Phage and the Origins of Molecular Genetics,* Cold Spring Harbor Laboratory Press.

De Chadarevian S. and Gaudilliere J.P., eds. 1996. *The Tools of the Discipline: Biochemists and Molecular Biologists.* Special issue of *J. Hist. Biol.* **29:** 327–462.

Lwoff A. and Ullman A., eds. 1979. *Origins of Molecular Biology. A Tribute to Jacques Monod,* Academic Press.

Monod J. and Borek E., eds. 1971. *Of Microbes and Life,* Cornell University Press.

Rich A. and Davidson N. 1968. *Structural Chemistry and Molecular Biology.* Freeman.

Sarkar S., ed. 1996. *The Philosophy and History of Molecular Biology: New Perspectives.* Kluwer Academic Publishers.

Soederqvist T., ed. 1997. *The Historiography of Contemporary Science and Technology.* Harwood Academic Publishers.

Other historical papers

Abir-Am P. 1992. *Osiris* **7:** 164–191.

Bunan R.M. 1993. *J. Hist. Biol.* **26:** 387–407.

Creager A.N.H. and Gaudilliere J.-P. 1996. *Hist. Stud. Phys. Biol. Sci.* **27:** 1–89.

Darden L. 1995. In *Concepts, Theories, and Rationality in the Biological Sciences* (ed. G. Wolters and J.G. Glennox), pp. 137–169, University of Pittsburgh Press.

Gaudilliere J.-P. 1992. *Hist. Phil. Life Sci.* **14:** 23–71.

Grmek M.D. and Fantini B. 1982. *Rev. Hist. Sci.* **35:** 193–215.

Keyes M.E. *Stud. Hist. Philos. Biol. Biomed. Sci.* (in press).

Thieffry D. and Bunan R.M. 1996. *Trends Biochem. Sci.* **21:** 114–117.

Witkowski J.A. 1988. *Trends Biochem. Sci.* **13:** 110–113.

Recollections by scientists

Baltimore D. 1995. *FASEB J.* **9:** 1660–1663.

Crick F.H.C. 1988. *What Mad Pursuit,* Basic Books.

Drake J.W. and Crow J.F. 1996. *Genetics* **144:** 1–6.

Hoagland M.B. 1990. *Towards the Habit of Truth: A Life in Science,* W.W. Norton.

Jacob F. 1987. *La Statue Interieure,* Editions Odile Jacob. (English edition: *The Statue Within: >An Autobiography,* Basic Books.)

Kornberg A. 1989. *For the Love of Enzymes,* Harvard University Press.

Luna S. 1984. *A Slot Machine, A Broken Test Tube: An Autobiography,* Harper and Row.

Nirenberg M.W. 1969. *Nobel Lecture* (Vol. I, Pt 21), Nobel Foundation.

Nomura M. 1990. *Trends Biochem. Sci.* **15:** 244–247.

Spiegelman S. 1978. *Tanpakushitsu Kakusan Koso* **23:** 7–43.

Watson J.D. 1968. *The Double Helix,* Weidenfeld and Nicholson.

Zamecnik P.C. 1979. *Ann. New York Acad. Sci.* **325:** 269–301.

A closely related question is that of the definition of and the relevance of the concept of information in molecular biology. This question[34–36] goes well beyond the scope of this short historical review. However, in the light of present biological knowledge, we would like to mention here the necessity of distinguishing between at least three types of molecular biological information (or specificity): the information associated with the sequence itself; the specificity of the spatial conformation of macromolecules; and, finally, the information that can be associated with regulatory mechanisms (including epigenetic specifications such as methylation). Of the three, only the first was addressed explicitly by Crick in 1958. The conformational specificity (of proteins) was supposed to be derived directly from the sequence but was ultimately shown to depend on various other factors, including other proteins (e.g. chaperonins). The various mechanisms of gene regulation already characterized add a third level of complexity that cannot be predicted solely on the basis of sequence or conformational information.

Curiously enough, the elucidation of the mechanisms of gene regulation in the late 1950s and early 1960s was rarely used in the criticisms of the Central Dogma or of the corresponding narrow conception of biological information. However, today, because regulatory mechanisms are known to involve almost any kind of biological molecule, either as targets or as effectors of regulation, a cell or an organism is increasingly considered to be a complex regulatory network. From this perspective, one might consider the Central Dogma effectively to be neglecting the feedback mechanisms present in the cell and thus lead to a simplified, but also much more accessible, scheme for the mechanisms controlling gene expression and protein synthesis.

ACKNOWLEDGEMENTS

We thank Richard Burian, Lily Kay, Evelyn Fox Keller, Michel Morange, Robert Olby, Hans-Jörg Rheinberger, Judy Johns Schloegel and Jan Witkowski for their critical reading of this paper and their many useful suggestions.

REFERENCES

1. Crick F.H.C. 1958. *Symp. Soc. Exp. Biol.* **XII:** 138–163.
2. Ephrussi B., Leopold U., Watson J.D., and Weigle J.J. 1953. *Nature* **171:** 701.
3. Mazia D. 1956. In *Enzymes: Units of Biological Structure and Function* (ed. E.H. Gaebler), pp. 261–278, Academic Press.
4. Spiegelman S. 1956. In *Enzymes: Units of Biological Structure and Function* (ed. E.H. Gaebler), pp. 67–89, Academic Press.
5. Lederberg J. 1956. In *Enzymes: Units of Biological Structure and Function* (ed. E.H. Gaebler), pp. 161–169, Academic Press.
6. Pollock M.R. 1953. In *Adaptation in Micro-organisms. 3rd Symposium of the Society for General Microbiology,* pp. 150–183, Cambridge University Press.
7. Dounce A.L. 1953. *Nature* **172:** 541–542.
8. Watson J.D. 1965. *Molecular Biology of the Gene,* W.A. Benjamin.
9. Olby R. 1974. *The Path to the Double Helix,* Macmillan Press.
10. Brachet J. 1952. *Le rôle des acides nucléiques dans la vie de la cellule et de l'embryon,* Masson & Cie.
11. Boivin A. and Vendrely R. 1947. *Experientia* **3:** 32–34.
12. Crick F.H.C. 1970. *Nature* **227:** 561–563.

13. Crick F.H.C. 1963. *Prog. Nucleic Acids Res. Mol. Biol.* **1:** 163–217.
14. Judson H.F. 1979. *The Eighth Day of Creation*, Simon & Schuster. (Expanded edition: Cold Spring Laboratory Press, 1996.)
15. Watson J.D. and Crick F.H.C. 1953. *Nature* **171:** 737–738.
16. Jacob F. and Monod F. 1961. *J. Mol. Biol.* **3:** 316–356.
17. Rheinberger H.-J. 1997. *Toward a History of Epistemic Things. Synthesizing Proteins in the Test Tube*, Stanford University Press.
18. Fraenkel-Conrat H. 1956. *J. Am. Chem. Soc.* **78:** 882–883.
19. Gierer A. and Schramm G. 1956. *Nature* **177:** 702–703.
20. Jeener R. 1956. *Adv. Enzymol.* **17:** 477–498.
21. Baltimore D. 1970. *Nature* **226:** 1209–1211.
22. Temin H.M. and Mizutani S. 1970. *Nature* **226:** 1211–1213.
23. Commoner B. 1964. *Nature* **203:** 486–491.
24. Commoner B. 1968. *Nature* **220:** 334–340.
25. Witkowski J.A. 1988. *Trends Biochem. Sci.* **13:** 110–113.
26. Morange M. 1994. *Histoire de la Biologie Moléculaire*, La Découverte.
27. Smith C.W., Patton J.G., and Nadal-Ginard B. 1989. *Annu. Rev. Genet.* **23:** 527–577.
28. Gilbert W. 1978. *Nature* **271:** 501.
29. Cattaneo R. 1991. *Annu. Rev. Genet.* **25:** 71–88.
30. Koslowski D.J. et al. 1990. *Cell* **62:** 901–911.
31. Prusiner S.B. 1982. *Science* **216:** 136–144.
32. Keyes M.E. *Stud. Hist. Philos. Biol. Biomed. Sci.* (in press).
33. Jablonka E. and Lamb M.J. 1995. *Epigenetic Inheritance and Evolution: The Lamarckian Dimension*, Oxford University Press.
34. Keller E.F. 1995. *Refiguring Life: Metaphors of Twentieth Century Biology*, Columbia University Press.
35. Sarkar S. 1996. In *The Philosophy and History of Molecular Biology: New Perspectives*, pp. 187–231, Kluwer Academic Publishers.
36. Kay L.E. *Who Wrote the Book of Life? A History of the Genetic Code*, Stanford University Press (in press).

Protein Sequencing and the Making of Molecular Genetics

SORAYA DE CHADAREVIAN

University of Cambridge,
Cambridge, UK

In the early history of molecular genetics, proteins—not nucleic acids—were at the centre of research. This was the case long after Watson and Crick had proposed their double-helical model of DNA. The reason was simple: proteins could be sequenced; nucleic acids could not.

Here, I present a brief history of protein sequencing, charting its development from a tool in protein chemistry to its uses in the study of gene function and the way it informed initial attempts at nucleic acid sequencing. I focus on Sanger's sequencing work in the Biochemistry Department in Cambridge (UK) and the use of his techniques by molecular geneticists in the nearby Physics Department. Sanger's work won him a double Nobel Prize, and his close interaction with molecular biologists culminated in common plans for a new Laboratory of Molecular Biology on the outskirts of Cambridge.

A NEW TOOL

Protein sequencing started as part of studies of protein structure and function. From the late 19th century onwards, proteins were associated with the basic functions of life, including heredity, and were the object of intense study and debate. Knowledge of protein structure was seen as the key to studies of protein function and as a step towards the synthetic production of proteins. Various theories regarding protein structure were proposed, and dis-

Reprinted from *Trends in Biochemical Sciences* May 1999, 24(5): 203–206.

carded, on the basis of sedimentation studies, amino acid analysis and X-ray studies. They invariably assumed that proteins exhibited a high degree of regularity.

Sanger's sequencing work developed from research into a new method for end-group determination in proteins, which he undertook as a postdoctoral student under Charles Chibnall in the Biochemistry Department in Cambridge during the war. End-group determination was an important tool for the estimation of the number and length of polypeptide chains in proteins, yielding basic information on protein structure. It could be used to identify proteins as well as to test their purity. Many different methods for end-group determination were described in the literature—a review article published in 1945 listed more than 20[1]—but none yielded reliable results.

Acting on Chibnall's suggestion, Sanger tried fluorodinitrobenzene as a reagent. Organic chemists had tended to steer clear of fluoroderivates because of the latter's toxicity, but during the war the chemicals were synthesized for research into chemical warfare. Sanger found that fluorodinitrobenzene reacted under much milder conditions than did the generally used chlorine compounds. Furthermore, the dinitrophenylamino acids were stable to the acid hydrolysis used to break down proteins and were bright yellow in solution. This made them amenable to the new method of partition chromatography.

Using his new technique, Sanger established that insulin consisted of two chains—not 18 chains, as Chibnall had postulated on the basis of its high free-amino-group content. Exploiting the same technique, Sanger identified short, 4–5-residue, sequences at the N-termini of the two chains of the molecule.[2] Extending the approach to peptides derived by partial acid hydrolysis, and later by enzymatic hydrolysis, Sanger and his co-workers,[3,4] in several years of painstaking work, were able to establish the complete sequence of the two insulin chains.

Sanger has suggested that end-group determination had already marked a change in the course of protein chemistry from an interest in amino acid analysis to one in the arrangement of amino acids in chains. Moving from there to sequencing did not require a great intellectual leap.[5] The decisive breakthrough, according to Sanger, was the development of new fractionation techniques by Richard Synge and Archer Martin in the context of research into the composition and characteristics of wool—research that was financed by the International Wool Secretariat.[6] Sanger's thesis finds confirmation in the fact that Synge and Martin themselves successfully applied their new fractionation techniques to determination of the structure of the pentapeptide GramicidinS. Their results were published before Sanger presented his first bit of sequence.[7]

Other researchers were active in the field. Pehr Edman[8] at the University of Lund in Sweden developed an elegant procedure that was based on the use of phenylisothiocyanate as a reagent and that allowed stepwise degradation of the protein. With the development of more reliable fraction collectors and of sensitive methods for detection of the colourless reaction products, Edman's method completely superseded Sanger's sequencing method. In the late 1950s, William Stein and Stanford Moore at the Rockefeller Institute in New York devised an automatic amino acid analyser that yielded quantitative results and facilitated the analytical work. Edman's procedure, in conjunction with the automatic analyser, allowed researchers to tackle larger proteins.

Sanger was interested in sequencing because he felt that it would provide insight into the mechanism of action of insulin. The expectation was that, once the mechanism of action of one protein was known, it would give clues to the functioning of protein hormones and enzymes more generally. When the full sequence of insulin, including the position of the three disulphide bridges, did not give any clue to the protein's function, Sanger explored new ways of using sequencing to achieve his aim. One avenue he pursued was to identify the "active centre" of insulin by determining and comparing the sequences of homologous proteins from different species. Another approach he followed was to label the active centre and to determine the sequence around it. In the course of their work, Sanger and his collaborators developed sensitive autoradiographic methods and "fingerprinting" techniques that allowed them to deduce the sequences of peptides without carrying out a complete amino acid analysis. Insulin's mechanism of action proved resilient to all these attacks, but the new methods became important analytical tools for structure determination (Fig. 1).

In a review article published in the late 1960s, Brian Hartley, who worked in Sanger's group, documented the exponential growth in both the speed and volume of sequencing work.[9] Despite the problems encountered with insulin, researchers expected that protein sequencing would yield insight into the structure and function of proteins. In combination with X-ray analysis, protein sequencing led to the first atomic model of a globular protein (myoglobin) and, some years later, to the first atomic model of an enzyme (lysozyme)[10,11]; a reaction mechanism for lysozyme was soon proposed. Sequencing also gave rise to evolutionary studies of proteins as a

FIGURE 1. Fred Sanger, holding an autoradiogramme, photographed in his laboratory in the Biochemistry Department in Cambridge in the late 1950s. (Photograph courtesy of F. Sanger.)

completely new area of research. Finally, protein sequencing was keenly seized upon by researchers concerned with the molecular mechanisms of gene function.

THE SEQUENCE HYPOTHESIS

Sanger's first sequencing results suggested to Crick—even before 1953—that genes determined the amino acid sequence of proteins (Ref. 12, pp. 34–36). Later, Crick expanded this insight in the sequence hypothesis, which stated that "the specificity of a piece of nucleic acid is expressed solely by the sequence of its bases, and that this sequence is a (simple) code for the amino acid sequence of a particular protein."[13]

The sequence hypothesis was a decisive step in early speculation on the genetic code. It boldly assumed that the amino acid sequence determined the folding of a protein. A few years before, Linus Pauling[14] had postulated the existence of a gene "responsible for the folding of polypeptide chains" to explain the different electrochemical charges of sickle-cell and normal haemoglobin, which appeared to have the same amino acid composition. In 1957, Crick still considered some possible exceptions to the rule, especially γ-globulins and adaptive enzymes. Heuristic reasons, however, made the sequence hypothesis attractive. In his celebrated lecture on protein synthesis, which he delivered before the Society for Experimental Biology, Crick conceded frankly: "Our basic handicap at the moment is that we have no easy and precise technique with which to study how proteins are folded, whereas we can at least make some experimental approach to amino acid sequences. For this reason, if for no other, I shall ignore folding in what follows and concentrate on the determination of sequences."[13] Thus, Sanger's analysis not only allowed the formulation of the hypothesis but also provided the experimental tool to test it.

It is worth noticing that, in the same lecture in which Crick for the first time explicitly stated the "central dogma" of molecular biology and defended an "informational" versus a biochemical view of the problem of protein synthesis, he stressed the central and unique importance of proteins in biology. Crick expected that, in contrast to the multiple and complex functions of proteins, nucleic acids acted in a "uniform and rather simple" way.[13]

Before testing the sequence hypothesis, it was necessary to show that an inherited defect was in fact laid down in the amino acid sequence of a protein. As is well known, Vernon Ingram's experiments on sickle-cell haemoglobin, which were performed in the Cavendish laboratory, provided such proof. Refining Sanger's fingerprinting techniques, Ingram succeeded in tracking down the difference beween normal and sickle-cell haemoglobin to a single amino acid residue (Fig. 2).[15]

Building on this first success, Crick, Brenner and Ingram, together with Seymour Benzer and George Streisinger, who gathered in Cambridge in 1957, tried to show that the order of mutations in a gene lined up with the order of changes in the amino acid sequence of the corresponding protein. The original plan was to use Benzer's finely mapped mutants of the rII region of bacteriophage T4 as a test case. When Benzer failed to isolate the corresponding protein, the group resorted to Streisinger's bacteriophage-T2 mutants, in which the tips of the tail fibers were affected. The key technique was again fingerprinting, which the group combined with the radioactive marking technique Sanger had pioneered.

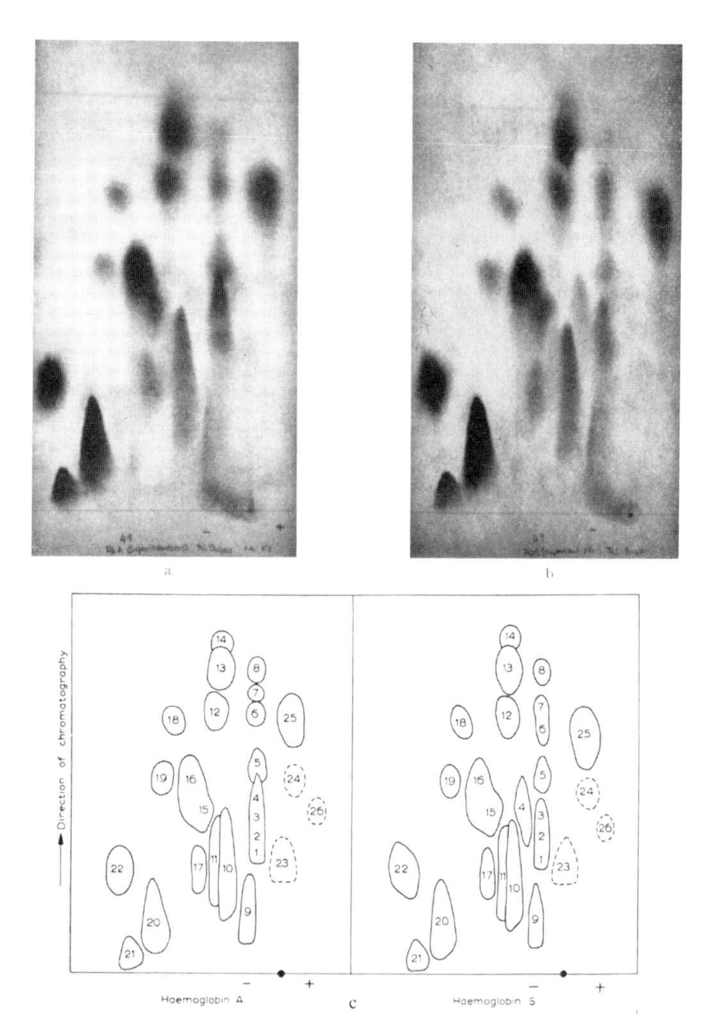

FIGURE 2. Fingerprints of normal and sickle-cell haemoglobin. Note the difference in peptide 4. (Figure reproduced, with permission, from Ref. 15.)

Radiographic techniques were much more sensitive than other chromatographic techniques. Sanger himself was experimenting with slices of oviducts, which he incubated with radioactive phosphate to get labelled ovalbumin. This was quite a lengthy and laborious procedure. As early as October 1956, Crick wrote to Brenner, who was still in Johannesburg: "I stressed to Fred [Sanger] how extremely favorable the phage system might be for this method . . . He seemed very interested."[16]

The experiments on the tail-fiber mutants did not yield conclusive results and were abandoned. Apparently, the group had not succeeded in isolating the right protein. However, Brenner continued to use the same techniques in work on the amber mutants of the phage T4. These mutants, which grew only on the *Escherichia coli* B strain, produced only fragments of the head protein—the protein that the affected gene encoded. By examining the fingerprints of the different mutants, Brenner and his collaborators were able to establish that the length of a fragment corresponded to the position of the mutation on the genetic map, thus proving collinearity. A few months earlier, Charles Yanofsky and his collaborators at Stanford University, using similar techniques, had proved the same point in studies on tryptophane synthetase mutants of *E. coli.*[17,18]

Besides proving collinearity, sequencing data were also used to establish some general features of the genetic code. On the basis of the few sequences then available, Crick disproved all possible versions of the first genetic code proposed by George Gamow (Ref. 12, p. 94). From published data on protein sequences and neighbor analysis, Brenner later deduced that an overlapping code was impossible.[19] A further handle on the problem of the genetic code came from analysis of the effects of chemical mutagens in combination with genetic and protein sequence analysis.[20,21]

The code itself was established by entirely different *in vitro* translation techniques, but the early experiments in which protein sequencing played a central role were nonetheless crucial for the formulation of the problem. Protein sequencing also remained an important tool for checking the validity of the code. Amino acid substitutions in hemoglobin variants established by fingerprinting, for instance, proved that the genetic code that had been established for bacteria and viruses was also valid for humans.[22]

Interestingly, Crick and Brenner not only used Sanger's sequencing work as a conceptual and practical tool for their own work in the newly defined field of molecular genetics, but also actively tried to interest Sanger in their work. They first approached him in the early 1950s, trying to convince him to move from the Biochemistry Department to the Cavendish Laboratory. Nothing came of this plan at the time. However, in 1957, the Cavendish group and Sanger joined forces, and together negotiated the creation of a new Laboratory of Molecular Biology.[23] To my knowledge, it was the first institution to carry that name. The combination of (both two- and three-dimensional) structural and genetic approaches became central to the definition of molecular biology at Cambridge. The creation of the new laboratory also had repercussions on the research agendas of those involved in the new venture.

FROM PROTEIN TO NUCLEIC ACID SEQUENCING

When trying to account for his "conversion" from protein to nucleic acid sequencing, Sanger referred to "the atmosphere" in the Laboratory of Molecular Biology and to the influence of his new colleagues. "With people like Francis Crick around," he reckoned, "it was difficult to ignore nucleic acids or to fail to realize the importance of sequencing them."[23] Originally, the main objective of nucleic acid sequencing was to try to "break the genetic code."[24] However, nucleic acid sequencing got going seriously only after the code was broken.

Initially, nucleic acid sequencing seemed an even-more daunting undertaking than protein sequencing had been. This was due to the lack of pure small substrates and to the composition of nucleic acids. Because nucleic acids possessed only four monomers, researchers expected the interpretation of results to be much more difficult. This expectation was based on the existing approaches to studying protein sequences, which required analysis of degradation products and the subsequent rearrangement of the pieces. New developments in sequencing techniques reversed the picture.

The first nucleic acid to be sequenced was alanine tRNA, the first small RNA to be isolated. The methods used were similar to those established for protein sequencing: enzymatic degradation followed by fractionation, analysis and interpretation of the degradation products.[25] These methods were too laborious to be applied to larger RNA or DNA molecules; however, the procedures for more rapid and reliable sequencing subsequently developed by Sanger and others continued to rely on methods pioneered with proteins or on information derived from protein sequencing. This is especially true of autoradiography and labelling techniques, which allowed one to "read off" the sequence from the autoradiogramme directly and therefore did not require complex interpretative procedures. This latter method became much more powerful in nucleic acid sequencing than it ever had been with proteins. A key development in nucleic acid sequencing was the introduction of copying techniques (instead of sequencing by degradation). But, again, the first primers employed to get the polymerases started were synthesized by using information provided by amino acid sequencing. Protein sequencing also served as an important check for the still unreliable DNA-sequencing data[26] (see also recent articles on DNA sequencing[27,28]).

CONCLUSIONS

In the debate on the role of biochemists in the history of molecular biology, which has been conducted heatedly since the late 1960s,[29–31] Sanger represents an interesting case. Despite joining the Laboratory of Molecular Biology in Cambridge, he never gave up his identity as a biochemist—or, more precisely, he never saw the necessity to draw a distinction between the two fields. Interestingly, too, protein sequencing is never mentioned among the techniques that biochemists introduced into molecular biology. My intention here, however, is not to fuel an old debate. The aim of my brief historical excursion is rather to show the important role of protein sequencing in the early history of molecular genetics. Today, ever faster and cheaper nucleic acid sequencing methods have overshadowed more-cumbersome methods of protein sequencing. However, this is only a fairly recent development. Long before nucleic acid sequencing techniques were at all conceivable, protein sequencing was at the forefront of research and offered a powerful tool for formulating and testing hypotheses about the functions of genes.

ACKNOWLEDGEMENT

I thank Sydney Brenner, Francis Crick, John Kendrew and Fred Sanger for extensive discussions, and Denis Thieffry for constructive comments on an earlier version of this paper.

REFERENCES

1. Fox S.W. 1945. *Adv. Protein Chem.* **2:** 155–177.
2. Sanger F. 1949. *Biochem. J.* **45:** 563–574.
3. Sanger F. and Tuppy H. 1951. *Biochem. J.* **49:** 481–490.
4. Sanger F. and Thompson E.O.P. 1953. *Biochem. J.* **53:** 353–374.
5. Sanger F. 1985. *Curr. Contents* **28:** 23.
6. Dowling L.M. and Sparrow L.G. 1991. *Trends Biochem. Sci.* **16:** 115–119.
7. Consden R., Gordon A.H., Martin A.J.P., and Synge R.L.M. 1947. *Biochem. J.* **41:** 596–602.

8. Edman P. 1950. *Acta Chem. Scand.,* **4:** 283–293.
9. Hartley B.S. 1970. In *British Biochemistry Past and Present. Biochemical Society Symposium No. 30* (ed. T.W. Goodwin), pp. 29–41, Academic Press.
10. Kendrew J.C. et al. 1960. *Nature* **185:** 422–427.
11. Blake C.C.F. et al. 1965. *Nature* **206:** 757–761.
12. Crick F. 1990. In *What Mad Pursuit: A Personal View of Scientific Discovery,* Penguin.
13. Crick F. 1958. In *The Biological Replication of Macromolecules. Symposia of the Society of Experimental Biology XII,* pp. 138–163, Cambridge University Press.
14. Pauling L. 1952. *Proc. Am. Philos. Soc.* **96:** 556–565
15. Ingram V.M. 1958. *Biochim. Biophys. Acta* **28:** 539–545.
16. Judson H. 1979. *Eighth Day of Creation. The Makers of the Revolution in Biology,* p. 331, Jonathan Cape.
17. Sarabhai A.S., Stretton A.O.W., Brenner S., and Bolle A. 1964. *Nature* **201:** 13–17.
18. Yanofsky C. et al. 1964. *Proc. Natl. Acad. Sci. U.S.A.* **51:** 266–272.
19. Brenner S. 1957. *Proc. Natl. Acad. Sci. U.S.A.* **43:** 687–694.
20. Crick F.H.C., Barnett L., Brenner S., and Watts-Tobin R. 1961. *Nature* **192:** 1227–1232.
21. Kay L. *Who Wrote the Book of Life? A History of the Genetic Code,* Stanford University Press (in press).
22. Beale D. and Lehmann H. 1965. *Nature* **207:** 259–262.
23. de Chadarevian S. 1996. *J. Hist. Biol.* **29:** 361–386.
24. Sanger F. 1988. *Annu. Rev. Biochem.* **57:** 1–28.
25. Holley R.W. et al. 1965. *Science* **147:** 1462–1465.
26. Sanger F. 1988. *Annu. Rev. Biochem.* **57:** 1–28.
27. Wu R. 1994. *Trends Biochem. Sci.* **19:** 429–433.
28. Sutcliffe J.G. 1995. *Trends Biochem. Sci.* **20:** 87–90.
29. Cohen S.C. 1984. *Trends Biochem. Sci.* **9:** 334–336.
30. Abir Am P.G. 1992. *Osiris* **7:** 210–237.
31. The Tools of the Discipline: Biochemists and Molecular Biologists (1996) [special issue] *J. Hist. Biol.* **29:** 327–462.

pBR322 and the Advent of Rapid DNA Sequencing

J. GREGOR SUTCLIFFE

The Scripps Research Institute,
La Jolla, California, USA

In these days of the Human Genome Project, it seems surprising in retrospect that after the double-helical nature of DNA and its protein-encoding mechanism had been revealed, there was no general perception of a need for DNA sequencing technology. Rather, a proteinocentric view emerged in the late 1960s, through the quests to understand the largely enzymological properties of replication, recombination, transcription and translation. Furthermore, in contrast to proteins and RNA, the featureless properties of DNA and the absence of enzymes that cleaved it with sequence specificity made it technologically recalcitrant as a means to obtain sequence information. The protein and RNA sequencing technologies of the day were laborious and produced relatively short fragments of sequence, so few thought it would be worth the effort to develop DNA sequencing technologies. At the time, genes were viewed as segregated blocks collinear with their RNA and protein counterparts; overlapping genes, introns, signal peptides and such had yet to be recognized. Nor had the possibility been widely considered that previously unrecognized genes might be discovered for which neither proteins nor RNAs were known.

THE GILBERT LABORATORY

I was fortunate to conduct my graduate research in the Harvard laboratory jointly led by Jim Watson, Wally Gilbert (my thesis advisor), Klaus Weber

Reprinted from *Trends in Biochemical Sciences* February 1995, 20(2): 87–90.

and David Dressler. Most of the enzymological issues mentioned above were being addressed, including the nature of interactions between proteins and the lactose operator/promoter. This was an active period in which the progress of members of the lab (a collection of talented, competitive and ambitious individuals) and their competitors was hilariously chronicled in the *Biolabs Midnight Hustler*, a non-politically-correct, equal-opportunity-offender parody that was published sporadically to blow off steam.

Many of the earliest DNA sequencing methods (recently reviewed by Ray Wu, *TIBS* 19, 429–433), used enzymatically derived complementary copies of DNA molecules, either transcripts or extended oligonucleotides, and were developed in the Gilbert laboratory and that of Fred Sanger.[1–3] The highly complex "plus–minus" method was the first rapid DNA sequencing method, and was the mainstay for determining the ϕX174 genome sequence, a tour de force that relied heavily upon information obtained by determining the sequences of ϕX174 proteins and RNA transcripts for its assembly. Despite several corroborating sets of data, the accuracy of the ϕX174 genome sequence was in doubt at its time of publication, with the authors estimating that their sequence had only a 99.3% confidence level, or approximately 30 errors in the 5375-nucleotide sequence they reported. Sanger and colleagues declared: "As with other methods of sequencing nucleic acids, the plus–minus technique used by itself cannot be regarded as a completely reliable system and occasionally errors occur. Such errors and uncertainties can only be eliminated by more laborious experiments and, although much of the [ϕX174] sequence has been so confirmed, it would probably be a long time before the complete sequence could be established. We are not certain that there is any scientific justification for establishing every detail . . .."[3] This comment summarized the state of the art in 1977. DNA sequence data could not yet stand on their own. In fact, the reality of highly accurate DNA sequences had yet to be achieved, and consequently DNA sequences were not conceived of as primary sources of protein structural information.

THE MAXAM-GILBERT METHOD

Meanwhile, Gilbert, in collaboration with Andrei Mirzabekov and Allan Maxam, had been investigating the direct use of fragments of DNA produced by restriction endonuclease cleavage and end-labeling for probing the interactions between the *lac* operator and the *lac* repressor. These studies[4] were the forerunners of what is now known as DNA footprinting. Maxam and Gilbert extended the technique of using partial chemical degradation with a panel of reagents with reactivities that were specific to subsets of nucleotides and more-or-less insensitive to the sequence context of those nucleotides.[5] Once the Maxam-Gilbert technology had been introduced, it was applied in laboratories around the world. Two important early applications contemporaneous with my studies, which are described below, serve to demonstrate the zeitgeist. Phil Farabaugh determined the sequence of the gene encoding the *lac* repressor protein by fitting his DNA sequence data to the known protein sequence and the hypothetical structure Jeffrey Miller had deduced from mutagenesis studies.[6] To the surprise of all, he found that a peptide fragment had gone undetected when the protein was sequenced, hence his study corrected the record. Sauer and Andregg conducted sequencing experiments with the purified phage repressor protein and its gene in parallel, and rationalized the two sets of data against one another.[7]

During this period, several members of Gilbert's lab were using the new method to characterize the sequences of various phage and bacterial promoters. I had been vocally skeptical of these studies because some of the promoters had been characterized by neither genetic nor biochemical experiments, but were merely sites to which 5′-ends of RNAs had been mapped by extremely difficult, and hence slightly suspect, experiments.

At the same time, Jeremy Knowles, a professor in the Chemistry Department who regularly lunched with Gilbert, was wondering whether the complete sequence of a protein could be obtained entirely from the structural analysis of its gene. Knowles worked with β-lactamase, the enzyme that renders bacteria resistant to penicillins and cephalosporins. He knew that Richard Ambler and his colleagues in Edinburgh had been working for a considerable time on the amino acid sequence of the β-lactamase from *Escherichia coli* and obtained their partial sequence data, which he kept locked in his desk.

SEQUENCING THE β-LACTAMASE GENE

Gilbert approached me in February 1977 with the suggestion that I undertake determination of the sequence of a β-lactamase gene, and mentioned that the cloning vector pBR322, recently constructed in Herb Boyer's laboratory by Francisco Bolivar and Ray Rodriguez,[8] carried ampicillin as well as tetracycline resistance. I searched the literature and traced the origin of the ampicillin resistance allele carried by pBR322 and that encoding the enzyme under study in the Ambler lab. Both were thought to encode identical 27 kDa proteins, classified serologically as RTEM1. I had independently learned of pBR322 from a postdoctoral fellow from Boyer's lab who visited our lab, and I had been impressed with its versatility, which suggested that it had great potential as a cloning vector. Thus, I accepted this challenging research project, both because it would serve as a rigorous test of the Maxam–Gilbert methodology and because the sequence of the β-lactamase itself would represent a worthwhile contribution to the field. Also, information for a portion of pBR322's estimated 4000 base pairs would enhance the utility of the vector.

During the first month of effort on the project, I learned the protocols for the Maxam–Gilbert methodology. In addition to library study, I interviewed all of the members of the group who had used the method, especially Allan Maxam, to pick up any tips they might have. There were as many variations as individuals that I consulted, and some of these were based upon rather mystical notions. I verified the unpublished observation of Bolivar that the β-lactamase gene spanned the single site in pBR322 for the restriction endonuclease *PstI* by digesting *PstI*-cleaved plasmid DNA with the single-strand-specific nuclease S1 and recircularizing the plasmid with DNA ligase. After transformation into *E. coli,* the majority of tetracycline-resistant isolates had lost the ability to confer ampicillin resistance, suggesting that the *PstI* site was within the amp^r gene.

I initiated the project knowing that I needed to solve a sequence containing a triplet open reading frame (ORF) of at least 700 nucleotides (large enough to encode a 27 kDa protein) overlapping the *PstI* site of pBR322. The method uses restriction endonucleases to generate fragments for sequence analysis, but only six useful enzymes were available in the lab at the time. These we isolated ourselves from bacterial cultures and used in barter with other labs. During the course of the project, I characterized the cleavage properties of two new restriction enzymes.[9,10] We also prepared our γ-labeled nucleotides weekly for kinase-labeling the fragments, since these were not commercially available —dreaded 25–50 mCi adventures (the responsibility rotated) shielded by 1/4 inch of plexiglass.

I began by mapping restriction cleavage fragments in the vicinity of the *PstI* site and isolating them for sequence analysis. In order to recognize the putative β-lactamase ORF, there could be no errors in the sequence that altered the phase of the frame which, of course, would have predicted either an imaginary protein sequence or possibly none at all. The endeavor was

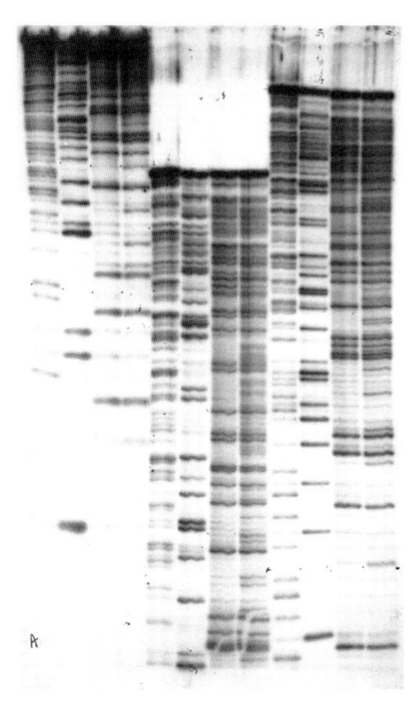

FIGURE 1. A typical autoradiogram from the era (lanes are 1.6 cm wide). These gels used wicks to transfer current and had evolved from those used to display restriction endonuclease digestion products. It took an embarrassingly long time before sleeker, wickless, thin gels with narrow lanes accommodating multiple fragments emerged, despite the fact that protein gel technology, usually performed in the same laboratories, utilized these tools.

daunting in that, until that time, no-one in the laboratory had attempted to apply the method to determine simultaneously the sequence of more than a single fragment; thus I instituted procedures for preparing and running ten or more gels per day. By today's standards, the gel technology was extremely primitive, with only single fragments applied to individual gels with 1.6 cm wide lanes (Fig. 1). Pipets for loading gels were pulled by hand.

I soon found that artifacts such as band compressions, caused by sequence-context-dependent intramolecular interactions during electrophoresis, were frequently encountered, so I devised procedures for minimizing these complexities. As the project proceeded, I came to realize that even carefully obtained sequences could be wrong if they were read from only one strand, so I decided that both complementary strands would need to be determined before a sequence could be considered certain. Additionally, it soon became evident that no two fragments could be abutted in a sequence unless their intersection had been sequenced across, as even carefully constructed maps of restriction fragments often lacked the resolution to guarantee that a ten-base-pair fragment had not been missed. It was necessary for me to follow these (for the time) stringent rules because this was to be the first protein sequence to be determined from its DNA sequence, and my career depended upon arriving at a correct sequence.

Although it might be obvious that these studies were carried out before the advent and convenience of the polymerase chain reaction, pipetmen and commercial kits, perhaps less obvious is that this work preceded word processing and computer programs for entering, compiling and analysing DNA sequences. Thus, a considerable portion of the effort went into the manual exercise of assembling the sequences, which was actually much like assembling a one-dimensional jigsaw puzzle. I read the sequences from the autoradiograms, assembled "contigs" manually, and strung together what eventually became 1100 contiguous nucleotide pairs surrounding the single *Pst*I site of pBR322. During the seven months it took me to assemble the contig, Gilbert consistently followed my progress so as to determine whether the test of his method was likely to work, to give advice, and to follow my findings about the structure of the gene and improvements in the application of the technology. He brought visitors to my bench to show off the absurd battery of ten gels running, wires dangling, with another ten waiting to serve. Wally Gilbert was a distant (but not ineffective) advisor to most of his students, and this was more attention than any of us expected or could appreciate.

As the work progressed, I periodically scanned the contigs for potential ORF fragments and their putative translation products. Alan Hall, a student in the Knowles laboratory, visited occasionally and shared my progress, which he passed on. On a few occasions Hall gave feedback from Knowles indicating that there was some (unspecified) agreement with the Edinburgh data. I posted a graph in the form of twin thermometers (Fig. 2), such as might be used to monitor a fund-raising campaign, on the laboratory door to advertise progress towards completing the β-lactamase sequence. The growing sequence was written on a roll of continuous chart paper with pencils of various colors, one for nucleotides, one for restriction enzyme recognition sites, one for deduced amino acids, and one for potential ribosome-binding sites. The antisense nucleotide chain was written upside down in yet another color so as to indicate its opposite chemical orientation, because at

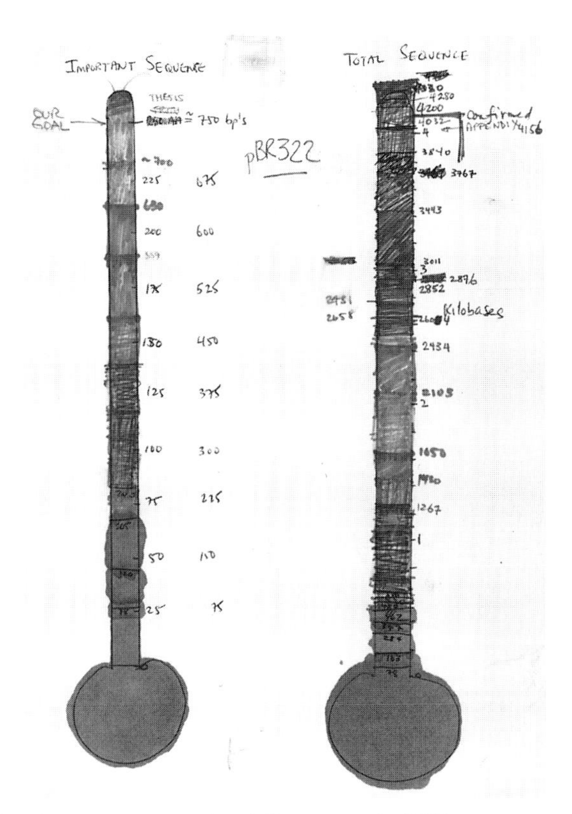

FIGURE 2. The thermometer on the left represents ORF sequence, given in both triplets and nucleotides; that on the right total nucleotides accumulated.

the beginning of the study I had no bias as to which strand might carry the message. Eventually, the length of the roll of paper exceeded 10 feet!

TEA WITH KNOWLES

When I had finally assembled a contig that I considered unambiguous, with a 286 triplet ORF bounded on both sides by termination triplets, Gilbert checked a subset of the autoradiograms and concurred with the sequence. On a September weekend afternoon, Wally drove me in his maroon Chevy convertible to visit Knowles at home. We were served tea on the patio, and proceeded amino acid by amino acid, comparing the sequence deduced from my DNA study with the partial sequence from Ambler's lab. The mood was tense, perhaps adversarial, at the beginning of the comparison. Our first finding was that the amino terminus of the purified, mature β-lactamase, which had been determined both in Edinburgh and by Hall, did not appear until residue 24 of my predicted protein. I had anticipated that its amino terminus might be processed because β-lactamase was known to be excreted into the periplasmic space of the bacterium. Thus, the amino-terminal 23 residues in the nascent protein were utilized as a signal peptide. β-lactamase became at that moment one of the first proteins to support the signal hypothesis of Blobel.

The Edinburgh data were most complete in the amino-terminal region, with long stretches of contiguous amino acids. At residue 37, my sequence predicted a glutamine, as opposed to a lysine in the Ambler sequence. Their lysine could not have been incorrect because it specified the end of a tryptic fragment. After a moment of despair, I realized that the codon encoding the predicted glutamine was in part specified by a sequence that was recognized by the restriction enzyme *Mbo*I. I had cleaved at this site with *Mbo*I during my sequencing studies, thus the pBR322 plasmid could not encode a lysine. It later turned out that the β-lactamase-producing bacteria supplied as RTEM1 to the research community had been misclassified by the reference laboratory in Porton, and they were actually sending out bacteria producing the RTEM2 enzyme isotype to Ambler and many others.

In the middle portion of the sequence, only the amino acid compositions of some of the ordered peptide fragments were available. However, we found that we could easily super-impose these on my predicted sequence. As our confidence grew, the mood changed. In the carboxy-terminal region, only a few of the fragments had been ordered or sequenced. We easily accounted for all of the data, except for one fragment in the Edinburgh collection that was of submolar yield and was later determined to be a contaminant.

I arrived home quite late for a dinner party my wife and I were hosting, but felt little guilt. To say that the comparison had been exhilarating is almost an understatement. Its result was highly gratifying given the effort and care that had gone into achieving it. The study and that from the Edinburgh group were published back-to-back in *Proceedings of the National Academy of Sciences.*[11,12] I could conclude that the hypothetical protein predicted by the DNA sequence was probably totally correct, potential errors being restricted to nucleotide positions for which degenerate codons could serve. As the project had required only seven months from initiation until completion, this suggested[12] "that direct DNA sequence analysis is capable of providing accurate results over extensive regions in a very short time relative to protein sequencing."

THE REST OF pBR322

Two circumstances led me to expand the sequence to include all of pBR322. One was my experience that maps ordering restriction-endonuclease-generated fragments were consistently unreliable, usually in subtle ways. Therefore, during the course of assembling the β-lactamase contig, a substantial amount of sequence outside of the contig was accumulated. Initially, these data were collected because of uncertainties in my maps. Later, the sequences of nearly all fragments produced by particular restriction enzymes were intentionally determined and the appropriate fragments selected for the sequence contig, as this proved more efficient than deliberate isolation of fragments that had predetermined relationships. This was for me the birth of random, or shotgun, sequencing, an approach still widely in use. The second reason was the tremendous interest my work elicited. When I began the lactamase work, my colleagues had been extremely supportive, but not necessarily very interested. The results made a powerful statement about the methodology and enhanced the "fast gun" reputation of the Gilbert lab (as reported in the *Biolabs Midnight Hustler*[13]).

There was also a response from researchers around the country. Many had begun using pBR322 for cloning experiments and wanted to know as much as possible about the vector. Some were interested in the transcriptional promoters carried by pBR322, the mechanism of tetracycline resistance, the mechanism of transposition of drug resistance genes, or the operation of the origin of plasmid replication. And there were those who simply wanted accurate DNA size markers for calibrating their gels.

Although completing the 4363 base-pair sequence required intense effort for the next six months and approximately 400 sequencing gels (it is sobering to realize that high school students now accomplish the same size of project in a few weeks), there was substantial reinforcement. Proof-reading was a major hurdle, and I was quite a pest about recruiting assistants. In the end, my careful checking procedures were assisted by nearly every member of the laboratory, although one fellow graduate student was an exception and told me (I thought ungraciously) to check my own (expletive deleted) films. The raw data as read from the autoradiograms were correlated directly with the manually typed sequence that was to be the figure for publication. In the end, it was only possible to obtain double-stranded sequence for approximately 75% of the plasmid because some areas were inaccessible owing to a scarcity of cleavage sites for available restriction enzymes. In subsequent years, only one sequence error has been reported (4 guanosines read as 3), although an embarrassing one as it caused a single base-pair deletion that affected the interpretation of the mechanism of tetracycline resistance, erroneously suggesting overlapping genes.

GOING PUBLIC

Most of the proof-reading was completed in time for the 1978 Cold Spring Harbor Symposium on DNA: *Replication and Repair.* I brought several hundred hand-outs of the plasmid sequence and derived restriction maps to the meeting to distribute, which went like hot cakes after my presentation. I had a similar experience at the Gordon Conference on nucleic acids the following month. Word spread rapidly that the pBR322 sequence was available and I received many phone requests for the hand-out. On a few occasions several years later, while visiting another institution, I have observed an age-yellowed copy of that hand-out taped to the wall next to an investigator's desk.

After all of the proof-reading was complete, I wrote two manuscripts, one cataloguing the restriction fragments of pBR322[14] and a second describing the new findings that I had learned from the plasmid's sequence and the sequencing project itself. I had been asked to contribute a manuscript for the proceedings of the Cold Spring Harbor Symposium, and decided to submit the second paper, which had been rejected by *Cell,* rather than shop around for another journal and write yet another paper for the Symposium volume. It is very gratifying to me that the paper[15] became by far the most cited paper from the Cold Spring Harbor Symposia series.

All three papers about the sequence[12,14,15] became citation classics.[16] The papers were cited because the data they contained had direct application to many studies. pBR322 became the cloning vector of choice, in part because of the availability of its sequence. Its sequence is the basis of the "dinosaur" DNA sequence in the Michael Crichton novel Jurassic Park (my parents received an autographed copy from the author). However, the greater value of the work, to my mind, was that it demonstrated the startling accuracy and speed with which kilo-bases of DNA sequence could be determined by the new technologies. For their accomplishments in developing these technologies, Gilbert and Sanger shared the 1980 Nobel Prize for Chemistry with Paul Berg.

Retrospectively, the pBR322 study represents a defining moment for several reasons. Other workers were having success with the Maxam–Gilbert chemical-degradation and Sanger chain-termination methods, but the pBR322 sequence was determined by a single researcher in an exceedingly short time for the era. It arrived on the scene abruptly and made its point dramatically. The study shifted the way in which DNA sequences were considered. From that time on, they became the primary source of amino acid sequence information, and now a majority of new proteins are first learned of through sequences of cDNA clones of their mRNAs. This lesson, that carefully determined sequences can provide solid foundations for experimental work, has led to the consideration of sequencing large genomes, including the human genome, as achievable goals.

ACKNOWLEDGEMENTS

I thank all of the members of the Watson–Gilbert–Weber–Dressler group, 1972–1978, for their collegiality and intellectual companionship, and for suffering the proof-reading chores. I especially thank Ron Ogata and Mike Wilson for their comments on the manuscript and historical perspective.

REFERENCES

1. Gilbert W. and Maxam A. 1973. *Proc. Natl. Acad. Sci. USA* **70:** 3581–3585.
2. Sanger F. et al. 1973. *Proc. Natl Acad. Sci. USA* **70:** 1209–1213.
3. Sanger F. et al. 1977. *Nature* **265:** 687–695.
4. Gilbert W., Maxam A., and Mirzabekov A. 1976. *Control of Ribosome Synthesis* (ed. N.C. Kjeldgaard and O. Maaloe, p. 139, Munksgaard.
5. Maxam A. and Gilbert W. 1977. *Proc. Natl. Acad. Sci. USA* **74:** 560–564.
6. Farabaugh P.J. 1978. *Nature* **274:** 765–769.
7. Sauer R.T. and Andregg R. 1978. *Biochemistry* **17:** 1092–1100.
8. Bolivar F. et al. 1977. *Gene* **2:** 95–113.
9. McConnell D.J., Searcy D.G., and Sutcliffe J.G. 1978. *Nucleic Acids Res.* **5:** 1729–1739.
10. Sutcliffe J.G. and Church G.M. 1978. *Nucleic Acids Res.* **5:** 2313–2319.
11. Ambler R.P. and Scott G.K. 1978. *Proc. Natl. Acad. Sci. USA* **75:** 3732–3736.

12. Sutcliffe J.G. 1978. *Proc. Natl. Acad. Sci. USA* **75:** 3737–3741.
13. Witkowski J.A. 1995. *Trends Biochem. Sci.* **20:** 163–168.
14. Sutcliffe J.G. 1978. *Nucleic Acids Res.* **5:** 2721–2728.
15. Sutcliffe J.G. 1979. *Cold Spring Harbor Symp. Quant. Biol.* **43:** 77–90.
16. Sutcliffe J.G. 1990. *Current Contents Life Sciences* **33:** 23.

Biochemistry Strikes Back

SYDNEY BRENNER

MRC Molecular Genetics Unit, Cambridge, UK

Many years ago I attended a meeting to discuss the contributions of the philosopher Karl Popper to the natural sciences. He claimed that science proceeds by what was called the hypotheticodeductive method, which involves the formulation and testing of hypotheses and not by the method of induction, that is, of collecting more and more instances and assembling a theory from these. He differed from most other writers on the philosophy of science by claiming that no hypothesis could be absolutely verified and that scientists seek, or should seek, to *disprove* the hypothesis, because one instance would be enough to eliminate it forever. For a long time Popper claimed that the theory of evolution by natural selection was not a proper scientific theory, because it was impossible to propose a way of disproving it. The argument at that meeting ranged—and raged—from the one extreme of hailing Popper as a genius in the class of thinkers such as Aristotle, Galileo, Newton and Einstein, to the other extreme of dismissing his work as completely trivial. One remark of a very eminent zoologist sticks in my mind. He thought it had all been done before, by a Frenchman called Claude Bernard, and it was called the experimental method, which Bernard had followed in his studies of physiology. The method of experiment, of intervening in the natural world, is at the heart of what we do; our experiments are designed to test causal hypotheses and discover how one event leads to another. Biochemistry inherited this tradition from physiology, and the discovery of the process governing the flow of energy and the biosynthesis of the molecular components of cells was the product of experiments aiming at the reconstitution of these processes in the test tube. In years gone by, biochemists viewed most of the rest of biology as a descriptive science and from the even loftier standpoint of physics, Rutherford dismissed it as stamp collecting.

Reprinted from *Trends in Biochemical Sciences* December 2000, 25(12): 584.

The early days of molecular biology were marked by what seemed to many to be an arrogant cleavage of the new science from biochemistry. People like myself, whose application for admission to the Cambridge Department of Biochemistry was ignored and who did not even receive the attention of a rejection letter, often expressed exaggerated views of our relationship to biochemistry. However, our argument was not concerned with the methods of biochemistry, but only with their blindness in ignoring the new field of the chemistry of information. The realization that one could intervene powerfully in biological processes by mutation provided a powerful addition to the experimental approach in biochemistry and physiology. As is well known, molecular biology, which I once defined as a way of life and not a subject, effected a fusion between biochemistry and genetics.

Over the past dozen years and particularly the past five years we have witnessed the rise of what can be called omic science. First, we had genomics, the study of the genome; actually a purely descriptive or observational science, focussing on the determination of the DNA sequences of genomes. This is perfectly reasonable as the tools are there and it is data that we can use. The genome has spawned further omes: the transcriptome, the proteome and although I have yet to encounter transcriptomics, proteomics is now well established as the next flavour of biological research.

If one surveys the so-called "new way of doing biology" that is omic science, it has several characteristics; it is based on high-throughput methods, on making observations on as much as possible at the same time, and on its reliance on technological improvements to enhance, improve and often automate many old methods. Thus arrays of oligonucleotide probes are used to measure mRNA expression rather than the old method of "dot blots." I am all for these technological advances but what dismays me about omic science is its departure from the hypothesis-generating-experiment basis of scientific investigation. I have even heard claims that it will liberate us from the domination of hypothesis, that is, thinking, in biology. Elaborate statistical methods are proposed to extract information from the millions of data points provided by arrays and by sequences.

Now there are sciences where it is not possible to do experiments. In astrophysics and cosmology we cannot directly intervene in the universe and in meteorology we cannot do experiments to alter the climate and see what happens. And in that other omic science, economics, we also cannot do experiments effectively; although one might argue that governments carry out experiments all the time but these, of course, are uncontrolled. It is no wonder then that people working in these fields, trying to determine the age of the universe, or predicting the weather, or the movements on the stock market, have to discover causation by other methods, and have to test hypotheses by observations alone, that is, by natural experiment. Nowhere else is it advocated that thinking should be abandoned and that we should rely on bioinformatics to provide us with the answers.

I am confident that much of this will vanish when we come to realize that we need to put everything into an evolutionary framework, simply because complexity arises in biological systems by accretion and modification and not by reinvention. Thus, the properties of many of the components in our cells, whether these are mRNAs or proteins, will be conditioned not only by processes of selection for specified activities and levels because these are positively required but may also take up any value because there are no negative consequences for the organism. This "don't care" condition will almost certainly be present because it is a cheap solution to the regulation problem of complex systems. Thus a 20% or a twofold increase, or indeed the very presence, of a protein may be very significant or totally irrelevant depending on whether it is following a "don't care" condition. Only experiment can decide that.

I once made the remark that two things disappeared in 1990: one was communism, the other was biochemistry and that only one of these should be allowed to come back. Of course, biochemistry never really went away but continued to flourish in the thousands of unread pages of biochemical journals. Protein interactions will not be solved by proteomics or protein chips but by protein biochemistry. The genome sequences tell us about the proteins we can expect to find in cells and give us the tools to make large amounts of the proteins for reconstitution studies and for detailed structural analysis. We do not have to resurrect biochemistry, and it will flourish because it provides the only experimental basis for causal understanding of biological mechanisms. That is why this article is not called "The return of biochemistry."

Further Reading

Berg P. and Singer M. 2003. *George Beadle: An Uncommon Farmer.* Cold Spring Harbor Laboratory Press.

Brenner S. 2001. *My Life in Science.* BioMed Central, 2001.

Cairns J., Stent G.S., and Watson J.D., eds. 1966. *Phage and the Origins of Molecular Biology.* Cold Spring Harbor Laboratory of Quantitative Biology (Expanded edition: Cold Spring Harbor Laboratory Press, 1992).

Chargaff E. 1978. *Heraclitean Fire: Sketches from a Life Before Nature.* Rockefeller University Press.

Crick F.H.C. 1988. *What Mad Pursuit: A Personal View of Scientific Discovery.* Basic Books.

de Chadarevian S. 2002. *Designs for Life: Molecular Biology after World War II.* Cambridge University Press.

Ferry G. 2000. *Dorothy Hodgkin: A Life.* Cold Spring Harbor Laboratory Press.

Hoagland M.B. 1990. *Toward the Habit of Truth: A Life in Science.* W.W. Norton.

Holmes F. 2001. *Meselson, Stahl, and the Replication of DNA: A History of 'The Most Beautiful Experiment in Biology'*. Yale University Press.

Inglis J., Sambrook J., and Witkowski J., eds. 2003. *Inspiring Science: Jim Watson and the Age of DNA.* Cold Spring Harbor Laboratory Press.

Jacob F. 1995. *The Statue Within: An Autobiography.* Cold Spring Harbor Laboratory Press.

Judson H.F. 1979. *The Eighth Day of Creation: Makers of the Revolution in Biology,* Simon & Schuster (Expanded edition: Cold Spring Laboratory Press, 1996).

Kay L.E. 1993. *The Molecular Vision of Life: Caltech, the Rockefeller Foundation, and the Rise of the New Biology.* Oxford University Press.

Kay L.E. 2000. *Who Wrote the Book of Life? A History of the Genetic Code,* Stanford University Press.

Kornberg A. 1989. *For the Love of Enzymes: The Odyssey of a Biochemist.* Harvard University Press.

Luria S.E. 1984. *A Slot Machine, A Broken Test Tube: An Autobiography,* Harper and Row.

Lwoff A. and Ullman A., eds. 1979. *Origins of Molecular Biology: A Tribute to Jacques Monod, Academic Press* (Revised edition: ASM Press, 2003).
Maddox B. 2002. *Rosalind Franklin: The Dark Lady of DNA.* HarperCollins.
McCarty M. 1985. *The Transforming Principle—Discovering That Genes Are Made of DNA.* W.W. Norton.
McElroy W.D. and Glass B., eds. 1957. *The Chemical Basis of Heredity.* The Johns Hopkins Press.
Morange M. 1998. *A History of Molecular Biology.* Harvard University Press.
Olby R. 1974. *The Path to the Double Helix.* University of Washington (Seattle); Macmillan Press (London) (New edition: Dover Publications, 1994).
Portugal F.H. and Cohen J.S. 1977. *A Century of DNA: History of the Discovery of the Structure and Function of the Genetic Substance.* MIT Press.
Rheinberger H.-J. 1997. *Toward a History of Epistemic Things: Synthesizing Proteins in the Test Tube.* Stanford University Press.
Sanger F. and Dowding M. 1996. *Selected Papers of Frederick Sanger.* World Scientific, Singapore.
Stahl F.W. 2000. *We Can Sleep Later: Alfred D. Hershey and the Origins of Molecular Biology.* Cold Spring Harbor Laboratory Press.
Taylor J.H., ed. 1965. *Selected Papers on Molecular Genetics: A Collection of Reprints, with Introductory Material by J. Herbert Taylor.* Academic Press.
Watson J.D. 1965. *Molecular Biology of the Gene.* W.A. Benjamin.
Watson J.D. 1968. *The Double Helix: A Personal Account of the Discovery of the Structure of DNA.* Atheneum (New York); Weidenfeld & Nicolson (London).
Watson J.D. 2001. *Genes, girls, and Gamow: After* The Double Helix. Oxford University Press (1st U.S. edition: Alfred A. Knopf, 2002).
Watson J.D., Baker T.A., Bell S.P, Gann A., Levine M. and Losick R. 2003. *Molecular Biology of the Gene, Fifth Edition.* Benjamin Cummings/Cold Spring Harbor Laboratory Press.
Wilkins M. 2003. *The Third Man of the Double Helix: The Autobiography of Maurice Wilkins.* Oxford University Press.

The following volumes of the Cold Spring Harbor Laboratory Symposia on Quantitative Biology provide contemporary accounts of the latest research on DNA, RNA, and protein, from 1938 to 2001. Many of the key findings were reported first at these Symposia. (The dates are of the meetings and not the publication dates of the volumes.)

Ponder E., ed. 1938. *Protein Chemistry.* Cold Spring Harbor Symposia on Quantitative Biology VI.
Demerec M., ed. 1941. *Genes and Chromosomes: Structure and Organization.* Cold Spring Harbor Symposia on Quantitative Biology IX.
Demerec M., ed. 1946. *Heredity and Variation in Microorganisms.* Cold Spring Harbor Symposia on Quantitative Biology XI.
Demerec M., ed. 1947. *Nucleic Acids and Nucleoproteins.* Cold Spring Harbor Symposia on Quantitative Biology XII.

Demerec M., ed. 1949. *Amino Acids and Proteins.* Cold Spring Harbor Symposia on Quantitative Biology XIV.
Demerec M., ed. 1951. *Genes and Mutations.* Cold Spring Harbor Symposia on Quantitative Biology XVI.
Demerec M., ed. 1953. *Viruses.* Cold Spring Harbor Symposia on Quantitative Biology XVIII.
Demerec M., ed. 1956. *Genetic Mechanisms: Structure and Function.* Cold Spring Harbor Symposia on Quantitative Biology XXI.
Demerec M., ed. 1958. *Exchange of Genetic Material: Mechanism and Consequences.* Cold Spring Harbor Symposia on Quantitative Biology XXIII.
Chovnik, A. 1961. *Cellular Regulatory Mechanisms.* Cold Spring Harbor Symposia on Quantitative Biology XXVI.
Umbarger H.E., ed. 1963. *Synthesis and Structure of Macromolecules.* Cold Spring Harbor Symposia on Quantitative Biology XXVIII.
Cairns J., ed. 1966. *The Genetic Code.* Cold Spring Harbor Symposia on Quantitative Biology XXXI.
Cairns J., ed. 1968. *Replication of DNA in Microorganisms.* Cold Spring Harbor Symposia on Quantitative Biology XXXIII.
Cairns J., ed. 1969. *The Mechanism of Protein Synthesis.* Cold Spring Harbor Symposia on Quantitative Biology XXXIV.
Watson J.D., ed. 1970. *Transcription of Genetic Material.* Cold Spring Harbor Symposia on Quantitative Biology XXXV.
Watson J.D., ed. 1971. *Structure and Function of Proteins at the Three-Dimensional Level.* Cold Spring Harbor Symposia on Quantitative Biology XXXVI.
Stillman B. 2001. *The Ribosome.* Cold Spring Harbor Symposia on Quantitative Biology XVI.

Index